Molecular Biology Intelligence Unit

Origin and Evolution of Telomeres

Jozef Nosek
Department of Biochemistry
Comenius University
Bratislava, Slovakia

Ľubomír Tomáška
Department of Genetics
Comenius University
Bratislava, Slovakia

CRC Press
Taylor & Francis Group
Boca Raton London New York

CRC Press is an imprint of the
Taylor & Francis Group, an **informa** business

Origin and Evolution of Telomeres

Molecular Biology Intelligence Unit

CRC Press
Taylor & Francis Group
6000 Broken Sound Parkway NW, Suite 300
Boca Raton, FL 33487-2742

First issued in hardback 2019

ISBN 13: 978-1-58706-309-1 (hbk)

Visit the Taylor & Francis Web site at
http://www.taylorandfrancis.com

and the CRC Press Web site at
http://www.crcpress.com

Library of Congress Cataloging-in-Publication Data

Origin and evolution of telomeres / [edited by] Jozef Nosek, Lubomír Tomáska.
p. ; cm.
Includes bibliographical references and index.
ISBN 978-1-58706-309-1
1. Telomere. I. Nosek, Jozef, Ph.D. II. Tomáska, Lubomír.
[DNLM: 1. Telomere--physiology. 2. Evolution, Molecular. 3. Telomerase--genetics.
4. Telomerase--physiology. 5. Telomere--genetics. QU 470 O695 2008]
QH600.3.O75 2008
572.8'7--dc22

About the Editors...

JOZEF NOSEK, PhD, DSc, is an Associate Professor at the Department of Biochemistry, Faculty of Natural Sciences at Comenius University in Bratislava, Slovakia. His research interests include biology and evolution of telomeres, replication of linear mitochondrial genomes and function of their terminal structures, biogenesis of organelles, and morphogenesis of eukaryotic cells. He received his PhD and DSc degrees in genetics and molecular biology from Comenius University and did his Postdoctoral at the Curie Institute in Orsay, France. Since 2001, he is an International Research Scholar of the Howard Hughes Medical Institute.

About the Editors...

ĽUBOMÍR TOMÁŠKA, PhD, DSc, is an Associate Professor at the Department of Genetics, Faculty of Natural Sciences at Comenius University in Bratislava, Slovakia. He is interested in nucleo-protein structure of nuclear and mitochondrial telomeres, evolutionary pathways leading to different means of telomere maintenance, mitochondrial morphogenesis and inheritance and interplay between the compartments of eukaryotic cells. He serves as the Editor of Current Genetics and FEMS Yeast Research. He received his PhD and DSc degrees in genetics and molecular biology from Comenius University and obtained additional training at Cornell University in Ithaca, New York and at the University of North Carolina in Chapel Hill, North Carolina.

Dedication

We would like to dedicate this book to Ladislav Kováč, our mentor and friend, on the occasion of his 75 birthday.

CONTENTS

EDITORS

Jozef Nosek
Department of Biochemistry
Comenius University
Bratislava, Slovakia
Email: nosek@fns.uniba.sk
Chapter 12

Ľubomír Tomáška
Department of Genetics
Comenius University
Bratislava, Slovakia
Email: tomaska@fns.uniba.sk
Chapter 12

CONTRIBUTORS

Note: Email addresses are provided for the corresponding authors of each chapter.

Chantal Autexier
Departments of Anatomy, Cell Biology and Medicine
McGill University
and
Bloomfield Centre for Research in Aging
Sir Mortimer B. Davis Jewish General Hospital
Montreal, Quebec, Canada
Email: chantal.autexier@mcgill.ca
Chapter 1

Marie-Eve Brault
Departments of Anatomy and Cell Biology
McGill University Bloomfield Centre for Research in Aging
Sir Mortimer B. Davis Jewish General Hospital
Montreal, Quebec, Canada
Chapter 1

Sherwood Casjens
Department of Pathology
University of Utah Medical School
Salt Lake City, Utah, USA
Email: sherwood.casjens@path.utah.edu
Chapter 11

Anthony J. Cesare
Cancer Research Unit
Children's Medical Research Institute
Westmead, New South Wales, Australia
Chapters 3, 4

Michael Chester
School of Biological and Chemical Sciences
Queen Mary University of London
London, UK
Chapter 8

Marita Cohn
Department of Cell and Organism Biology
Lund University
Lund, Sweden
Email: marita.cohn@cob.lu.se
Chapter 5

Sarah A. Compton
Lineberger Comprehensive Cancer Center
University of North Carolina
Chapel Hill, North Carolina, USA
Chapter 4

P. Gregory DeBaryshe
Department of Biology
Massachusetts Institute of Technology
Cambridge, Massachsettes, USA
Chapter 2

Yasmin D'Souza
Departments of Anatomy
and Cell Biology
McGill University
and
Bloomfield Centre for Research
in Aging
Sir Mortimer B. Davis Jewish General
Hospital
Montreal, Quebec, Canada
Chapter 1

Jiří Fajkus
Department of Functional Gemonics
and Protemics
Masaryk University
and
Institute of Biophysics
Czech Academy of Science
Brno, Czech Republic
Email: fajkus@sci.muni.cz
Chapter 8

Nicole Fouche
Lineberger Comprehensive Cancer Center
University of North Carolina
Chapel Hill, North Carolina, USA
Email: nicole.fouche@cxi.net
Chapter 4

Eric Gilson
Laboratoire de Biologie Moléculaire
de la Cellule
Ecole Normale Supérieure de Lyon
Lyon, France
Email: eric.gilson@ens-lyon.fr
Chapter 9

Jack D. Griffith
Lineberger Comprehensive Cancer Center
University of North Carolina
Chapel Hill, North Carolina, USA
Email: jdg@med.unc.edu
Chapter 4

Martin P. Horvath
Department of Biology
University of Utah
Salt Lake City Utah, USA
Email: horvath@biology.utah.edu
Chapter 6

Wai Mun Huang
Department of Pathology
University of Utah Medical School
Salt Lake City, Utah, USA
Chapter 11

Andrew R. Leitch
School of Biological and Chemical
Sciences
Queen Mary University of London
London, UK
Chapter 8

José Arturo Londoño-Vallejo
Telomeres & Cancer Lab
Institut Curie
Paris, France
Email: arturo.londono@curie.fr
Chapter 10

Frederique Magdinier
Laboratoire de Biologie Moléculaire
de la Cellule
Ecole Normale Supérieure de Lyon
Lyon, France
Chapter 9

Michael J. McEachern
Department of Genetics
Fred C. Davison Life Sciences Complex
University of Georgia
Athens, Georgia, USA
Email: mjm@arches.uga.edu
Chapter 7

Alexandre Ottaviani
Laboratoire de Biologie Moléculaire de la Cellule
Ecole Normale Supérieure de Lyon
Lyon, France
Chapter 9

Sezgin Ozgur
Lineberger Comprehensive Cancer Center
University of North Carolina
Chapel Hill, North Carolina, USA
Chapter 4

Mary-Lou Pardue
Department of Biology
Massachusetts Institute of Technology
Cambridge, Massachsettes, USA
Email: mlpardue@mit.edu
Chapter 2

Roger R. Reddel
Cancer Research Unit
Children's Medical Research Institute
Westmead, New South Wales, Australia
Email: rreddel@cmri.usyd.edu.au
Chapter 3

Eva Sýkorová
Department of Functional Gemonics and Protemics
Masaryk University
and
Institute of Biophysics
Czech Academy of Science
Brno, Czech Republic
Chapter 8

PREFACE

There are two principal forms of chromosomes, circular and linear. The circular form is more frequently (albeit not exclusively) associated with simpler, prokaryotic organisms, whereas the linear form is commonly found in eukaryotic cells. Although some bacteria contain linear DNA molecules and circular DNAs occur in several eukaryotes, the general trend is that linearity of chromosomal DNA is associated with nuclei of more complex eukaryotic organisms. Both forms, however, bring about a number of problems related to DNA replication, stabilization and inheritance. Cells overcome these problems by using intricate strategies and complex mechanisms. In the case of linear DNA, architecture-specific problems result from distinct features of replication and maintenance of the two ends of each molecule. Most notable is the fact that standard DNA polymerases are unable to finalize replication at these ends; a phenomenon referred to as the 'end-replication problem' defined independently in the early 1970s by Alexei Olovnikov and James D. Watson. Secondly, the ends of linear DNA molecules represent natural double-strand breaks (DSB). When a DSB occurs within DNA, it is recognized and fixed by robust DNA-repair machinery. In contrast, chromosomal ends must be ignored or they would undergo end-to-end fusions, leading to the formation of dicentric chromosomes and genomic instability. To overcome these problems, cells containing linear DNA genomes have evolved complex terminal nucleoprotein structures called telomeres.

The term telomere (derived from two Greek words telos—terminus and meros—part) was originally introduced by Hermann J. Muller in 1938 to distinguish the termini of eukaryotic chromosomes. Muller defined telomere as a 'terminal gene' that ensures sealing of the chromosomal ends. Currently, telomeres are defined as special complexes located at the ends of eukaryotic chromosomes essential for preserving genome integrity and ultimately cell survival. At the same time, the linearity of chromosomes is crucial for proper pairing during meiosis and sexual reproduction of eukaryotic organisms. In addition to providing solutions to the end-replication problem and protection of chromosomal termini, telomeres are involved in a number of interactions within eukaryotic nuclei, including interactions between chromosomes and specific nuclear regions. These interactions are implicated in the control of expression of loci located in subterminal chromosomal regions; a phenomenon called telomere position effect. Equally as important and perhaps more intriguing is the fact that telomeres are thought to function as molecular clocks controlling cellular life span and are therefore involved in complex biological phenomena such as cell senescence, carcinogenesis, and immortalization. Several lines of evidence indicate that telomere shortening plays a key role in the process of replicative senescence and represents a potent tumor suppression mechanism in human cells. As a result, pathways involved in telomere maintenance represent the Achille's heel of most cancers—a specific molecular target suitable for rational drug design and therapeutic intervention. On the other hand, directed manipulation of telomere maintenance pathways provides an opportunity to extend cellular longevity with potential implications for treatment of various degenerative diseases and tissue engineering.

Several recent reviews in scientific journals and books have extensively covered the field of telomere biology. However, questions related to the origin and evolution of telomeres remain unanswered. In this book, inspired by F. Theodosius Dobzhansky's idea that 'nothing in biology makes sense except in the light of evolution', we intended to fill this gap in the literature and shed light on the origin and evolution of telomeres, their functions and the consequences of eukaryotic chromosome linearity. As indicated above, the linearity of nuclear chromosomes seems to be an essential prerequisite for meiotic cell division and, thus, sexual reproduction. Selective pressure toward linearization must have been associated with the emergence of robust and redundant mechanisms for maintenance of telomeric structures. On the other hand, linearity per se, the presence of telomeres and/or a specific component of their replication machinery, may provide a selective advantage. Segmentation of the genome into multiple linear chromosomes is a typical feature of eukaryotes, but is rare among bacterial and archaeal species. Therefore, it may represent an evolutionary innovation associated with the origin of eukaryotic cells. This raises questions of how linear chromosomes and primordial pathways for maintenance of their terminal structures emerged in early eukaryotes and how primordial telomeres transformed into their modern robust form backed up by alternative pathways for their maintenance.

Jozef Nosek
Ľubomír Tomáška

Acknowledgements

The idea to publish a book on the origin of telomeres came from Ronald G. Landes (Landes Bioscience), whom we would like to thank for his generous proposal. Moreover, we wish to thank to Cynthia Conomos and other staff members from Landes Bioscience for help with administrative and technical problems, and to all the contributors who participated in this exciting project as well as the many scientists in the field of telomere biology for stimulating discussions. In addition, we thank all our colleagues with whom we collaborate in the field of telomere biology: Ladislav Kováč, Hiroshi Fukuhara, Monique Bolotin-Fukuhara, Jack Griffith, Jiří Fajkus, Mike McEachern, Ed Louis, Mundy Wellinger, Dudy Tzfati, and Peter Griač. In addition, we wish to thank our previous and current PhD students, Ľubica Adamíková, Roman Szabo, Blanka Kucejová, Martin Kucej, Adriana Ryčovská, Júlia Zemanová, Peter Kosa, Ľubomír Lanátor, Silvia Petrezselyová, Judita Slezáková, Stanislava Gunišová, Zuzana Holešová, Slavomír Kinský, Juraj Kramara, Lenka Abelovská, Dominika Fričová, and Matúš Valach, and numerous undergraduate students for their invaluable experimental and intellectual inputs and many inspirations. Last but not least, many of our colleagues at the Departments of Biochemistry and Genetics are highly appreciated for their effort to generate a stimulating and friendly atmosphere. Over the years the research in our laboratory has been kindly supported by grants from the Howard Hughes Medical Institute, Fogarty International Collaboration Awards, Slovak grant agencies APVV and VEGA.

Acknowledgements

Chapter 1

Telomerase:
Evolution, Structure and Function

Marie-Eve Brault,[†] Yasmin D'Souza[†] and Chantal Autexier*

Abstract

Telomerase is a unique ribonucleoprotein reverse transcriptase that uses an integral RNA template to catalyze the addition of telomeric repeats at telomeres. This mechanism is required for the maintenance of chromosome termini, as the structure and integrity of telomeres are essential for genome stability. Although the catalytic subunit of the enzyme shares several features with other reverse transcriptases, it differs in having telomerase-specific structures and functions. Structurally, the telomerase reverse transcriptase protein component contains unique amino–and carboxy–terminal domains that flank centrally-located reverse transcriptase motifs. Functionally, unlike reverse transcriptases, telomerase contains an integral RNA component and is able to synthesize telomeric repeats. Here we discuss the evolutionary relationship of telomerase to other nucleic acid polymerases and reverse transcriptases and the various functions that regulate the elongation of chromosome ends. Such functions include interaction with the DNA substrate, nucleolytic activity, multimerization and processivity. In addition, we propose telomere-lengthening independent functions ascribed to telomerase.

Introduction

Telomeres are nucleoprotein complexes located at the ends of eukaryotic chromosomes. In most organisms, they are composed of tandem repeats of G-rich DNA sequences that terminate in a 3' single-strand G-rich overhang.[1] Telomeric DNA sequences assemble with various proteins to create protective structures or "caps" at chromosome termini which are essential for the maintenance of genomic stability and integrity.[2] Telomere capping prevents chromosome ends from being recognized as DNA breaks, thus inhibiting the activation of the DNA damage response in undamaged cells and subsequent chromosome-to-chromosome fusions.[2,3] Telomeres also protect the ends of linear chromosomes from inappropriate nuclease degradation and recombination.[4] In addition to their "capping" function, telomeres are essential for promoting the complete replication of chromosome ends. The conventional semi-conservative replication machinery is unable to replicate the ends of linear DNA molecules, resulting in telomere attrition with each cell division, a process known as the "end replication problem".[5] Extensive telomere erosion to a critical length triggers cell cycle arrest and entry into an irreversible nondividing state termed "replicative senescence".[6,7] Telomere shortening can, however, be counteracted by elongation mechanisms, most commonly by a specialized cellular reverse transcriptase (RT) named telomerase.[8-10] Considerable efforts in the last two decades have been directed towards understanding the structure and function of telomerase components and the evolutionary relationship between this unique enzyme and other RTs. In this chapter, we discuss the evolution and structure of telomerase and the various

[†]These authors contributed equally to the chapter.
*Corresponding Author: Chantal Autexier—Departments of Anatomy, Cell Biology and Medicene, McGill University, Bloomfield Centre for Research in Aging, Sir Mortimer B. Davis Jewish General Hospital, 3755 Cote Ste. Catherine Road, Montreal, Quebec, Canada, H3T 1E2. Email: chantal.autexier@mcgill.ca

Origin and Evolution of Telomeres, edited by Jozef Nosek and Ľubomír Tomáška.

functions ascribed to each telomerase component which contribute to the efficient maintenance of telomeres. Such functions include DNA substrate interaction, nucleolytic activity, multimerization and processivity. In addition, we propose extracurricular functions of telomerase that are independent of its role in telomere maintenance.

Telomere-Lengthening Dependent and Independent Functions of Telomerase

Telomerase is a ribonucleoprotein (RNP) composed of a catalytic subunit named telomerase reverse transcriptase (TERT) and an RNA component known as telomerase RNA (TR, TER or TERC) that serves as a template for the synthesis of telomeric repeats. Because coexpression of TERT and TR in a transcription and translation system suffices to reconstitute enzymatic activity in vitro, these two subunits comprise the minimum functional core of the enzyme complex.[11] Similarly, telomerase activity can be reconstituted using recombinant baculovirus-expressed TERT and in vitro transcribed telomerase RNA.[12,13] Telomerase is active in unicellular organisms such as yeast and ciliated protozoa and maintains telomere length to ensure the long-term proliferation of the cell population. However, in multicellular organisms such as human and chicken, telomerase activity is not detected in the vast majority of somatic cells except for highly proliferative cells or renewal tissues.[14] Primary human cells in culture undergo telomere shortening-dependent replicative senescence, which is observed in parallel with increasing passage.[15] However, ectopic expression of human TERT (hTERT) in these cells reconstitutes telomerase activity, lengthens telomeres and extends replicative life span.[9,16,17] Strikingly, telomerase is active in more than 85% of cancer cells, a prerequisite for their acquisition of infinite life span.[18] Telomere elongation may also occur by the alternative lengthening of telomeres (ALT) pathway, a mechanism based on homologous recombination.[19] However, elongation of chromosome ends by telomerase remains the most widespread mechanism to maintain telomeres.

Telomerase elongates the 3' end of the chromosome, thus enabling other polymerases to synthesize the complementary strand.[20] In vitro telomerase assays have been valuable in elucidating the mechanism by which this enzyme catalyzes nucleotide addition. During telomere extension, telomerase repeatedly uses the same short region within its RNA moiety as a template for DNA synthesis. This mechanism entails a cyclic reaction, described as follows:[20,21] the single-stranded DNA substrate first base pairs with the telomerase RNA template (Fig. 1). The DNA substrate is then extended using this template. Nucleotide addition processivity (NAP) is defined as the addition of nucleotides, one at a time, until the 5' boundary of the RNA template is encountered. Subsequently, the template-DNA hybrid is disrupted and the telomerase complex either dissociates from the DNA substrate, or the enzyme translocates to the new 3' end of the DNA substrate, which becomes available for another round of elongation by telomerase. Successive rounds of nucleotide addition and enzyme translocation, which allows reiterative addition of telomeric repeats onto the DNA 3' end, is known as repeat addition processivity (RAP). While telomerase isolated from yeast, fungi and rodents are nonprocessive, predominantly generating DNA products consisting of only one repeat,[22-25] human and ciliate telomerases can translocate more efficiently, adding hundreds of nucleotides to a DNA substrate in vitro.[26,27]

Telomerase activity has been reported in almost all eukaryotes with the exception of some organisms such as *Drosophila melanogaster* and other insects of the order Diptera.[28] Enzyme activity was first identified in the ciliated protozoa *Tetrahymena thermophila*,[8] and subsequently in the hypotrichous ciliate *Euplotes crassus*,[29] human,[26] mouse,[22] several yeasts[23-25] and plants.[30-33] Although telomerase is the most widespread mechanism used to maintain the ends of eukaryotic chromosomes, other methods are used by viruses and bacteria for the same purpose, such as protein priming, terminal hairpins and recombination.[34] It has been speculated that the prevalence of telomerase in eukaryotes could be explained by the enzyme's capacity to mediate functions that extend beyond its classical role in telomere length regulation. For example, TERT plays an active role in neuroprotection by preventing neuronal apoptosis induced by a variety of stresses, including hypoxia and ischemia.[35,36] Anti-apoptotic functions of TERT have also been reported in immune,[37]

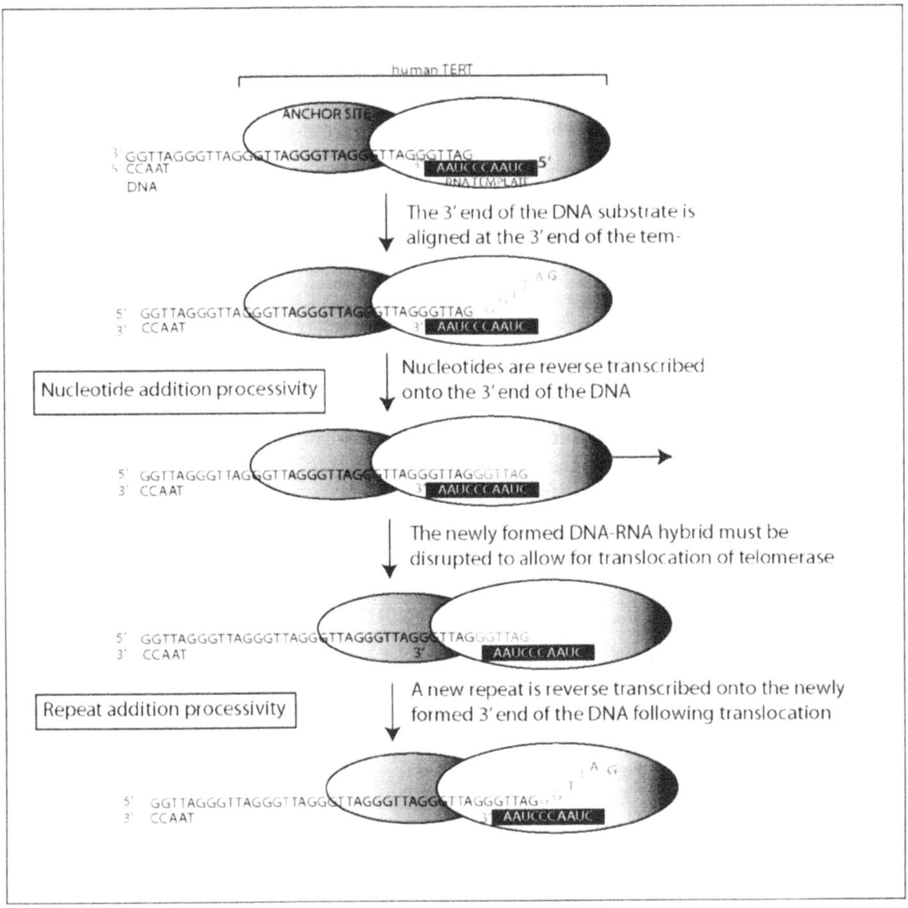

Figure 1. Model for processive elongation by telomerase. The 3′ end of the DNA substrate forms a hybrid with the 3′ end of the RNA template, while nucleotides upstream of the DNA 3′ end are postulated to interact with an anchor site. Next, template-directed addition of nucleotides onto the 3′ end of the DNA occurs until the 5′ end of the template is encountered. This process is known as nucleotide addition processivity (NAP). The active site then translocates to reposition itself and the 3′ end of the template at the newly formed 3′ end of the DNA substrate. Another round of nucleotide addition is then initiated. Reiterative translocation and nucleotide addition, resulting in the formation of multiple repeats is known as repeat addition processivity (RAP). (Adapted from Autexier and Lue 2006; Lue 2004.)

muscle[38] and some cancer cells.[39,40] In addition, evidence of a role for TERT in the promotion of tumorigenesis has been described. Telomerase-negative ALT cells fail to promote tumorigenesis when overexpressing oncogenic H-Ras alone while the same cells expressing H-Ras in combination with TERT reveal a strong tumorigenic phenotype. The expression of an HA-tagged TERT protein (TERT-HA), defective in maintaining telomere length in normal cells, also reveals a tumorigenic phenotype when coexpressed with H-Ras in ALT cells, suggesting a role for TERT in tumorigenesis independent of telomere lengthening.[41] Moreover, TERT expression in mice, which possess long telomeres, promotes breast cancer and papillomas.[42,43] A less characterized function for TERT is its role as a transcriptional regulator of cell growth by upregulating growth-promoting genes[44] and downregulating apoptotic genes.[45] The involvement of TERT in regulating genomic stability, chromatin structure and response to DNA damage has also been shown and appears to rely on its association with telomeres rather than its catalytic activity.[46,47] In addition, the ability of TERT expression to promote the proliferation of epidermal stem cells was found in at least one study to be independent of the telomerase RNA component.[48,49] Further experiments will be needed to confirm the extracurricular functions of telomerase, which still remain incompletely characterized. However, the existence of such moonlighting functions might explain why telomerase, rather than other mechanisms of telomere maintenance, has been favoured evolutionarily in eukaryotes.

How Ancient Is Telomerase?

It has been proposed that telomerase is an ancient mechanism of telomere maintenance, perhaps originating with early eukaryotes.[50] In support of this hypothesis, the most ancient evidence of telomerase may be in the parasitic protozoan *Giardia*, which is thought to be one of the most primitive eukaryotic species. Though telomerase activity from *Giardia lamblia* has not been reported, a putative *G. lamblia* TERT was recently identified.[51] The ancient origin of telomerase is further supported by the nature of the enzyme, which has been conserved in the form of a ribonucleoprotein (RNP) throughout evolution. The presence of an RNA component led to the speculation that telomerase is a remnant from the time of the RNA- to DNA-world transition.[52] According to the RNA-world hypothesis, DNA evolved from RNA.[53,54] In the RNA world, RNA molecules were able to store genetic information and catalyze all the reactions required for the survival of the earliest forms of life. In our modern world however, DNA and protein appropriated these functions as DNA is a more stable nucleic acid and protein is a more efficient catalyst than RNA.[55] Presumably, the transition from the RNA-world to the current DNA-protein-world occurred in two steps: first with the addition of protein components to RNA to form a RNP world and second with the addition of DNA to form the world we know today. According to this, it has been speculated that the telomerase RNA component evolved from an ancient ribozyme whose catalytic activity was subsequently acquired by a more efficient protein component.[56]

While almost all essential functions in eukaryotes are performed by DNA-dependent DNA polymerases (DdDP), telomere maintenance is achieved by an RNA-dependent DNA polymerase (RdDP) that carries its own RNA component. Interestingly, telomerase can also function as a RNA-dependent RNA polymerase(RdRP).[57,58] RdRPs are enzymes that perform viral RNA replication and are the presumed ancestor of actual polymerases.[59] Similar to telomerase, RdRPs utilize an RNA template to synthesize a single-stranded nucleic acid: DNA for telomerase and RNA for RdRPs. Hence, it is possible that telomerase evolved from an ancestral RdRP following modifications in enzyme specificity.[57]

Protein Component (TERT)

It is the identification of an RNA subunit acting as a template for the addition of telomeric repeats that first led to the assumption that the catalytic subunit of telomerase could share similarities with RTs.[60,61] When the first catalytic subunits of telomerase were identified, characterization of the sequences encoding the protein component (p123) of *Euplotes aediculatus* telomerase and its *Saccharomyces cerevisiae* homologue Est2 indeed revealed the presence of universally conserved RT motifs.[62-64] Telomerase catalytic subunits were subsequently identified in *Schizosaccharomyces pombe* and humans by sequence alignment of RT motifs.[65-68]

Reverse Transcriptase (RT) Domain of TERT

Alignment of TERT proteins from evolutionarily close or distant organisms reveals a conserved structural organization composed of hallmark RT motifs, referred to as 1, 2, A, B', C, D and E (Fig. 2B). Alteration of conserved sequences in these motifs leads to the inactivation of telomerase activity and reduced telomere length.[11,63,65,66] For example, deletions within motifs B', C, D and E in the RT domain of *Schizosaccharomyces pombe* TERT results in progressive telomere shortening in cells and eventual senescence.[65] Importantly, motifs A and C contain three essential aspartic acid residues which catalyze phosphoryl transfer and are conserved between TERTs and other RTs.[65]

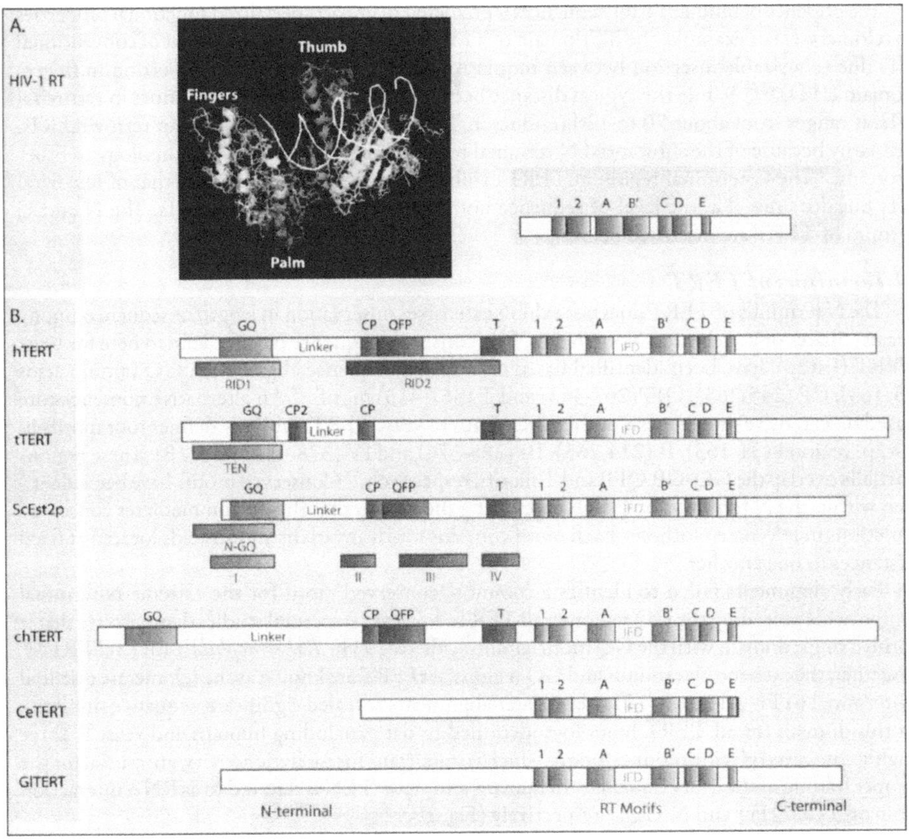

Figure 2. Structure of TERTs. A) The structure of the HIV-1 RT p66 subunit N-terminal polymerase domain is shown with a bound DNA duplex. The polymerase domain of p66 resembles a human "right hand" composed of a thumb, a palm and fingers. The connection domain is shown but the RNaseH domain has been omitted. This figure was created with PyMOL using an available HIV-1 RT sequence.[107] B) *Homo sapiens* (hTERT), *Tetrahymena thermophila* (tTERT), *Saccharomyces cerevisiae* (ScEst2p), chicken (chTERT), *Giardia lamblia* (GlTERT) and *Caenorhabditis elegans* (CeTERT) telomerase RTs are compared with HIV-1 RT. Alignment of TERTs reveals seven conserved RT motifs (1, 2, A, B, C, D and E) and a telomerase-specific insertion in fingers domain (IFD) between motifs A and B'. The N-teminus of TERTs is variable but demonstrates four conserved motifs (GQ, CP, QFP and T or I, II, III and IV in Est2p). Although highly conserved in TERTs, the T motif is absent in *G. lamblia* and *C. elegans*. The C terminus of TERTs reveals only low sequence homology. The linker region is the most variable region of TERTs in terms of sequence and length. Reprinted, with permission, from the *Annual Review of Biochemistry*, Volume 75 ©2006 by Annual Reviews www.annualreviews.org.

Mutagenesis of any of these three invariant residues abolishes telomerase activity[63,69] and telomere length maintenance.[64,65,69] Residues can also diverge within similar motifs between TERTs and RTs. To date, alignments of motif E demonstrate the presence of the consensus sequence WxGx among several TERT proteins, (except in *Giardia lamblia* and several yeast TERTs) whereas hLGx is the characteristic sequence present in other RTs[65,70] ("h" refers to hydrophobic amino acids and "x" refers to any amino acid). Similar to the role of RT motif E, TERT motif E is implicated in interactions with the DNA substrate.[70,71]

The presence of unique TERT sequences is predicted to impart specialized functional properties to telomerase. For example, the RT domain of TERTs is distinguishable from that of conventional RTs due to a sizable insertion between motifs A and B', referred to as the "insertion in fingers domain (IFD).[63,65] While the typical distance between the two motifs is 20 residues in retroviral RTs, it ranges from about 70 to 120 residues in TERTs.[72] TERTs are larger than retroviral RTs, primarily because of the substantial N-terminal region which has no obvious homology to other proteins.[73] The C-terminal region in TERT exhibits no sequence homology to that of retroviral RTs but does reveal a low level of sequence homology among different TERTs. The N and C termini of TERT are discussed below.

N Terminus of TERT

The N terminus of TERT does not exhibit extensive conservation in length or sequence among yeasts, ciliates or vertebrates. Nonetheless, four conserved sequence motifs, defined here for yeast TERT (Est2p), have been identified based on multiple sequence alignments: GQ (amino acids 45-163), CP (245-265), QFP(267-343) and T (367-413) motifs.[74] An alternative nomenclature based on extensive mutagenesis of the N terminus of yeast TERT, similarly defines four motifs in Est2p: region I (31-163), II (214-265), III (285-374) and IV (378-432) (Fig. 2B). These regions partially overlap the GQ, CP, QFP and T motifs, respectively.[73] Conserved motifs have been identified within the N terminus of TERT, supporting the notion that this region mediates conserved functions in telomere synthesis. Each motif contains nearly invariant amino acids located at fixed distances to one another.[74]

Early alignments failed to identify a common, conserved motif for the extreme N-terminal amino acids preceding the GQ region in all TERTs, however structural studies demonstrate that it forms a single domain with the GQ motif, amino acids 72-193 in *T. thermophila* TERT (tTERT).[75] Together, the extreme N terminus and GQ regions of tTERT are known as the telomerase essential N-terminal (TEN) domain.[75] However, later alignments revealed significant sequence similarity in this domain for all TERT homologs identified to date, including humans and yeast.[75] Three highly conserved glycine residues, one of which is important for catalytic activity, are crucial for the proper folding of the TEN domain.[75] In humans and yeast, TEN is referred to as RNA interaction domain 1 (RID1)[76] and N-GQ,[74] respectively (Fig. 2B).

A linker region physically and functionally separates the RID-1 or N-GQ domain from the rest of the protein, composed of the remaining portion of the N terminus, the RT region and the C terminus (Fig. 2B).[74,76,77] This region of TERT is the most diverse in terms of sequence and ranges in size from ~20 amino acids in *Encephalitozoan cuniculi*[78] to ~500 amino acids in several *Plasmodium* species.[79] Notably, among vertebrates, chicken TERT (chTERT) possesses the longest flexible linker at 298 amino acids.[80]

The telomerase-specific (T) motif is a region of high homology among all TERTs,[65] with the exception of *G. lamblia* and *C. elegans* TERTs, in which this motif is missing.[51] Also, the CP motif is a region of homology in TERTs from ciliated protozoa. It contains conserved cysteine and leucine residues that are weakly conserved in TERTs of other organisms.[81]

The TERT gene of higher eukaryotes contains a putative mitochondrial leader sequence at its N terminus, which is absent from yeast and ciliate TERTs. Active telomerase is detected in mitochondrial extracts from human cells,[82] which is surprising since mitochondrial DNA is circular and lacks telomeric structures.[83] Oxidative stress induces tyrosine phosphorylation of hTERT by the Src kinase family,[84] triggering nuclear export of hTERT and amplifying cellular sensitivity of cells to reactive oxygen species (ROS)-induced mitochondrial DNA (mtDNA) damage and apoptosis.[82,84]Abolishing mitochondrial targeting through mutational disruption of the leader sequence leads to a decrease in ROS-induced mtDNA damage and subsequently protects cells from apoptosis.[83] Therefore nuclear hTERT is thought to function as an endogenous inhibitor of the mitochondrial pathway of apoptosis,[85] independent of its role in telomere elongation. In the mitochondria, telomerase might have a "pruning role", by targeting cells with damaged mtDNA to apoptose.[83]

C Terminus of TERT

The C terminus of TERT exhibits a low level of conservation among yeasts, ciliates and vertebrates[70] (Fig. 2B). Similar although not identical changes to the C termini of yeast, ciliate and human TERTs elicit different phenotypes. Sequence divergence could underlie differences in the biological function of telomerase in higher versus lower eukaryotes as described below. In yeast, Est2p C-terminal deletions exhibit significantly reduced growth and shortened telomeres[70] however, such deletions do not compromise cell survival.[73] Conversely, modification or truncation of part or all of the C terminus of tTERT and hTERT severely impacts telomerase activity in vitro.[86-89] In addition, such alterations compromise telomere length maintenance and lifespan of human cells.[86,89] These results indicate that the C terminus of ciliate and vertebrate TERT may encode functions that are not conserved in other eukaryotes.[89]

Alignment of the TERT C terminus among vertebrates, including mice and humans, shows two blocks of strong conservation throughout the length of the C-terminal domain (C1, amino acids 985-1083; C2, amino acids 1083-1132 of hTERT).[90] Sequences within C1 are responsible for regulating TERT protein accumulation because exchange of these sequences between mouse TERT (mTERT) and hTERT reverses their relative protein levels. Furthermore C1 sequences of hTERT are absolutely essential for immortalization of human cells because a chimeric hTERT containing a substitution of the C1 domain with the mTERT C1 region cannot extend the proliferative lifespan of human cells, nor maintain these cells' telomere lengths.[90] These results indicate that certain hTERT C-terminal sequences are required for immortalization and telomere maintenance in vivo.[90]

An alternative nomenclature for C-terminal domains of TERT based on mutational analysis reveals four regions critical for enzymatic activity, denoted E-I to E-IV.[89] They are separated by spacer regions in which substitutions are less deleterious to telomerase function. E-I and E-II (~hTERT residues 973-1052) are binding sites for CRM1 and 14-3-3 proteins respectively in vitro. These regions regulate the intracellular localization of telomerase.[91] EI and EII, as well as the remaining portion of the C terminus in humans, are also determinants of processivity.[92] Many hTERT C-terminal mutants exhibit significant defects in processivity and overall DNA

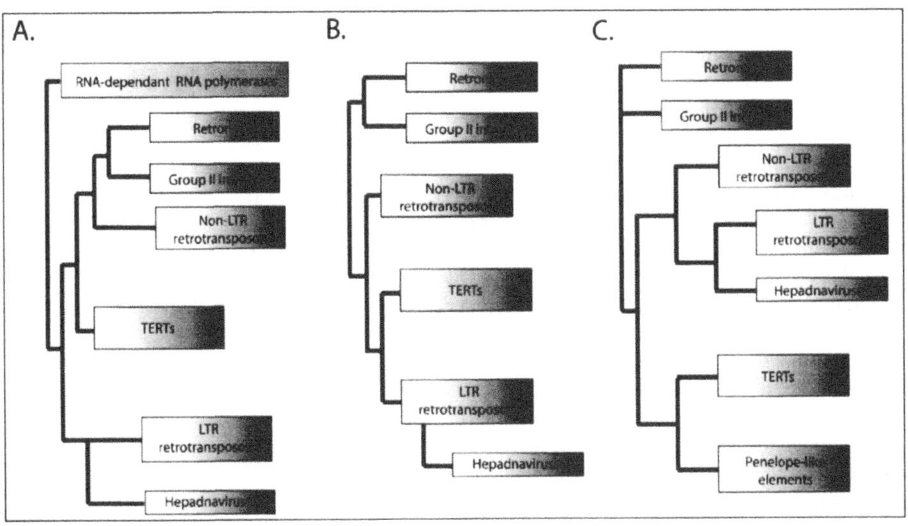

Figure 3. Phylogenetic relationships of retroelements. (Adapted from Eickbush 1997; Arkhipova et al 2003). A) Tree rooted with RNA-dependent RNA polymerases. B) Tree rooted with bacterial retrons. C) Addition of the new Penelope-like elements class of retroelements. Tree is rooted with bacterial retrons.

synthesis.[92] Similarly in yeast, alterations of the C terminus also reduce DNA synthesis levels[70] but do not compromise cell viability.[73]

As shown above, the C terminus of TERT is clearly important for several functions in vivo. Interestingly, several roundworm members, such as *C. elegans,* appear to lack this structure completely.[51] Loss of the C terminus is hypothesized to be compensated by the acquisition or optimization of other domains.[21]

Phylogenetic Analyses of Reverse Transcriptases

In addition to telomerase, RTs are encoded by a wide variety of elements which together form the large group of RT encoding elements or retroelements. Molecular phylogenetic analyses of retroelements based on the seven conserved RT motifs indicate that TERT is not just an RT structurally but also evolutionarily. Assuming that the current world was preceded by an RNA world, previous phylogenetic studies divided retroelements into two groups using RdRPs to root the phylogenetic tree; one group contained retroviruses and LTR-retrotransposons and the other contained non-LTR retrotransposons, group II introns and retrons. Sequence comparisons placed telomerase in the second group with group II introns and non-LTR retrotransposons (Fig. 3A).[93]

Interestingly, the catalytic subunit of telomerase is functionally analogous to non-LTR retrotransposons and group II introns. Non-LTR retroelements, also known as long interspersed nuclear elements (LINE)-like elements are mobile genetic elements that are inserted in the genome by reverse transcription of an RNA intermediate. Non-LTR elements use an encoded endonuclease to create a nick at a target site, thereby generating a 3'-OH group to prime a reverse transcription reaction, a mechanism known as target-primed reverse transcription (TPRT).[94] During catalysis, TERT utilizes the 3'-OH group at the end of chromosomes to prime the addition of telomeric repeats. Surprisingly, it has been recently shown that LINE-1 elements, a human-specific family of non-LTR retrotransposons, can also utilize this 3'-OH end to retrotranspose to telomeres after inactivation of their endonuclease domain. However, this recognition of telomeres by nuclease-deficient LINE-1 only occurs in cells defective in both telomere capping and nonhomologous end joining.[95] Group II introns also use a variation of the TPRT mechanism.[96] These retroelements are large catalytic RNAs found in the organelles of many organisms such as bacteria, plants, fungi, yeasts and algae and in the genome of bacteria. These introns must be removed for host gene expression; they are able to catalyze their own excision from their precursor premRNA. After splicing, the excised intron RNA lariat remains associated with an intron-encoded protein, forming an RNP able to transfer the RNA lariat site-specifically into an intronless allele. This process is termed retrohoming or intron homing.[96] The catalytic activity of the RT requires stable interaction between the RNA lariat and the intron-encoded protein.[97] This interaction is similar to the association formed by TERT and TR in telomerase. The relationship between telomerase and non-LTR retrotransposons is further supported by the telomeric addition mechanism in insects of the order Diptera, such as *Drosophila,* where non-LTR retrotransposons are used as an alternative to telomerase. In *Drosophila,* instead of reverse transcribing only a short segment of an RNA template, an RT synthesizes two entire retrotransposons, TART and HeT-A, at the ends of chromosomes.[28]

Although it seems apparent that telomerase shares phylogenetic traits with other retroelements, the precise relationship between them is not obvious. Retroelements possess diverse origins and share little or no homology, with the exception of the RT region that is common to all of them.[98] Thus, phylogenetic analyses are often restricted to RT sequence comparisons, which also reveal low levels of identity and hence contain small numbers of aligned amino acids. In addition, there are two different approaches to root the phylogenetic tree of retroelements, depending on which element is chosen as the ancestor of RTs.[50,93] The two approaches imply that each retroelement evolved in different orders. Instead of using RdRPs to root the tree, an alternative way has been proposed. If the assumption is made that prokaryotes evolved into more complex eukaryotes, then prokaryotic and organellar retroelements can be employed to root the tree (Fig. 3B).[50] This rooting implies that non-LTR retrotransposons came first and that telomerase and LTR-retroelements diverged from this branch. The tree rooted with RdRPs proposes an earlier origin for telomerase which precedes non-LTR retrotransposons and retrons.

Recently, homology between telomerase and the RT encoded by the newly described Penelope-like element (PLE) class has been reported, highlighting a potential evolutionary link between those two classes of retroelements. RT domains have been identified in the *Penelope* transposon of *Drosophila virilis*, the first identified member of the PLE class.[99] However, analyses revealed only a weak sequence homology between Penelope and other RTs, suggesting that this element could belong to any of the previously described classes of retroelements.[99] Penelope shares structural traits with LTR and non-LTR retrotransposons, introns II and retrons. Screening of diverse organisms such as crustaceans, fish and amphibians has led to the detection of PLEs.[100,101] Phylogenetic analysis of these elements based on the seven RT motifs places PLEs and telomerase together in a third branch, distinct from non-LTR and LTR retrotransposons (Fig. 3C).[102,103] Interestingly, terminal PLEs have been identified in the Athena group of PLEs, a distinct group of PLEs present in the asexual invertebrate bdelloid rotifers. These terminal PLEs are endonuclease-deficient and can be recruited to telomeres. It has been proposed that they perform essential functions in telomere maintenance, perhaps by supporting the telomerase-based system.[104] These recent findings add new facets to the debate concerning the evolution and origin of telomerase.

All phylogenetic analyses support an ancient origin for telomerase and emphasize the relationship between telomerase and other retroelements. However, it is not clear if telomerase is the ancestor of other retroelements or if it was derived from a more ancient family of retroelements. The first scenario implies that telomerase is an ancient RNA-entity that later acquired a protein component. In the second situation, telomerase could be derived from a RT that later associated with an RNA component to acquire a specific activity to maintain telomeres.

Structural Comparison of TERT to HIV-1 RT

The crystal structure of the entire TERT molecule has not been resolved. However, modeling of TERT's structure based on that of HIV-1 RT has been predictive of TERT domain function. The three-dimensional (3-D) structure of HIV-RT is typically described as a right hand (Fig. 2A). The core polymerase is composed of catalytically essential "palm", "fingers" and "thumb" subdomains.[105] Motif C is located at the catalytic center within the "palm", while motif E is located near the interface between the "palm" and "thumb". The C terminus of RT, comprised of a bundle of three alpha-helices, constitutes the "thumb".[106,107] According to this model, TERT would have longer fingers due to the presence of the unique IFD in this domain. In the 3-D model for HIV1-RT, the fingers are predicted to be in close proximity to the nucleotide triphosphates, the 5' end of the template and the 3' end of the DNA substrate. In TERT, however, the longer fingers may make additional contacts with the RNA-DNA duplex to enhance stability.[72] In fact, it is possible that the tip of the fingers may extend sufficiently to make contact with the putative thumb domain of TERT, thus fully encircling the substrate. This type of full closure may augment enzyme-substrate stability.[72] While the basic RT catalytic mechanism is likely conserved in telomerase, there are several features of telomerase that distinguish it from other RTs, including the stable association of the RNA component and its ability to catalyze the reiterative addition of a short DNA sequence. The domains that are unique to telomerase, including the IFD and N and C termini, contribute to these differences.[72,88,92,108]

Telomerase RNA Component

RTs are best known as viral proteins that copy an RNA genome into DNA.[109] However, unlike viral RTs, telomerase contains an intrinsic RNA molecule that serves as the template for the addition of telomeric sequences.[60,61] The RNA subunit contains sequences complementary to its cognate G-rich telomeric repeat to catalyze the synthesis of telomeric repeats at the 3' end of single-stranded DNA.[110] Curiously, repeat synthesis in some organisms, such as the ciliate *Paramecium tetraurelia*, is not strictly determined by the template, which contains nucleotides consistent with the synthesis of G_4T_2 repeats.[111,112] This organism's telomeres consist of variable repeats, composed primarily of a random mixture of G_4T_2 and G_3T_3.[111,113,114] This variability is due to the misincorporation of dT residues. It is postulated that a posttranscriptional modification of one of the template nucleotides promotes pairing to dTTP rather than dGTP, or that there may be a greater binding efficiency for dTTP rather than the other dNTPs (excluding the cognate dGTP) at that position in the polymerization cycle.[112] Template infidelity has also been observed in *S. cerevisiae*, due primarily to premature dissociation of the template.[23]

Nonetheless, the RNA component modulates enzyme activity and acts as a scaffold for the assembly of the telomerase RNP.[109,115] Telomerase RNAs from different organisms range in size considerably, depending on the species and have been identified in twenty-four ciliate species (148-209 nts in length), forty vertebrates (including mouse, 397 nts and human, 451 nts) and yeasts, for example *Kluyveromyces lactis* (~1300 nts) and *Saccharomyces cerevisiae* (~1160 nts).[60,115-125] TRs from diverse species share little sequence homology, but do appear to share common secondary structures. Phylogenetic comparative analysis has proven to be the most powerful approach for inferring higher-order RNA structures.[115,117,118,123,124,126,127] To construct a secondary structure model for telomerase RNA based on phylogenetic analysis, the aligned sequences are evaluated for nucleotide covariation between species. For example, an A/U base pair in the RNA of one species may be a G/C, C/G or U/A base pair in the same RNA of other species. Covariation of a nucleotide pair that maintains base pairing among species is considered supportive evidence of a paired structural element (P), such as a helix. In some cases, unusual base pairs, such as G/U, G/A and C/A are observed; these base pairs also maintain a stable helix because they allow possible hydrogen-bonding interactions.[123,128] Mutational analyses can also support the presence of putative paired structural elements. Disrupting important base pairings on either side of a helix results in alteration of telomerase function, while mutations that restore base pairings, but do not necessarily re-establish the original sequence, reconstitute telomerase function, thus reinforcing the existence of such base pairing.[124,129,130]

Vertebrate Telomerase RNA

Vertebrate TRs contain structural elements that are universally conserved, including the template, the conserved region CR4-CR5 domain, the pseudoknot domain, the H/ACA box and the CR7 domain (Fig. 4A). The template region is a region of high homology; it has been shown that the consensus sequence 3'-UCCCAAUC-5' is universally conserved among all vertebrates.[123] The template region can be subdivided in two regions: the elongation domain which determines the telomeric repeat GGTTAG and the alignment domain that is implicated in the positioning of the DNA substrate during elongation.[123] While the elongation domain CCAAUC is invariant in vertebrates, the alignment domain is less conserved and can vary from 2 nucleotides (nt) for rodents to 5 nt for humans. Because some vertebrates possess a template region of 8 nt, it is postulated that 8 nt is the minimum length required for telomerase function.[131]

The pseudoknot domain (Fig. 4A) contains both a pseudoknot structural element and the template region and is established by paired elements P2a.1, P2a, P2b and P3, which form helices. The junction J2b/3 is conserved in sequence and in length,[123] and may be required for maintenance of a stable pseudoknot conformation. Nucleotides upstream of the template base pair with nucleotides downstream of the pseudoknot to form helix P1 in several vertebrate TRs, including human. However, rodent TRs lack P1 because the 5' end begins 2 nt upstream of the template.[123] The CR4-CR5 domain (Fig. 4A) is found downstream of the pseudoknot and consists of a stem-loop structure. The loop contains a paired region (P6.1) within itself.[129] The domain also contains the J6 loop.[132] P4, P7a, P7b, Box H and Box ACA (Fig. 4A) form the Box H/ACA domain, a conserved structure similar to that found in small nucleolar RNAs (snoRNAs).[123,133] The H/ACA box is responsible for RNA accumulation and nucleolar targeting.[133-135] Finally, the highly conserved CR7 domain contains helices P8a and P8b and loop L8.[123]

Ciliate Telomerase RNA

Phylogenetic comparisons, enzymatic and chemical probing and mutational analyses suggest that the secondary structure of telomerase RNAs from various divergent groups of ciliates share

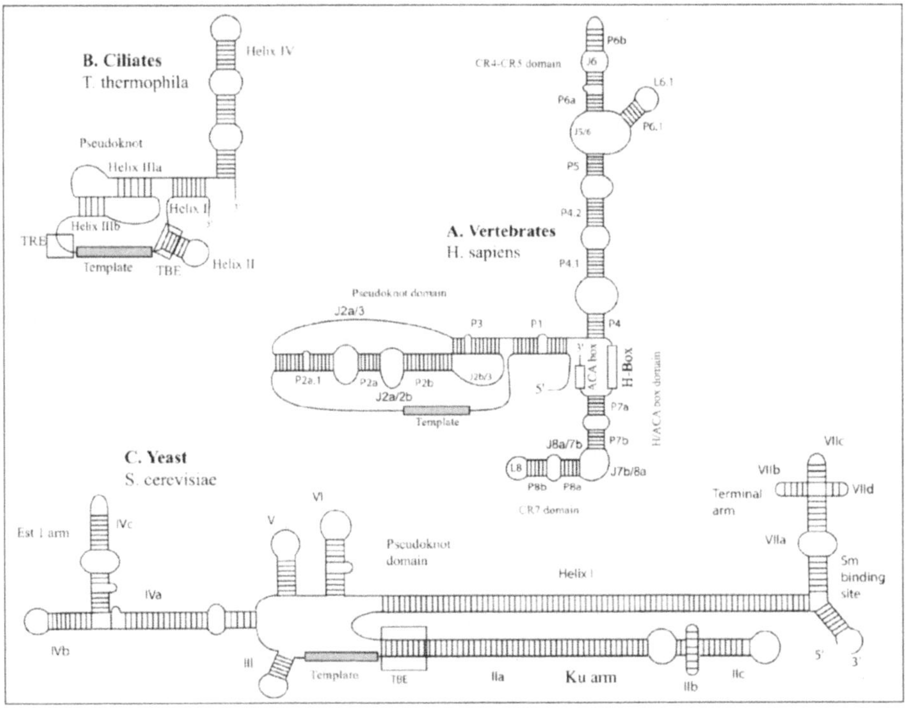

Figure 4. Examples of predicted secondary structures for vertebrate, ciliate and yeast telomerase RNA. A) *H. sapiens* telomerase RNA is composed of a pseudoknot-template domain, a CR4-CR5 domain, a H/ACA box domain and a CR7 domain. B) *T. thermophila* telomerase RNA contains a pseudoknot-template domain and helix IV. In addition, the template recognition element (TRE) and the template boundary element (TBE) are shown. C) *S. cerevisiae* RNA contains a template and putative pseudoknot domain, which are considered the central core of the molecule. Three arms, required for binding different cellular proteins, extend from the core. Stem loops V and VI have been proposed to exist in the pseudoknot domain. (Adapted from Autexier and Lue 2006; Theimer and Feigon 2006; Dandjinou et al 2004.)

similarities with that of vertebrate TRs (Fig. 4B).[117,118,121,123,136-140] The core structure of ciliate TR comprises a template region followed by a pseudoknot domain, named Helix III. Nucleotides 5' to the template base pair with a region downstream of the pseudoknot to form Helix I, presumably an analog of the P1 helix in vertebrate TRs.[117,121,140,141] Finally, the structure of stem loop IV, which was determined by NMR spectroscopy, forms a kinked structure which ends in a loop.[142,143] Stem loop IV of ciliate TR resembles the CR4-CR5 domain in vertebrate TR.[144] Box H/ACA and the CR7 domain appear to be vertebrate TR-specific elements because the shorter ciliate TRs do not possess these structures.

Yeast Telomerase RNA

Determining the structure of yeast telomerase RNA (TLC1) has been a daunting task due to its large size, however four labs simultaneously reported its secondary structure using a combination of phylogenetic comparative analysis, computer prediction and RNase H probing to confirm single-stranded regions.[115,124,126,130] A model for the secondary structure of TLC1 (Fig. 4C) consists of a central hub with three major arms emanating from the core, each associating with a different protein in vivo. A putative pseudoknot found adjacent to the template[125] is required for association with Est2p.[124,130] Alternate models have been suggested for the pseudoknot domain.[115,124,130]

One model postulates the existence of two hairpins, V and VII in this domain (Fig. 4C).[115] The template boundary element (TBE)[145,146] and a stem-loop mediating binding to the Ku70/80 heterodimer[147,148] are readily identifiable upstream of the template.[126] One bulgy stem-loop structure downstream of the template interacts with Est1p.[115,124,126,149] A long range base pairing, termed helix 1, brings the template in close proximity to the Est2p binding site and is structurally analogous to the ciliate helix I and vertebrate P1 structures.[115,124,126] However in TLC1, this region is unusually long (86 bp); a structure at the distal most 3' end of this arm binds Sm proteins[126] and is responsible for efficient biogenesis of TLC1.[150]

Viral Telomerase RNA

Interestingly, the etiological agent of Marek's disease, Marek's disease alphaherpesvirus (MDV), encodes a viral telomerase RNA (vTR) gene which is 88% homologous to the chicken TR (cTR) gene.[151] The disease is characterized by the vTR-dependent development of T-cell lymphomas in chickens and turkeys.[151,152] vTR contains several point mutations and deletions compared with cTR, mainly within the pseudoknot domain.[151,153] Although vTR reconstitutes a more active telomerase complex than cTR when expressed with chTERT in vitro, it is unknown if complex formation is required for the tumor-promoting function of vTR.[153,152]

Interaction between the Protein and RNA Components of Telomerase

Because telomerase requires both TERT and TR for synthesis of telomeric repeats, assembly of these two components is essential for enzymatic function. Mutational analyses suggest that the telomerase RNA binding domain of tTERT encompasses motif CP, which is conserved among ciliate TERTs and motif T, which is present in most TERTs (Fig. 2B).[88,154,155] Amino acid deletions within either motif inhibit telomerase RNA association.[88,154,155] In tTR, the 5' template boundary element (TBE) (Fig. 4B) functions as the high affinity TERT binding structure. TR variants with substitutions in the TBE interact less efficiently with tTERT.[88] Two additional regions of tTR appear to contribute to lower affinity interactions with tTERT. One of these tTR regions corresponds to the template recognition element (TRE) 3' of the template (Fig. 4B), proposed to position the template 3' end in the active site.[156,157] Competitive binding experiments show that tTR variants with TBE or TRE substitutions do not inhibit tTR-tTERT interactions as effectively as wild-type TR.[156] The second region, the distal loop of stem IV, is required for efficient use of the entire template and a high level of activity.[156] Crosslinking and co-immunoprecipitation experiments demonstrate that this loop makes direct contact with tTERT.[144,157] In addition, substitution of the entire stem IV loop strongly reduces binding to the N terminus of TERT.[156] These results imply that the N terminus of tTERT could recognize a structure formed by the distal loop of stem IV and TRE together, or the contact regions could be distributed over residues in both of these motifs.[156] These results suggest a model for tTERT in which N terminus-TR interaction serves to position stem IV distal loop and TRE residues into the catalytic site. Binding between the TBE and the N terminus should constitute a high affinity interaction because the TBE must prevent residues 5' of the template from entering the active site.[156] The lower affinity interactions may be required for structural rearrangement during translocation.[156]

Because the CP motif is not highly conserved in TERTs of nonciliates, it was originally thought that the RNA binding domains of TERTs might vary from organism to organism.[88,158] However, in yeast, the QFP, T and CP motifs are required for efficient telomerase RNP formation and telomere maintenance in vivo,[78] while the GQ motif contributes minimally to binding.[73] The T motif is absent in *G. lamblia* and *C. elegans*.[51] The absence of this motif in *C. elegans* could indicate an impaired ability to recognize its cognate TR or that a different TERT region mediates binding to TR. Notably, the TR from *C. elegans* remains unidentified to date.[51]

In humans, TERT interacts with TR via two regions. RID1, composed of extreme N-terminal residues and the GQ motif, is a low affinity interaction domain. RID2, which encompasses the CP, QFP and T motifs, is a high affinity interaction domain.[76,77] Human TERT containing substitutions within motifs CP and QFP of the N terminus has a reduced ability to bind human TR (hTR) and fails to rescue telomerase-negative primary cells from crisis.[159] RID2 interacts with the CR4-CR5 domain of hTR.[76,129,160,161] Requirement of the CR4-CR5 domain has been demonstrated by electrophoretic mobility shift assay (EMSA), where hTR lacking the CR4-CR5 domain is unable to inhibit the interaction of wild-type hTR with hTERT in vitro.[158] Specifically, the P6.1 helix and the J6 internal loop in CR4-CR5 are directly involved in binding RID2; a mouse TR (mTR) containing mutations in P6.1 fails to bind mTERT in vitro and deletion of the J6 loop in hTR also abolishes hTERT association.[88,129] The J6 internal loop may introduce a twist in the RNA structure that positions the entire CR4-CR5 domain onto the catalytic site.[132] The additional hTERT-hTR interaction mediated by RID1 involves binding to the hTR pseudoknot-template domain (nts 1-209 of hTR).[77,88]

Interaction of TERT with Telomeric DNA

The mechanism whereby telomerase recognizes the ends of chromosomes in vivo is not clearly understood. In vitro, telomerase efficiently elongates single-stranded DNA primers resembling the natural G-rich 3' overhangs of chromosomal termini.[162] Evidence suggests that telomerase interacts with oligonucleotide primers in a bipartite manner: with the 3'-end of the primer positioned at the template domain and with nucleotides upstream of the 3' end of the primer. The latter interaction is mediated by a TERT region referred to as the "anchor site".[71,163-171] These interaction sites are thought to act in concert to facilitate polymerization. Following base-pairing between the DNA primer 3' terminus and the RNA template, nucleotides are reverse transcribed onto the 3' end of the DNA in accordance with the RNA template. Once the 5' end of the template is reached, the new 3' primer end is repositioned at the 3' end of the template, simultaneously threading the primer 5' end further into the anchor site.[163-168] The anchor site mediates processive elongation by preventing dissociation of the enzyme during translocation (Fig. 1).

Longer oligonucleotides sustain more processive elongation by *Tetrahymena* telomerase, while shorter oligonucleotides, presumed to lack sequences required for a DNA-tTERT interaction, are only extended by one repeat.[165,166] Also, the sequence of the DNA primer 5' to the RNA-DNA hybrid contributes significantly to the overall binding affinity for the primer.[165,167,169,172] These results support the existence of an anchor site. Furthermore, the RNA-DNA duplex contributes weakly to stable binding, perhaps to allow for a transient interaction required for translocation of the enzyme during repeat DNA synthesis.[169] Indeed, hybridization of the telomerase RNA to the 3' end of an oligonucleotide is not required for efficient initiation of polymerization since oligonucleotides with nontelomeric sequences at their 3' ends, which lack complementarity to the template, can be elongated.[164]

The N-GQ domain of Est2p binds DNA in a nonsequence-specific manner, with a slight preference for single-stranded DNA in vitro.[74] Similarly, altered primer utilization and processivity has been demonstrated for human telomerase variants with mutations in RID1,[87,170,173] suggesting that this region may be a constituent of the hypothesized anchor site.[171] These mutants may be impaired in their ability to rebind the DNA substrate at the 3'-region of the RNA template following one round of synthesis, thus preventing the synthesis of additional repeats (Fig. 1).[171]

While the GQ motif is conserved in evolution, the extreme N-region is not highly conserved, raising the possibility that different anchor sites may have evolved slightly different recognition properties for species-specific functions.[78] For example, ciliate telomerases recognize nontelomeric DNA during macronuclear development,[174] characterized by genome fragmentation and amplification.[175] Extension of new chromosome ends entails recognition of nontelomeric sequences at break sites.[164,174,176] Interactions with nontelomeric DNA by ciliate TERT may be accomplished with the aid of the postulated anchor site.

In humans, the C terminus of TERT may also be required for anchor site interactions with DNA, perhaps by cooperative interaction with or regulation of the RID1 domain.[76,170,173] Furthermore, the postulated anchor site in Est2p, the N-GQ domain, and the C terminus can physically interact and regulate affinity for DNA substrates.[70,74,177] Interestingly yeast does not require the Est2p C terminus for cell survival,[73] nor is it highly processive.[178] Perhaps the C terminus, which normally modulates RAP by interacting with TEN/RID1 in ciliates and humans, has a different function in yeast. In yeast, the C terminus may not necessarily interact with N-GQ, leading to an inability to synthesize long telomeric repeats. Deletion of this domain may be inconsequential to cell survival because yeast does not require a processive telomerase for survival.

Nucleolytic Activity

It is postulated that cleavage of chromosome ends by telomerase acts as a proofreading function.[162,166,179] Telomerase may encounter nontelomeric sequences in vivo, for example in ciliates during macronuclear development.[174] Telomerase uses two pathways for processing nontelomeric 3' ends. One pathway results in the addition of telomeric repeats directly onto a nontelomeric 3' end.[162] Alternatively, 3' end nucleotides can be removed prior to elongation.[162] This cleavage reaction has been observed with chimeric oligonucleotides. These chimeric primers consist of 3'nontelomeric DNA which is noncomplementary to the RNA template, flanked by 5' telomeric DNA. In ciliates and humans, nontelomeric DNA is cleaved before extension by telomerase.[162,166,180] Furthermore, replacement of the 3' terminal nucleotide with a chain terminator in a chimeric primer does not affect the formation of the elongation products. This result indicates that the 3' terminal nontelomeric sequence, including the chain terminator, is removed before elongation.[180]

Cleavage of nontelomeric DNA from the 3' end of a primer occurs preferentially at the junction between telomeric and nontelomeric DNA.[181] If telomerase were to initiate polymerization on 3' nontelomeric DNA, these nontelomeric nucleotides would become sealed within the telomere.[162,166,179] However, incubation of telomerase with a cognate telomeric primer can also result in the formation of cleavage products.[166,182,183] For example, *Euplotes* telomerase synthesizes perfect 5'-TTTTGGGG-3' repeats in a mechanism involving successive rounds of primer elongation, translocation and realignment of the newly synthesized 3' DNA terminus on the RNA template (Fig. 1).[29,116] The fidelity of these three events is essential. If the primer fails to translocate at the proper location and nucleotides in the telomerase RNA beyond the 5' end of the template domain are reverse transcribed into DNA, the repeat sequence is altered.[29,116] Alteration of the telomere sequence leads to changes in telomere length, senescence or death.[61,184-186] Therefore, a DNA substrate that aligns at the 3' end of the template will not be cleaved while one that aligns at or beyond the 5' end of the template is cleaved, possibly to ensure that only nucleotides within the RNA template are utilized during telomere synthesis. This mechanism is also consistent with a proofreading function.[162,183]

Both DNA products generated by cleavage can be extended by the polymerization activity of yeast telomerase (Fig. 5).[181] Though *Tetrahymena* telomerase exists as a monomer and is capable of nuclease activity, it is difficult to rationalize the extension of both cleavage products if the enzyme is monomeric. Nonetheless, it has been hypothesized that two active sites could each have distinct roles: one in endonucleolytic cleavage, the other in elongation (Fig. 5A).[181] Alternatively, it has been postulated that efficient extension of the two cleavage products might require telomerase multimerization.[166,181,187] Multimerization could lead to a telomerase with subunits that are each able to perform both elongation and endonucleolytic cleavage (Fig. 5B).

Multimerization of Telomerase Components

One of the major structural questions regarding the telomerase holoenzyme is the oligomerization state of the complex. Both TERT and TR are capable of forming dimeric/multimeric complexes. Biochemical data demonstrate that the human,[13,188] yeast[189] and *Euplotes*[190] telomerase complexes exist in dimeric forms. In fact, telomerase from *E. crassus* can exist in both dimeric and higher order forms.[191] Interestingly, recombinant telomerase from *T. thermophila* functions as a monomer, suggesting that the inherent biochemical activity of telomerase does not necessarily require a dimeric complex.[187]

Early evidence for the multimerization of telomerase extends from studies using mutant telomerase yeast TRs.[189] Activity can be restored to a telomerase RNP containing a mutant TR template by co-expression of wild-type TR. The telomeric DNA synthesized in vivo in the presence of both

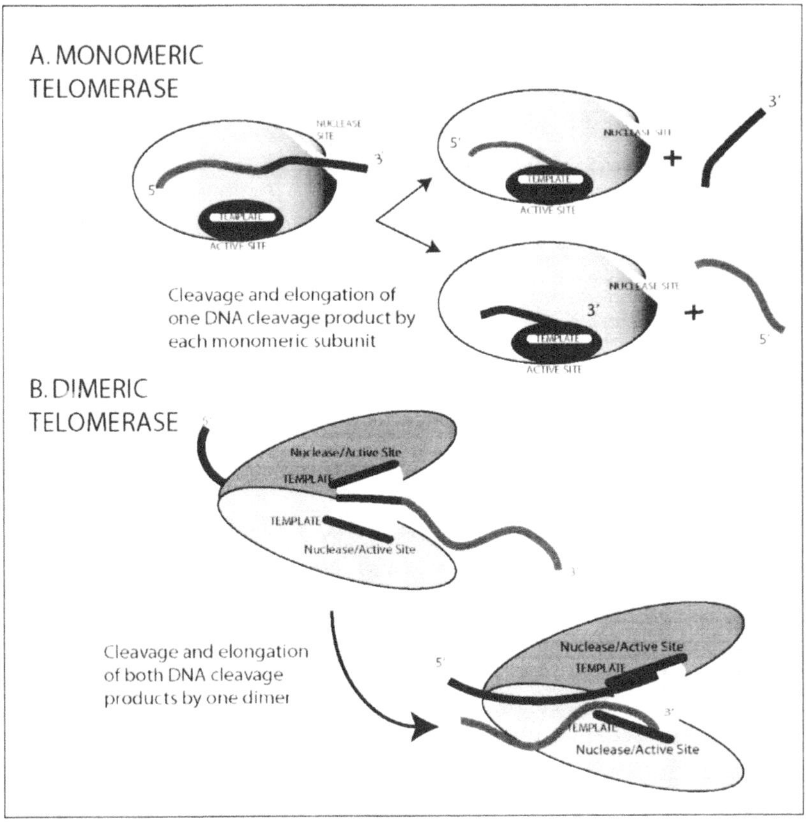

Figure 5. Cleavage, followed by elongation of DNA by monomeric and dimeric telomerase. Two models illustrate telomerase-mediated cleavage, followed by elongation of the two resulting cleavage products. A) In the first model, telomerase is shown to possess two distinctive active sites for nuclease and elongation activity. Following cleavage of the primer, the active site captures either fragment for elongation. B) Telomerase is shown in dimeric form. Each subunit contains a bifunctional active site capable of both cleavage and elongation. Following cleavage of the primer, the resulting cleavage products can be elongated by separate active sites on each individual subunit. (Adapted from Niu et al 2000.)

enzymes consists of tracts of closely interspersed mutant and wild-type repeats.[189] The clustering of these repeats is consistent with synthesis by a heterodimeric telomerase containing two active sites and two TRs that can functionally interact.[189] These results have also been observed with human telomerase assembled with mutant and wild-type templates.[13]

Co-immunoprecipitation experiments reveal that hTERT and *Euplotes crassus* telomerase (ecTERT) form multimers through N-terminal and C-terminal protein-protein contacts that are independent of telomerase RNA.[77,191-193] Interactions were found to occur between hTERT N- and C-terminal fragments (N-C) in vivo, but not between two N-terminal (N-N) or two C-terminal fragments (C-C).[193] However, in *E. crassus*, N-N, C-C and N-C interactions occur, indicating that ecTERT is capable of head-to-tail, head-to-head and tail-to-tail oligomers in vitro.[191]

Despite the observation that telomerase can multimerize without TR, data indicate that TR is also capable of dimerization. Full-length hTR and a truncated hTR consisting of the template, pseudoknot helices P2 and P3, as well as helix P1, form a dimer when examined by nondenaturing gel electrophoresis.[194] Mutations in helices P1 and P3 cause marked reductions in dimerization.[76,194]

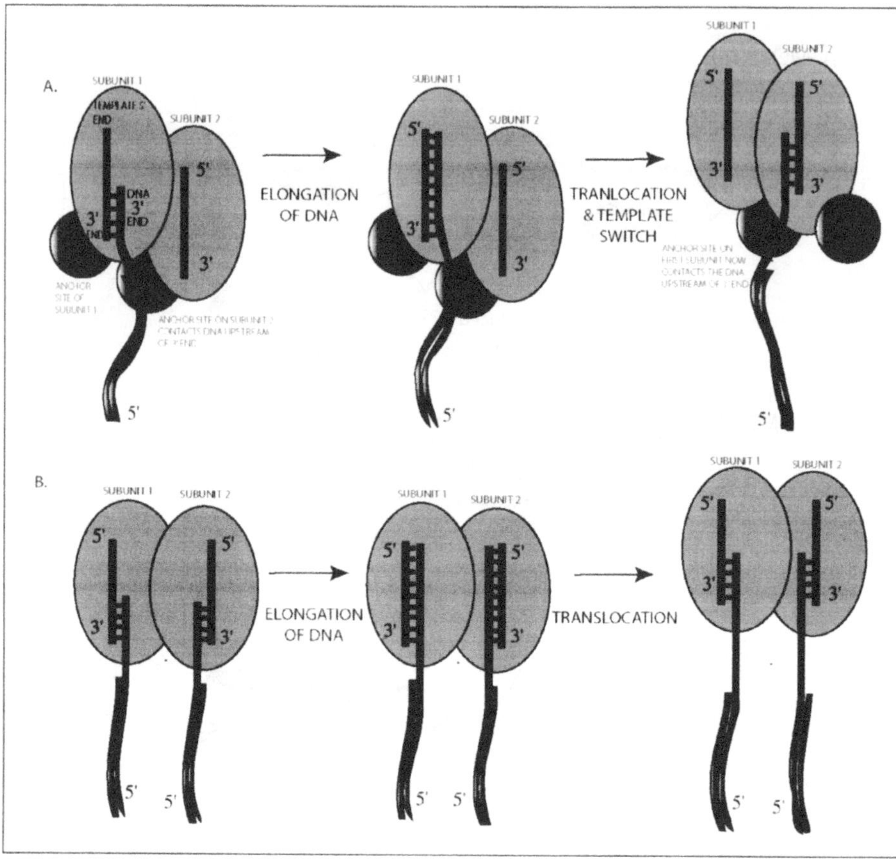

Figure 6. Cooperative elongation of DNA substrates by a dimeric telomerase complex. A) The template switching model postulates that the dimeric telomerase complex alternates between two templates for elongation of the DNA. Once the 5' boundary of one template has been reached, the 3' end of the DNA is shifted or "switched" over to the alternate template. B) In the parallel extension model, two DNA 3' ends are extended cooperatively within one dimeric telomerase complex. (Adapted from Kelleher et al 2002; Wenz et al 2001.)

Furthermore, a chimeric human-*Tetrahymena* TR (htTR) containing the tTR pseudoknot but lacking the conserved region required for the formation of the P3 helix is able to dimerize in vitro, despite the observation that this helix is a determinant of TR dimerization.[76,194,195] This finding suggests that another region in the chimera can mediate dimerization. Indeed, mutational experiments determined that the hTR J7b/8a region of the htTR chimera, found at the junction of the H/ACA box and CR7 domains, is also involved in dimerization.[195,196] *S. cerevisiae* TLC1 also dimerizes in vitro. This dimerization is dependent on a 6-base palindromic sequence similar to that present in HIV-1 and other retroviruses,[197] suggesting that RNA dimerization is shared by both viral and cellular RTs. The importance of dimerization in HIV-1 function and telomere shortening observed in yeast cells expressing dimerization-defective TLC1 alleles support the hypothesis that telomerase RNA dimerization is necessary for telomerase function.[198,197]

Presently, the mechanistic basis for cooperation between two telomerase subunits is uncertain. The extension of telomeric 3' ends during processive synthesis could be enhanced by oligomerization. Two models may account for this functional interaction.[13,189] A template switching model postulates that a dimeric telomerase, containing two active sites, extends one DNA substrate in a cooperative fashion. Substrate extension by one subunit until the template 5' end is encountered is followed by realignment of the substrate with the RNA template of the other subunit, allowing addition of the next repeat (Fig. 6A).[13] The template switching model supports the existence of an anchor site, where one TERT subunit is postulated to extend the 3' end of the substrate while the other subunit contacts the substrate at a site upstream of the 3' end from the previous elongation cycle, allowing for processive elongation.[13,168] However, it is postulated that if this anchor site were formed as a result of multimerization, for example of yeast and human telomerase, its molecular nature may have diverged from that in monomeric complexes, such as *Tetrahymena* telomerase.[199] Alternatively, the parallel synthesis model suggests that a dimeric telomerase, which contains two active sites, catalyzes the addition of repeats onto two substrates, such as two sister chromatids, simultaneously in vivo (Fig. 6B).[13,189]

Processivity

A unique feature of telomerase is its ability to processively synthesize long stretches of DNA on telomeric ends, despite the presence of a short RNA template.[20] This section focuses on the structural elements within TERT and TR that mediate telomerase processivity.

Nucleotide Addition Processivity

There appears to be a mechanistic conservation between TERT and prototypical RTs with regard to the determinants of nucleotide addition processivity (NAP). Substitutions within conserved RT motifs that are predicted to affect dNTP binding and primer positioning, namely motifs 1, 2 and E, dramatically reduce NAP.[21,70,200,201]

Analysis of the catalytic center of telomerase, motif C, has proven to be most interesting. In all retroviral RTs, the amino acid preceding the two catalytic aspartates in motif C is a tyrosine, while it is a leucine in tTERT. This amino acid is predicted to form several contacts with the template and primer to position it relative to the active site.[106,108] Substitution of the *Tetrahymena* TERT leucine with a tyrosine results in increased NAP, creating an enzyme which more closely resembles the highly processive classical RT.[108] This improvement in enzyme action suggests a lack of evolutionary advantage for a highly processive telomerase in vivo.

Substitutions in the C terminus can also compromise NAP. Both yeast and vertebrate telomerase C-terminal mutants exhibit NAP defects despite weak conservation in this region, suggesting the presence of a conserved functional motif between their C-terminal domains.[70,92,177]

Repeat Addition Processivity

Repeat addition processivity is a mechanism unique to telomerase and appears to be regulated by telomerase-specific structures not shared by conventional RTs. Several structural elements of telomerase modulate RAP, including the TERT IFD and RID1 (N-GQ) domains and several elements in TR, including the hTR P1 helix, the template alignment region and the pseudoknot domain.

While yeast telomerase is generally considered nonprocessive, elongation products beyond the first repeat can sometimes be observed in vitro.[178] Est2p variants containing mutations in the IFD generate only one telomeric repeat in vitro, indicating defects in RAP. This same mutant is also less active on primers that form short hybrids with the RNA template, suggesting that the IFD stabilizes enzyme-DNA interactions.[72] This IFD mutation may cause reduced protein-DNA interactions, i.e., an IFD mutant with a "loosened grip" may bind DNA too weakly to perform RAP.[72]

RAP may be partly regulated by telomerase-DNA interactions at the anchor site, minimally constituted by the RID1 region[74,76,87,170,171,173] Interestingly, RID1 deletion mutants can only synthesize one DNA repeat, suggesting that RID1 is a RAP determinant[76] Furthermore, altering hTR P1 sequences reduces processivity.[76] Notably, all vertebrate TRs described to date, except those of hamster, mouse and rat, contain a P1 helix.[123] An analogous structure, helix I, is also found in the TRs of ciliates.[117,121,140] Interestingly, human and *Tetrahymena* telomerases, which contain a P1 helix, or helix I, are more processive in vitro than mouse telomerase, which lacks a P1 helix.[22,27,202,203]

The insertion of an artifical P1 helix in mTR confers increased levels of activity compared to an mTR without P1.[203] These results support the notion that the P1 helix is a RAP determinant.

The primer alignment region of TR is also proposed to account for the low processivity of rodent telomerase.[120,141,202] This region is short in the rodent template. Alteration of two residues in this region, resulting in an increased number of base pairs between the primer and template, produces an enzyme with significantly increased processivity.[202] However, this increase in processivity with increased complementarity has a limit. It has been proposed that long RNA templates, with the potential for a high number of base pairings with the DNA,[145,172] leads to a decreased ability of telomerase to translocate.[202] For example, the yeast *Kluyveromyces lactis* has a 30-nucleotide template domain. In vitro, *K. lactis* telomerase catalyzes one round of repeat synthesis.[178] Increasing the number of complementary primer DNA-template RNA base pairs results in stalling of the enzyme, suggesting that the RNA-DNA hybrid is not sufficiently unpaired during elongation in vitro.[25] Formation of a long, presumably rigid RNA-DNA hybrid duplex during elongation is likely to interfere with the constrained movement of the template through the catalytic center. Such a duplex may also prevent the active site of the RNP from assuming the correct conformation necessary for polymerization. Hence, as the hybrid lengthens, steric restriction may become sufficient to prevent further polymerization.[25]

Dynamic rearrangements of the template RNA and/or DNA substrate within the catalytic active site are likely to affect RAP. It is postulated that distortion of the tTR pseudoknot, mediated by RNA template positioning in the active site, could create a conformational change in the RNP, leading to dissociation of the DNA product 3' end from the template.[156] Once the template is freed from hybridization, it can return to its default position. Pseudoknot refolding could facilitate repositioning of the product 3' end in the active site and formation of a new RNA-DNA hybrid.[156] In tTR, pseudoknot element stem III is important in conjunction with stem IV for RAP.[118,138,156] Comparably, hTR P3 helix mutants are defective in RAP as well.[76]

Finally, it is thought that oligomerization may contribute to processivity. End-to-end pairing of two DNA molecules by multimerized *E. crassus* telomerase has been observed by electron microscopy.[190] This observation is consistent with a model of processive repeat synthesis consisting of two cooperating telomerases that bind to and extend a single DNA substrate,[190] perhaps by TERT-TERT[77] or TR-TR[76] interactions.

TERT-TERT interactions are speculated to contribute to processivity.[77] The C termini of human[192,193] and *E. crassus*[191] TERTs are implicated in functional multimerization in vitro. Several hTERT C-terminal mutants exhibit RAP defects, suggesting that a determinant of human telomerase RAP resides in the C terminus. These defects in RAP could perhaps result from an inability of TERT to multimerize due to the absence of a functional C terminus, leading to improper alignment of the substrate with the template, defects in primer or dNTP binding or defects in translocation.[92]

Reductions in hTR dimerization also coincide with processivity defects. Impairment of the P1 helix or helices in the pseudoknot weakens dimerization and reduces processivity.[76] Although the mechanistic contribution of these structures to RAP remains unclear, it is postulated that hTR-hTR interactions could stabilize the dimer, permitting one TR to allosterically influence the function or conformation of the other TR.[194]

Conclusions

Telomerase is the most widespread mechanism used to maintain eukaryotic telomeres. The prevalence of telomerase in phylogenetically diverse organisms suggests that telomere maintenance by telomerase is an ancient mechanism. Evidence supports the existence of telomerase in early eukaryotes, however the exact origin of telomerase remains uncertain. Is telomerase derived from an ancient RT that later acquired an RNA component specific for the maintenance of eukaryotic telomeres or is it derived from an ancient RNA-based enzyme that was replaced by a more efficient retroelement? Additional information is required to determine how telomerase evolved to its current form. However it now seems clear that telomerase is an evolutionarily old enzyme whose origin can be traced back to other retroelements which were present during the RNA-DNA transition.

Telomerase activity appears to depend not only on the presence of conserved RT domains essential for catalytic activity, but also on a network of protein-nucleic acid interactions formed by the conserved secondary structures in TR, TERT's unique IFD, N- and C-terminal domains and its DNA substrate. These interactions orchestrate the proper positioning of the template and primer in the active site.[109] The most unique feature of telomerase is its ability to polymerize repetitive sequences despite the use of a relatively short template.[20] Multimerization is predicted to contribute to increased processivity, through, for example, the optimal use of an anchor site. Human and *Tetrahymena* telomerases are extremely processive in vitro while yeast telomerase is not, despite the observations that yeast telomerase can multimerize and *T. thermophila* telomerase functions as a monomer. Why is this scenario the case when multimerization is predicted to be a determinant of processivity?

Furthermore, what is the functional relevance of these characteristics in vivo? It is thought that the maintenance of physiologic telomere lengths is dependent upon the natural processivity of telomerase.[70] Yeast and mouse telomerase are both nonprocessive in vitro, therefore one would suspect that telomere maintenance would only require a minimal level of processive elongation in these organisms, preferentially while telomeres are still long. Indeed, yeast has a low telomere attrition rate (3-4 bp/cell division), however the number of nucleotides added to a telomere in a single cell cycle in yeast varies between a few to more than 100 nucleotides.[204,205] In addition, analysis of telomere attrition rates in mouse cells lacking telomerase suggests that the enzyme would need to add at least 50-100 nts of DNA per cell division in order to counter the loss incurred by incomplete replication.[15,20,206]

The characteristic differences between telomerase from diverse species and the discrepancies between the in vitro and in vivo observations highlighted above illustrate that many questions remain concerning the evolutionary divergence of telomerase, as well as the in vivo requirements for telomere maintenance. Nonetheless, the conservation of telomerase in these species indicates its importance in various cellular functions. Appreciation of the various functions of telomerase and how they relate to genomic instability is critical to our understanding of cancer. An increased comprehension of telomerase regulation is currently leading to the validation of methods for the early and accurate diagnosis of cancer and of novel anti-telomerase cancer therapeutics.[207]

Acknowledgements

We thank D.T. Marie-Egyptienne, J. Fakhoury and G.A.M. Nimmo for critical reading of the manuscript. Work in the laboratory of C. Autexier is funded by the Canadian Institutes of Health Research (CIHR). M.E. Brault is supported by a Cole Foundation Doctoral Award. Y. D'Souza is supported by a CIHR Doctoral Research Award. C. Autexier is a Chercheur-Boursier of Le Fond de la Recherche en Santé du Québec.

References

1. Makarov VL, Hirose Y, Langmore JP. Long G tails at both ends of human chromosomes suggest a C strand degradation mechanism for telomere shortening. Cell 1997; 88(5):657-666.
2. de Lange T. Shelterin: the protein complex that shapes and safeguards human telomeres. Genes Dev 2005; 19(18):2100-2110.
3. van Steensel B, Smogorzewska A, de Lange T. TRF2 protects human telomeres from end-to-end fusions. Cell 1998; 92:401-413.
4. Sandell LL, Zakian VA. Loss of a yeast telomere: arrest, recovery and chromosome loss. Cell 1993; 75(4):729-739.
5. Olovnikov AM. A theory of marginotomy. The incomplete copying of template margin in enzymic synthesis of polynucleotides and biological significance of the phenomenon. J Theor Biol 1973; 41(1):181-190.
6. Allsopp RC, Vaziri H, Patterson C et al. Telomere length predicts replicative capacity of human fibroblasts. Proc Natl Acad Sci USA 1992; 89:10114-10118.
7. Kipling D, Wynford-Thomas D, Jones CJ et al. Telomere-dependent senescence. Nat Biotechnol 1999; 17:313-314.

8. Greider CW, Blackburn EH. Identification of a specific telomere terminal transferase activity in Tetrahymena extracts. Cell 1985; 43:405-413.
9. Bodnar AG, Ouellette M, Frolkis M et al. Extension of life-span by introduction of telomerase into normal human cells. Science 1998; 279(5349):349-352.
10. de Lange T. Lasker Laurels for telomerase. Cell 2006; 126(6):1017-1020.
11. Weinrich SL, Pruzan R, Ma L et al. Reconstitution of human telomerase with the template RNA component hTR and the catalytic protein subunit hTRT. Nat Genet 1997; 17:498-502.
12. Masutomi K, Kaneko S, Hayashi N et al. Telomerase activity reconstituted in vitro with purified human telomerase reverse transcriptase and human telomerase RNA component. J Biol Chem 2000; 275:22568-22573.
13. Wenz C, Enenkel B, Amacker M et al. Human telomerase contains two cooperating telomerase RNA molecules. EMBO J 2001; 20(13):3526-3534.
14. Forsyth NR, Wright WE, Shay JW. Telomerase and differentiation in multicellular organisms: turn it off, turn it on and turn it off again. Differentiation 2002; 69(4-5):188-197.
15. Harley CB, Futcher AB, Greider CW. Telomeres shorten during ageing of human fibroblasts. Nature 1990; 345:458-460.
16. Counter CM, Meyerson M, Eaton EN et al. Telomerase activity is restored in human cells by ectopic expression of hTERT (hEST2), the catalytic subunit of telomerase. Oncogene 1998; 16:1217-1222.
17. Vaziri H, Benchimol S. Reconstitution of telomerase activity in normal human cells leads to elongation of telomeres and extended replicative life span. Curr Biol 1998; 8:279-282.
18. Kim NW, Piatyszek MA, Prowse KR et al. Specific association of human telomerase activity with immortal cells and cancer. Science 1994; 266(5193):2011-2015.
19. Muntoni A, Reddel RR. The first molecular details of ALT in human tumor cells. Hum Mol Genet 2005; 14(Spec No. 2):R191-196.
20. Lue NF. Adding to the ends: what makes telomerase processive and how important is it? Bioessays 2004; 26(9):955-962.
21. Autexier C, Lue NF. The structure and function of telomerase reverse transcriptase. Annu Rev Biochem 2006; 75:493-517.
22. Prowse KR, Avilion AA, Greider CW. Identification of a nonprocessive telomerase activity from mouse cells. Proc Natl Acad Sci USA 1993; 90:1493-1497.
23. Cohn M, Blackburn EH. Telomerase in yeast. Science 1995; 269:396-400.
24. Lue NF, Peng Y. Identification and characterization of a telomerase activity from Schizosaccharomyces pombe. Nucleic Acids Res 1997; 25(21):4331-4337.
25. Fulton TB, Blackburn EH. Identification of Kluyveromyces lactis telomerase: discontinuous synthesis along the 30-nucleotide-long templating domain. Mol Cell Biol 1998; 18(9):4961-4970.
26. Morin GB. The human telomere terminal transferase enzyme is a ribonucleoprotein that synthesizes TTAGGG repeats. Cell 1989; 59:521-529.
27. Greider CW. Telomerase is processive. Mol Cell Biol 1991; 11:4572-4580.
28. Pardue ML, Rashkova S, Casacuberta E et al. Two retrotransposons maintain telomeres in Drosophila. Chromosome Res 2005; 13(5):443-453.
29. Shippen-Lentz D, Blackburn EH. Telomere terminal transferase activity from Euplotes crassus adds large numbers of TTTTGGGG repeats onto telomeric primers. Mol Cell Biol 1989; 9:2761-2764.
30. Fajkus J, Kovarik A, Kralovics R. Telomerase activity in plant cells. FEBS Lett 1996; 391(3):307-309.
31. Fitzgerald MS, McKnight TD, Shippen DE. Characterization and developmental patterns of telomerase expression in plants. Proc Natl Acad Sci USA 1996; 93(25):14422-14427.
32. Heller K, Kilian A, Piatyszek MA et al. Telomerase activity in plant extracts. Mol Gen Genet 1996; 252(3):342-345.
33. Fajkus J, Fulneckova J, Hulanova M et al. Plant cells express telomerase activity upon transfer to callus culture, without extensively changing telomere lengths. Mol Gen Genet 1998; 260(5):470-474.
34. Nosek J, Kosa P, Tomaska L. On the origin of telomeres: a glimpse at the pretelomerase world. Bioessays 2006; 28(2):182-190.
35. Kang HJ, Choi YS, Hong SB et al. Ectopic expression of the catalytic subunit of telomerase protects against brain injury resulting from ischemia and NMDA-induced neurotoxicity. J Neurosci 2004; 24(6):1280-1287.
36. Baek S, Bu Y, Kim H et al. Telomerase induction in astrocytes of Sprague-Dawley rat after ischemic brain injury. Neurosci Lett 2004; 363(1):94-96.
37. Roth A, Yssel H, Pene J et al. Telomerase levels control the lifespan of human T-lymphocytes. Blood 2003; 102(3):849-857.
38. Cao Y, Li H, Mu FT et al. Telomerase activation causes vascular smooth muscle cell proliferation in genetic hypertension. FASEB J 2002; 16(1):96-98.
39. Cao Y, Li H, Deb S et al. TERT regulates cell survival independent of telomerase enzymatic activity. Oncogene 2002; 21(20):3130-3138.

40. Rahman R, Latonen L, Wiman KG. hTERT antagonizes p53-induced apoptosis independently of telomerase activity. Oncogene 2005; 24(8):1320-1327.
41. Stewart SA, Hahn WC, O'Connor BF et al. Telomerase contributes to tumorigenesis by a telomere length-independent mechanism. Proc Natl Acad Sci USA 2002; 99(20):12606-12611.
42. Artandi SE, Alson S, Tietze MK et al. Constitutive telomerase expression promotes mammary carcinomas in aging mice. Proc Natl Acad Sci USA 2002; 99(12):8191-8196.
43. Gonzalez-Suarez E, Samper E, Ramirez A et al. Increased epidermal tumors and increased skin wound healing in transgenic mice overexpressing the catalytic subunit of telomerase, mTERT, in basal keratinocytes. EMBO J 2001; 20(11):2619-2630.
44. Smith LL, Coller HA, Roberts JM. Telomerase modulates expression of growth-controlling genes and enhances cell proliferation. Nat Cell Biol 2003; 5(5):474-479.
45. Dudognon C, Pendino F, Hillion J et al. Death receptor signaling regulatory function for telomerase: hTERT abolishes TRAIL-induced apoptosis, independently of telomere maintenance. Oncogene 2004; 23(45):7469-7474.
46. Sharma GG, Gupta A, Wang H et al. hTERT associates with human telomeres and enhances genomic stability and DNA repair. Oncogene 2003; 22(1):131-146.
47. Masutomi K, Possemato R, Wong JM et al. The telomerase reverse transcriptase regulates chromatin state and DNA damage responses. Proc Natl Acad Sci USA 2005; 102(23):8222-8227.
48. Sarin KY, Cheung P, Gilison D et al. Conditional telomerase induction causes proliferation of hair follicle stem cells. Nature 2005; 436(7053):1048-1052.
49. Flores I, Cayuela ML, Blasco MA. Effects of telomerase and telomere length on epidermal stem cell behavior. Science 2005; 309(5738):1253-1256.
50. Nakamura TM, Cech TR. Reversing time: origin of telomerase. Cell 1998; 92(5):587-590.
51. Malik HS, Burke WD, Eickbush TH. Putative telomerase catalytic subunits from Giardia lamblia and Caenorhabditis elegans. Gene 2000; 251(2):101-108.
52. Blackburn EH. Telomeres: structure and synthesis. J Biol Chem 1990; 265(11):5919-5921.
53. Pace NR, Marsh TL. RNA catalysis and the origin of life. Orig Life Evol Biosph 1985; 16(2):97-116.
54. Cech TR. A model for the RNA-catalyzed replication of RNA. Proc Natl Acad Sci USA 1986; 83(12):4360-4363.
55. Gilbert W. Origin of Life—The RNA World. Nature 1986; 319(6055):618-618.
56. Blackburn EH. Telomerase. In: Gestland RF, Atkins JF, eds. The RNA world: Cold Spring Harbor Laboratory Press, 1993:557-576.
57. Collins K, Greider CW. Utilization of ribonucleotides and RNA primers by Tetrahymena telomerase. EMBO J 1995; 14(21):5422-5432.
58. Collins K, Kobayashi R, Greider CW. Purification of Tetrahymena telomerase and cloning of genes encoding the two protein components of the enzyme. Cell 1995; 81(5):677-686.
59. Poch O, Sauvaget I, Delarue M et al. Identification of four conserved motifs among the RNA-dependent polymerase encoding elements. EMBO J 1989; 8:3867-3874.
60. Greider CW, Blackburn EH. A telomeric sequence in the RNA of Tetrahymena telomerase required for telomere repeat synthesis. Nature 1989; 337:331-337.
61. Yu G-L, Bradley JD, Attardi LD et al. In vivo alteration of telomere sequences and senescence caused by mutated Tetrahymena telomerase RNAs. Nature 1990; 344:126-132.
62. Lendvay TS, Morris DK, Sah J et al. Senescence Mutants of Saccharomyces cerevisiae With a Defect in Telomere Replication Identify Three Additional EST genes. Genetics 1996; 144:1399-1412.
63. Lingner J, Hughes TR, Shevchenko A et al. Reverse Transcriptase Motifs in the Catalytic Subunit of Telomerase. Science 1997; 276:561-567.
64. Counter CM, Meyerson M, Eaton EN et al. The catalytic subunit of yeast telomerase. Proc Natl Acad Sci USA 1997; 94:9202-9207.
65. Nakamura TM, Morin GB, Chapman KB et al. Telomerase Catalytic Subunit Homologs from Fission Yeast and Human. Science 1997; 277:955-959.
66. Harrington L, Zhou W, McPhail T et al. Human telomerase contains evolutionarily conserved catalytic and structural subunits. Genes Dev 1997; 11(23):3109-3115.
67. Meyerson M, Counter CM, Eaton EN et al. hEST2, the putative human telomerase catalytic subunit gene, is up-regulated in tumor cells and during immortalization. Cell 1997; 90(4):785-795.
68. Kilian A, Bowtell DDL, Abud HE et al. Isolation of a candidate human telomerase catalytic subunit gene, which reveals complex splicing patterns in different cell types. Hum Mol Gen 1997; 6:2011-2019.
69. Ouellette MM, Aisner DL, Savre-Train I et al. Telomerase activity does not always imply telomere maintenance. Biochem Biophys Res Commun 1999; 254:795-803.
70. Peng Y, Mian IS, Lue NF. Analysis of telomerase processivity: mechanistic similarity to HIV-1 reverse transcriptase and role in telomere maintenance. Mol Cell 2001; 7(6):1201-1211.

71. Wyatt HD, Lobb DA, Beattie TL. Characterization of physical and functional anchor site interactions in human telomerase. Mol Cell Biol 2007; 27(8):3226-3240.
72. Lue NF, Lin YC, Mian IS. A conserved telomerase motif within the catalytic domain of telomerase reverse transcriptase is specifically required for repeat addition processivity. Mol Cell Biol 2003; 23(23):8440-8449.
73. Friedman KL, Cech TR. Essential functions of amino-terminal domains in the yeast telomerase catalytic subunit revealed by selection for viable mutants. Genes Dev 1999; 13:2863-2874.
74. Xia J, Peng Y, Mian I et al. Identification of functionally important domains in the N-terminal region of telomerase reverse transcriptase. Mol Cell Biol 2000; 20:5196-5207.
75. Jacobs SA, Podell ER, Cech TR. Crystal structure of the essential N-terminal domain of telomerase reverse transcriptase. Nat Struct Mol Biol 2006; 13(3):218-225.
76. Moriarty TJ, Marie-Egyptienne DT, Autexier C. Functional organization of repeat addition processivity and DNA synthesis determinants in the human telomerase multimer. Mol Cell Biol 2004; 24(9):3720-3733.
77. Moriarty TJ, Huard S, Dupuis S et al. Functional multimerization of human telomerase requires an RNA interaction domain in the N terminus of the catalytic subunit. Mol Cell Biol 2002; 22(4):1253-1265.
78. Bosoy D, Peng Y, Mian IS et al. Conserved N-terminal motifs of telomerase reverse transcriptase required for ribonucleoprotein assembly in vivo. J Biol Chem 2003; 278(6):3882-3890.
79. Figueiredo LM, Rocha EP, Mancio-Silva L et al. The unusually large Plasmodium telomerase reverse-transcriptase localizes in a discrete compartment associated with the nucleolus. Nucleic Acids Res 2005; 33(3):1111-1122.
80. Delany ME, Daniels LM. The chicken telomerase reverse transcriptase (chTERT): molecular and cytogenetic characterization with a comparative analysis. Gene 2004; 339:61-69.
81. Bryan TM, Sperger JM, Chapman KB et al. Telomerase reverse transcriptase genes identified in Tetrahymena thermophila and Oxytricha trifallax. Proc Natl Acad Sci USA 1998; 95:8479-8484.
82. Santos JH, Meyer JN, Skorvaga M et al. Mitochondrial hTERT exacerbates free-radical-mediated mtDNA damage. Aging Cell 2004; 3(6):399-411.
83. Santos JH, Meyer JN, Van Houten B. Mitochondrial localization of telomerase as a determinant for hydrogen peroxide-induced mitochondrial DNA damage and apoptosis. Hum Mol Genet 2006; 15(11):1757-1768.
84. Haendeler J, Hoffmann J, Brandes RP et al. Hydrogen peroxide triggers nuclear export of telomerase reverse transcriptase via Src kinase family-dependent phosphorylation of tyrosine 707. Mol Cell Biol 2003; 23(13):4598-4610.
85. Massard C, Zermati Y, Pauleau AL et al. hTERT: a novel endogenous inhibitor of the mitochondrial cell death pathway. Oncogene 2006; 25(33):4505-4514.
86. Counter CM, Hahn WC, Wei W et al. Dissociation among in vitro telomerase activity, telomere maintenance and cellular immortalization. Proc Natl Acad Sci USA 1998; 95(25):14723-14728.
87. Beattie TL, Zhou W, Robinson MO et al. Polymerization defects within human telomerase are distinct from telomerase RNA and TEP1 binding. Mol Biol Cell 2000; 11(10):3329-3340.
88. Lai CK, Mitchell JR, Collins K. RNA binding domain of telomerase reverse transcriptase. Mol Cell Biol 2001; 21(4):990-1000.
89. Banik SS, Guo C, Smith AC et al. C-terminal regions of the human telomerase catalytic subunit essential for in vivo enzyme activity. Mol Cell Biol 2002; 22(17):6234-6246.
90. Middleman EJ, Choi J, Venteicher AS et al. Regulation of cellular immortalization and steady-state levels of the telomerase reverse transcriptase through its carboxy-terminal domain. Mol Cell Biol 2006; 26(6):2146-2159.
91. Seimiya H, Sawada H, Muramatsu Y et al. Involvement of 14-3-3 proteins in nuclear localization of telomerase. EMBO J 2000; 19(11):2652-2661.
92. Huard S, Moriarty TJ, Autexier C. The C terminus of the human telomerase reverse transcriptase is a determinant of enzyme processivity. Nucleic Acids Res 2003; 31(14):4059-4070.
93. Eickbush TH. Telomerase and retrotransposons: which came first? Science 1997; 277(5328):911-912.
94. Luan DD, Korman MH, Jakubczak JL et al. Reverse transcription of R2Bm RNA is primed by a nick at the chromosomal target site: a mechanism for non-LTR retrotransposition. Cell 1993; 72(4):595-605.
95. Morrish TA, Garcia-Perez JL, Stamato TD et al. Endonuclease-independent LINE-1 retrotransposition at mammalian telomeres. Nature 2007; 446(7132):208-212.
96. Zimmerly S, Guo H, Perlman PS et al. Group II intron mobility occurs by target DNA-primed reverse transcription. Cell 1995; 82(4):545-554.
97. Saldanha R, Chen B, Wank H et al. RNA and protein catalysis in group II intron splicing and mobility reactions using purified components. Biochemistry 1999; 38(28):9069-9083.
98. Xiong Y, Eickbush TH. Origin and evolution of retroelements based upon their reverse transcriptase sequences. EMBO J 1990; 9(10):3353-3362.

99. Evgen'ev MB, Zelentsova H, Shostak N et al. Penelope, a new family of transposable elements and its possible role in hybrid dysgenesis in Drosophila virilis. Proc Natl Acad Sci USA 1997; 94(1):196-201.
100. Lyozin GT, Makarova KS, Velikodvorskaja VV et al. The structure and evolution of Penelope in the virilis species group of Drosophila: an ancient lineage of retroelements. J Mol Evol 2001; 52(5):445-456.
101. Volff JN, Hornung U, Schartl M. Fish retroposons related to the Penelope element of Drosophila virilis define a new group of retrotransposable elements. Mol Genet Genom 2001; 265(4):711-720.
102. Arkhipova IR, Pyatkov KI, Meselson M et al. Retroelements containing introns in diverse invertebrate taxa. Nat Genet 2003; 33(2):123-124.
103. Pyatkov KI, Arkhipova IR, Malkova NV et al. Reverse transcriptase and endonuclease activities encoded by Penelope-like retroelements. Proc Natl Acad Sci USA 2004; 101(41):14719-14724.
104. Gladyshev EA, Arkhipova IR. From the Cover: Telomere-associated endonuclease-deficient Penelope-like retroelements in diverse eukaryotes. Proc Natl Acad Sci USA 2007; 104(22):9352-9357.
105. Sousa R. Structural and mechanistic relationships between nucleic acid polymerases. Trends Biochem Sci 1996; 21:186-190.
106. Ding J, Das K, Hsiou Y et al. Structure and functional implications of the polymerase active site region in a complex of HIV-1 RT with a double-stranded DNA template-primer and an antibody Fab fragment at 2.8 A resolution. J Mol Biol 1998; 284(4):1095-1111.
107. Huang H, Chopra R, Verdine GL et al. Structure of a Covalently Trapped Catalytic Complex of HIV-1 Reverse Transcriptase: Implications for Drug Resistance. Science 1998; 282:1669-1675.
108. Bryan T, Goodrich K, Cech T. A mutant of Tetrahymena telomerase reverse transcriptase with increased processivity. J Biol Chem 2000; 275:24199-24207.
109. Collins K. The biogenesis and regulation of telomerase holoenzymes. Nat Rev Mol Cell Biol 2006; 7(7):484-494.
110. Greider CW. Telomere length regulation. Annu Rev Biochem 1996; 65:337-365.
111. McCormick-Graham M, Romero DP. A single telomerase RNA is sufficient for the synthesis of variable telomeric DNA repeats in ciliates of the genus Paramecium. Mol Cell Biol 1996; 16:1871-1879.
112. McCormick-Graham M, Haynes WJ, Romero DP. Variable telomeric repeat synthesis in Paramecium tetraurelia is consistent with misincorporation by telomerase. EMBO J 1997; 16(11):3233-3242.
113. Baroin A, Prat A, Caron F. Telomeric site position heterogeneity in macronuclear DNA of Paramecium primaurelia. Nucleic Acids Res 1987; 15:1717-1728.
114. Forney JD, Blackburn EH. Developmentally controlled telomere addition in wild type and mutant paramecia. Mol Cell Biol 1988; 8:251-258.
115. Dandjinou AT, Levesque N, Larose S et al. A phylogenetically based secondary structure for the yeast telomerase RNA. Curr Biol 2004; 14(13):1148-1158.
116. Shippen-Lentz D, Blackburn EH. Functional evidence for an RNA template in telomerase. Science 1990; 247:546-552.
117. Romero DP, Blackburn EH. A conserved secondary structure for telomerase RNA. Cell 1991; 67:343-353.
118. Lingner J, Hendrick LL, Cech TR. Telomerase RNAs of different ciliates have a common secondary structure and a permuted template. Genes Dev 1994; 8:1984-1998.
119. Singer MS, Gottschling DE. TLC1: Template RNA component of Saccharomyces cerevisiae telomerase. Science 1994; 266:404-409.
120. Blasco M, Funk W, Villeponteau B et al. Functional characterization and developmental regulation of mouse telomerase RNA. Science 1995; 269:1267-1270.
121. McCormick-Graham M, Romero DP. Ciliate telomerase RNA structural features. Nucleic Acids Res 1995; 23:1091-1097.
122. Hinkley C, Blasco M, Funk W et al. The mouse telomerase RNA 5'-end lies just upstream of the telomerase template sequence. Nucleic Acids Res 1998; 26:532-536.
123. Chen JL, Blasco MA, Greider CW. Secondary structure of vertebrate telomerase RNA. Cell 2000; 100(5):503-514.
124. Lin J, Ly H, Hussain A et al. A universal telomerase RNA core structure includes structured motifs required for binding the telomerase reverse transcriptase protein. Proc Natl Acad Sci USA 2004; 101(41):14713-14718.
125. Tzfati Y, Knight Z, Roy J et al. A novel pseudoknot element is essential for the action of a yeast telomerase. Genes Dev 2003; 17(14):1779-1788.
126. Zappulla DC, Cech TR. Yeast telomerase RNA: a flexible scaffold for protein subunits. Proc Natl Acad Sci USA 2004; 101(27):10024-10029.
127. Shefer K, Brown Y, Gorkovoy V et al. A triple helix within a pseudoknot is a conserved and essential element of telomerase RNA. Mol Cell Biol 2007; 27(6):2130-2143.
128. Chen JL, Greider CW. An emerging consensus for telomerase RNA structure. Proc Natl Acad Sci USA 2004; 101(41):14683-14684.

129. Chen JL, Opperman KK, Greider CW. A critical stem-loop structure in the CR4-CR5 domain of mammalian telomerase RNA. Nucleic Acids Res 2002; 30(2):592-597.
130. Chappell AS, Lundblad V. Structural elements required for association of the Saccharomyces cerevisiae telomerase RNA with the Est2 reverse transcriptase. Mol Cell Biol 2004; 24(17):7720-7736.
131. Gavory G, Farrow M, Balasubramanian S. Minimum length requirement of the alignment domain of human telomerase RNA to sustain catalytic activity in vitro. Nucleic Acids Res 2002; 30(20):4470-4480.
132. Leeper TC, Varani G. The structure of an enzyme-activating fragment of human telomerase RNA. RNA 2005; 11(4):394-403.
133. Mitchell JR, Cheng J, Collins K. A box H/ACA small nucleolar RNA-like domain at the human telomerase RNA 3' end. Mol Cell Biol 1999; 19(1):567-576.
134. Narayanan A, Lukowiak A, Jady B et al. Nucleolar localization signals of box H/ACA small nucleolar RNAs. EMBO J 1999; 18:5120-5130.
135. Jady BE, Bertrand E, Kiss T. Human telomerase RNA and box H/ACA scaRNAs share a common Cajal body-specific localization signal. J Cell Biol 2004; 164(5):647-652.
136. Bhattacharyya A, Blackburn EH. Architecture of telomerase RNA. EMBO J 1994; 13:5521-5531.
137. Zaug AJ, Cech TR. Analysis of the structure of Tetrahymena nuclear RNAs in vivo: telomerase RNA, self-splicing rRNA and U2 snRNA. RNA 1995; 1:363-374.
138. ten Dam E, Van Belkum A, Pleij K. A conserved pseudoknot in telomerase RNA. Nucleic Acids Res 1991; 19:6951.
139. Autexier C, Greider CW. Mutational analysis of the Tetrahymena telomerase RNA: identification of residues affecting telomerase activity in vitro. Nucleic Acids Res 1998; 26:787-795.
140. Sperger JM, Cech TR. A stem-loop of Tetrahymena telomerase RNA distant from the template potentiates RNA folding and telomerase activity. Biochemistry 2001; 40(24):7005-7016.
141. Autexier C, Greider CW. Boundary elements of the Tetrahymena telomerase RNA template and alignment domains. Genes Dev 1995; 15:2227-2239.
142. Richards RJ, Wu H, Trantirek L et al. Structural study of elements of Tetrahymena telomerase RNA stem-loop IV domain important for function. RNA 2006; 12(8):1475-1485.
143. Chen Y, Fender J, Legassie JD et al. Structure of stem-loop IV of Tetrahymena telomerase RNA. EMBO J 2006; 25(13):3156-3166.
144. Mason DX, Goneska E, Greider CW. Stem-loop IV of tetrahymena telomerase RNA stimulates processivity in trans. Mol Cell Biol 2003; 23(16):5606-5613.
145. Tzfati Y, Fulton T, Roy J et al. Template boundary in a yeast telomerase specified by RNA structure. Science 2000; 288:863-867.
146. Seto AG, Umansky K, Tzfati Y et al. A template-proximal RNA paired element contributes to Saccharomyces cerevisiae telomerase activity. RNA 2003; 9(11):1323-1332.
147. Peterson SE, Stellwagen AE, Diede SJ et al. The function of a stem-loop in telomerase RNA is linked to the DNA repair protein Ku. Nat Genet 2001; 27(1):64-67.
148. Stellwagen AE, Haimberger ZW, Veatch JR et al. Ku interacts with telomerase RNA to promote telomere addition at native and broken chromosome ends. Genes Dev 2003; 17(19):2384-2395.
149. Lustig AJ. Telomerase RNA: a flexible RNA scaffold for telomerase biosynthesis. Curr Biol 2004; 14(14): R565-567.
150. Seto AG, Zaug AJ, Sobel SG et al. Saccharomyces cerevisiae telomerase is an Sm small nuclear ribonucleoprotein particle. Nature 1999; 401:177-180.
151. Fragnet L, Blasco MA, Klapper W et al. The RNA subunit of telomerase is encoded by Marek's disease virus. J Virol 2003; 77(10):5985-5996.
152. Trapp S, Parcells MS, Kamil JP et al. A virus-encoded telomerase RNA promotes malignant T-cell lymphomagenesis. J Exp Med 2006; 203(5):1307-1317.
153. Fragnet L, Kut E, Rasschaert D. Comparative functional study of the viral telomerase RNA based on natural mutations. J Biol Chem 2005; 280(25):23502-23515.
154. Bryan T, Goodrich K, Cech T. Telomerase RNA bound by protein motifs specific to telomerase reverse transcriptase. Mol Cell 2000; 6:493-499.
155. O'Connor CM, Lai CK, Collins K. Two purified domains of telomerase reverse transcriptase reconstitute sequence-specific interactions with RNA. J Biol Chem 2005; 280(17):17533-17539.
156. Lai CK, Miller MC, Collins K. Roles for RNA in telomerase nucleotide and repeat addition processivity. Mol Cell 2003; 11(6):1673-1683.
157. Miller MC, Collins K. Telomerase recognizes its template by using an adjacent RNA motif. Proc Natl Acad Sci USA 2002; 99(10):6585-6590.
158. Bachand F, Triki I, Autexier C. Human telomerase RNA-protein interactions. Nucleic Acids Res 2001; 29(16):3385-3393.
159. Armbruster BN, Banik SS, Guo C et al. N-terminal domains of the human telomerase catalytic subunit required for enzyme activity in vivo. Mol Cell Biol 2001; 21(22):7775-7786.

160. Mitchell J, Collins K. Human telomerase activation requires two independent interactions between telomerase RNA and telomerase reverse transcriptase. Mol Cell 2000; 6:361-371.
161. Bachand F, Autexier C. Functional regions of human telomerase reverse transcriptase and human telomerase RNA required for telomerase activity and RNA-protein interactions. Mol Cell Biol 2001; 21:1888-1897.
162. Melek M, Greene EC, Shippen DE. Processing of nontelomeric 3▯ ends by telomerase: default template alignment and endonucleolytic cleavage. Mol Cell Biol 1996; 16:3437-3445.
163. Morin GB. Recognition of a chromosome truncation site associated with α-thalassaemia by human telomerase. Nature 1991; 353:454-456.
164. Harrington LA, Greider CW. Telomerase primer specificity and chromosome healing. Nature 1991; 353:451-454.
165. Lee MS, Blackburn EH. Sequence-specific DNA primer effects on telomerase polymerization activity. Mol Cell Biol 1993; 13:6586-6599.
166. Collins K, Greider CW. Tetrahymena telomerase catalyzes nucleolytic cleavage and nonprocessive elongation. Genes Dev 1993; 7:1364-1376.
167. Melek M, Davis BT, Shippen DE. Oligonucleotides complementary to the Oxytricha nova telomerase RNA delineate the template domain and uncover a novel mode of primer utilization. Mol Cell Biol 1994; 14:7827-7838.
168. Hammond PW, Lively TN, Cech TR. The Anchor Site of Telomerase from Euplotes aediculatus Revealed by Photo-Cross-Linking to Single- and Double-Stranded DNA Primers. Mol Cell Biol 1997; 17:296-308.
169. Lue NF, Peng Y. Negative regulation of yeast telomerase activity through an interaction with an upstream region of the DNA primer. Nucleic Acids Res 1998; 26(6):1487-1494.
170. Moriarty TJ, Ward RJ, Taboski MA et al. An anchor site-type defect in human telomerase that disrupts telomere length maintenance and cellular immortalization. Mol Biol Cell 2005; 16(7):3152-3161.
171. Lue NF. A physical and functional constituent of telomerase anchor site. J Biol Chem 2005; 280(28):26586-26591.
172. Prescott J, Blackburn EH. Telomerase RNA mutations in Saccharomyces cerevisiae alter telomerase action and reveal nonprocessivity in vivo and in vitro. Gene Develop 1997; 11:528-540.
173. Lee SR, Wong JM, Collins K. Human telomerase reverse transcriptase motifs required for elongation of a telomeric substrate. J Biol Chem 2003; 278(52):52531-52536.
174. Bednenko J, Melek M, Greene EC et al. Developmentally regulated initiation of DNA synthesis by telomerase: evidence for factor-assisted de novo telomere formation. EMBO J 1997; 16(9):2507-2518.
175. Prescott DM. The DNA of Ciliated Protozoa. Microbiol Rev 1994; 58:233-267.
176. Karamysheva Z, Wang L, Shrode T et al. Developmentally programmed gene elimination in Euplotes crassus facilitates a switch in the telomerase catalytic subunit. Cell 2003; 113(5):565-576.
177. Hossain S, Singh S, Lue NF. Functional analysis of the C-terminal extension of telomerase reverse transcriptase. A putative "thumb" domain. J Biol Chem 2002; 277(39):36174-36180.
178. Bosoy D, Lue NF. Yeast telomerase is capable of limited repeat addition processivity. Nucleic Acids Res 2004; 32(1):93-101.
179. Greene EC, Bednenko J, Shippen DE. Flexible positioning of the telomerase-associated nuclease leads to preferential elimination of nontelomeric DNA. Mol Cell Biol 1998; 18(3):1544-1552.
180. Oulton R, Harrington L. A human telomerase-associated nuclease. Mol Biol Cell 2004; 15(7):3244-3256.
181. Niu H, Xia J, Lue NF. Characterization of the interaction between the nuclease and reverse transcriptase activity of the yeast telomerase complex. Mol Cell Biol 2000; 20(18):6806-6815.
182. Collins K, Gandhi L. The reverse transcriptase component of the Tetrahymena telomerase ribonucleoprotein complex. Proc Natl Acad Sci USA 1998; 95:8485-8490.
183. Huard S, Autexier C. Human telomerase catalyzes nucleolytic primer cleavage. Nucleic Acids Res 2004; 32(7):2171-2180.
184. Kirk KE, Harmon BP, Reichardt IK et al. Block in Anaphase Chromosome Separation Caused by a Telomerase Template Mutation. Science 1997; 275:1478-1481.
185. Prescott JC, Blackburn EH. Telomerase RNA template mutations reveal sequence-specific requirements for the activation and repression of telomerase action at telomeres. Mol Cell Biol 2000; 20(8):2941-2948.
186. McEachern MJ, Iyer S, Fulton TB et al. Telomere fusions caused by mutating the terminal region of telomeric DNA. Proc Natl Acad Sci USA 2000; 97(21):11409-11414.
187. Bryan TM, Goodrich KJ, Cech TR. Tetrahymena telomerase is active as a monomer. Mol Biol Cell 2003; 14(12):4794-4804.
188. Cohen SB, Graham ME, Lovrecz GO et al. Protein composition of catalytically active human telomerase from immortal cells. Science 2007; 315(5820):1850-1853.
189. Prescott J, Blackburn EH. Functionally interacting telomerase RNAs in the yeast telomerase complex. Genes Dev 1997; 11:2790-2800.

190. Fouche N, Moon IK, Keppler BR et al. Electron microscopic visualization of telomerase from Euplotes aediculatus bound to a model telomere DNA. Biochemistry 2006; 45(31):9624-9631.
191. Wang L, Dean SR, Shippen DE. Oligomerization of the telomerase reverse transcriptase from Euplotes crassus. Nucleic Acids Res 2002; 30(18):4032-4039.
192. Beattie TL, Zhou W, Robinson MO et al. Functional multimerization of the human telomerase reverse transcriptase. Mol Cell Biol 2001; 21(18):6151-6160.
193. Arai K, Masutomi K, Khurts S et al. Two independent regions of human telomerase reverse transcriptase are important for its oligomerization and telomerase activity. J Biol Chem 2002; 277(10):8538-8544.
194. Ly H, Xu L, Rivera MA et al. A role for a novel 'transpseudoknot' RNA-RNA interaction in the functional dimerization of human telomerase. Genes Dev 2003; 17(9):1078-1083.
195. Marie-Egyptienne DT, Cerone MA, Londono-Vallejo JA et al. A human-Tetrahymena pseudoknot chimeric telomerase RNA reconstitutes a nonprocessive enzyme in vitro that is defective in telomere elongation. Nucleic Acids Res 2005; 33(17):5446-5457.
196. Ren X, Gavory G, Li H et al. Identification of a new RNA. RNA interaction site for human telomerase RNA (hTR): structural implications for hTR accumulation and a dyskeratosis congenita point mutation. Nucleic Acids Res 2003; 31(22):6509-6515.
197. Paillart JC, Shehu-Xhilaga M, Marquet R et al. Dimerization of retroviral RNA genomes: an inseparable pair. Nat Rev Microbiol 2004; 2(6):461-472.
198. Gipson CL, Xin ZT, Danzy SC et al. Functional characterization of yeast telomerase RNA dimerization. J Biol Chem 2007 (in press).
199. Kelleher C, Teixeira MT, Forstemann K et al. Telomerase: biochemical considerations for enzyme and substrate. Trends Biochem Sci 2002; 27(11):572-579.
200. Miller M, Liu J, Collins K. Template definition by Tetrahymena telomerase reverse transcriptase. EMBO J 2000; 19:4412-4422.
201. Bosoy D, Lue NF. Functional analysis of conserved residues in the putative "finger" domain of telomerase reverse transcriptase. J Biol Chem 2001; 276(49):46305-46312.
202. Chen JL, Greider CW. Determinants in mammalian telomerase RNA that mediate enzyme processivity and cross-species incompatibility. EMBO J 2003; 22(2):304-314.
203. Garforth SJ, Wu YY, Prasad VR. Structural features of mouse telomerase RNA are responsible for the lower activity of mouse telomerase versus human telomerase. Biochem J 2006; 397(3):399-406.
204. Forstemann K, Hoss M, Lingner J. Telomerase-dependent repeat divergence at the 3' ends of yeast telomeres. Nucleic Acids Res 2000; 28(14):2690-2694.
205. Teixeira MT, Arneric M, Sperisen P et al. Telomere length homeostasis is achieved via a switch between telomerase- extendible and -nonextendible states. Cell 2004; 117(3):323-335.
206. Blasco MA, Lee HW, Hande MP et al. Telomere shortening and tumor formation by mouse cells lacking telomerase RNA. Cell 1997; 91(1):25-34.
207. Shay JW. Meeting report: the role of telomeres and telomerase in cancer. Cancer Res 2005; 65(9):3513-3517.

CHAPTER 2

Drosophila Telomeres:
A Variation on the Telomerase Theme

Mary-Lou Pardue* and P. Gregory DeBaryshe

Abstract

In *Drosophila*, the role of telomerase is carried out by three specialized retrotransposable elements, *HeT-A*, *TART* and *Tahre*. Telomeres contain long tandem head-to-tail arrays of these elements. Within each array, the three elements occur in random, but polarized, order. Some are truncated at the 5' end, giving the telomere an enriched content of the large 3' untranslated regions which distinguish these telomeric elements from other retrotransposons. Thus, *Drosophila* telomeres resemble other telomeres because they are long arrays of repeated sequences, albeit more irregular arrays than those produced by telomerase. The telomeric retrotransposons are reverse-transcribed directly onto the end of the chromosome, extending the end by successive transpositions. Their transposition uses exactly the same method by which telomerase extends chromosome ends—copying an RNA template. In addition to these similarities in structure and maintenance, *Drosophila* telomeres have strong functional similarities to other telomeres and, as variants, provide an important model for understanding general principles of telomere function and evolution.

Introduction: There Appear to Be Only a Few Ways to Build a Eukaryotic Telomere

The concept of the telomere was derived from analysis of early studies on *Drosophila* chromosomes. In 1938 Herman Muller noted that these chromosomes could survive many kinds of breakage, exchange and rejoining, but simple terminal deletions were never found.[1] He concluded that chromosome ends are capped by special structures and that broken chromosomes could not survive if they did not acquire a cap from another chromosome. He named the caps telomeres and noted that these regions had heterochromatic morphology.[1,2] We now understand that this cap is what distinguishes a chromosome end from a break in the chromosome. Breaks in chromosomes activate a checkpoint response that prevents the cell from proceeding through the cell cycle. Telomere caps allow cells to pass the checkpoint but we do not understand the mechanism involved.

Also in 1938, Barbara McClintock[3] reported that ends of broken chromosomes in corn tended to fuse with ends of other broken chromosomes, forming dicentric chromosomes. Dicentrics broke again when the two centromeres tried to enter different daughter nuclei at metaphase. These early studies suggested that telomeres in flies and corn were very similar, as might be expected if the basic structure of telomeres arose early in the evolution of linear nuclear chromosomes and had been conserved.

This evolutionary conservation is now strongly supported by studies of the molecular structure of the ends of eukaryotic chromosomes. Beginning with the work of Elizabeth Blackburn and

*Corresponding Author: Mary-Lou Pardue—Department of Biology, Massachusetts Institute of Technology, Cambridge, MA 02139, USA. Email: mlpardue@mit.edu

Origin and Evolution of Telomeres, edited by Jozef Nosek and Ľubomír Tomáška.

colleagues on Tetrahymena,[4] chromosome ends in all animals, plants and unicellular eukaryotes studied have been found to consist of long arrays of tandem repeat sequences[5] (which we will call telomere repeats) and, frequently, repeated telomere-associated sequences (TAS).[6]

The large amount of DNA found in the telomere arrays was unexpected because DNA viruses have much more economical telomere mechanisms and need only a few nucleotides to prevent erosion of the ends of their linear genomes (see ref. 7). In contrast, multicellular eukaryotes tend to have ten or more kilobases of these sequences per end and even unicellular eukaryotes have a few hundred base pairs of telomere repeats on each chromosome. This large investment of cellular resources is less surprising now that we know that eukaryotic telomeres have many roles beyond simply capping the end of the DNA. For example, the arrays are important in cell maintenance, senescence, genomic stability and oncogenesis in ways that are not understood but that are related to the length of the arrays.[8-10] The significance of array length and the mechanisms by which it is regulated are major questions in the field today.

For the vast majority of eukaryotes, telomere arrays are composed of 5-10 bp repeats added to the chromosome end by the enzyme complex, telomerase. The repeat sequence for each organism is determined by the RNA template used by its telomerase. Organisms with telomerase are also able to extend telomere arrays by a recombination-based mechanism. Recombination provides a back-up mechanism when telomerase is lost but seems to be of minor importance in cells with active telomerase. After loss of telomerase in budding yeast populations undergoing senescence, rare spontaneous survivors use recombination to elongate telomeres.[11,12] A significant fraction of human tumors and immortalized cell lines lack active telomerase and instead use recombination-based Alternative Lengthening of Telomeres (ALT) to extend telomeres.[13,14] Telomeres maintained by recombination in both yeast and human cells are very heterogeneous in length but return to normal when telomerase activity is restored experimentally, although vestiges of the recombination system remain.[12,15,16]

In most organisms the telomerase catalytic subunit and its template RNA seem to be encoded by single copy genes. This is a boon for biological research because knocking out either gene eliminates telomerase activity. It is also puzzling that the telomerase mechanism has persisted with very little change throughout the evolution of eukaryotes despite being so easily susceptible to experimental knockout.

Telomerase activity has been found almost everywhere in the Eukaryota. Indeed, we think it likely that the primary mechanism for maintenance of all eukaryotic telomeres utilizes reverse transcription of RNA templates. Unfortunately, it is difficult to test this possibility because of problems in identifying telomere sequences in new organisms. Telomeres, like all complex repeat sequences, present nearly insurmountable technical problems both for cloning and for correct sequence assembly. As a result, sequence databases contain little, if any, of these sequences, even for organisms with complete assembly of the non-heterochromatic genome. Thus it is not possible to do informatic surveys for telomere sequences. However, the sequence of the telomerase RNA template has been so strongly conserved that in situ hybridization has identified telomere repeats on the chromosomes of many organisms. Surveys of the insects have been especially interesting.

Many insects have TTAGG telomerase repeats, one nucleotide different from the vertebrate TTAGGG. This TTAGG sequence hybridizes with telomeres of many different species but neither TTAGG nor TTAGGG hybridizes to chromosomes in species in several branches of the insect phylogenetic tree (Fig. 1), including branches of the most successful lineage of insects, the superorder Endopterygota. This lack of hybridization suggests that telomerase has been lost, or at least modified, several times in insect evolution.[17,18]

It is of considerable interest to learn what has happened to telomeres in organisms that do not appear to use telomerase, but only a few of these species have been studied. In one case the change has been relatively minor; the flour beetle, *Triboleum castaneum*, uses a telomerase template, TCAGG, which does not cross-hybridize with TTAGG.[19] It is not impossible that recombination has completely replaced telomerase in other organisms. For example, telomeres in the midge, *Chironomus*, end in repeats of 176, 340 or 350 bp, depending on the species.[20] There

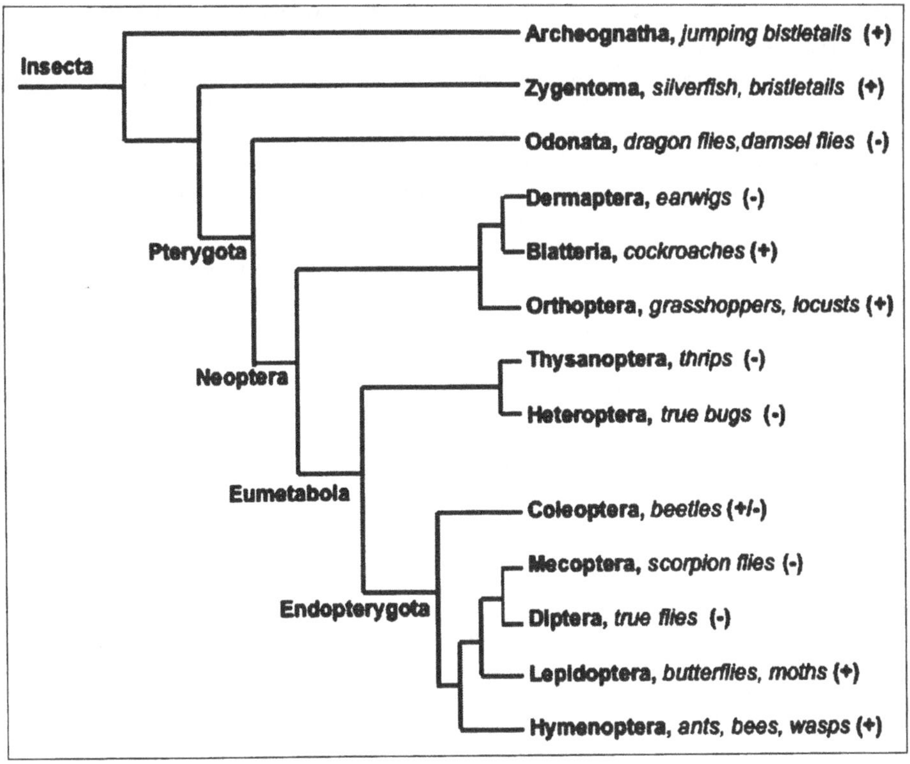

Figure 1. Some species in the Insecta do not use telomerase to maintain telomeres. Phylogeny of insect orders where one or more species have been analyzed for the presence of telomeric TTAGG repeats. (+): all species studied have TTAGG; (-): all species studied lack TTAGG; (+/-): some species studied have TTAGG, others do not. Only species from 3 genera without TTAGG have been studied further, *Drosophila, Chironomus* and *Anopheles*. All are Diptera (see text for discussion). At least one non-insect, the spider *Tegenaria ferrugenea* also lacks TTAGG.[76] Tree based on Frydrychova, et al.[17]

is clear evidence that these repeats maintain their homogeneity by recombination, although it is not clear that recombination also compensates for sequence erosion. Because extrachromosomal RNA-DNA complexes containing long runs of the telomere repeats have been found in all three studied species of *Chironomus,*[21] *Chironomus* may also use an RNA template to extend its telomeres. In the mosquito, *Anopheles gambiae,* analysis of a transgene on the end of a broken chromosome showed that chromosomes can be elongated by unequal recombination;[22] however the endogenous telomere sequences in this organism have not yet been characterized.

A strikingly different mechanism of telomere maintenance has been found in the genus *Drosophila*. In all studied species of this genus, specialized retrotransposons extend the telomeres. These retrotransposons provide a robust mechanism which may well be used by other members of the Endopterygota. The *Drosophila* telomere-specific retrotransposons also provide an unexpected link between chromosome structure and transposable elements, raising important questions about the evolution of both.

Drosophila Telomeres Are Maintained by Specialized Non-LTR Retrotransposons

The three retrotransposons that maintain *Drosophila* telomeres are all non-LTR (Long Terminal Repeat) retrotransposons (Fig. 2). Non-LTR elements differ from their relatives, LTR retrotransposons and retroviruses, in the way in which the RNA transposition intermediate is converted to chromosomal DNA. Of course, like all transposons, they also differ from retroviruses by lacking the viral envelope gene. LTR retrotransposons and retroviruses are reverse transcribed in the cytoplasm, transported to the nucleus and inserted into the chromosome as double stranded DNA.[23]

In contrast, non-LTR elements enter the nucleus as RNA, the 3' end associates with a nick in the chromosome and reverse transcriptase uses the 3' OH on the nicked DNA to initiate reverse transcription of the RNA.[24] Thus, new DNA is linked to the chromosome. The 5' end of the reverse-transcribed DNA is then joined to the other side of the nick in the chromosome, to complete the insertion. At least for some elements, the insertion is then converted to double-stranded DNA by the same reverse transcriptase,[24-26] although it is also possible that the second strand synthesis might be accomplished by normal DNA synthesis.

The mechanism by which non-LTR elements are incorporated into chromosomes is significant because it is basically equivalent to that used by telomerase (Fig. 3). We have postulated that telomeric retrotransposon RNA associates with the end of the DNA, rather than with an internal nick like other elements. Thus each transposition of a telomeric element adds a new end to the DNA, extending the chromosome (Fig. 3). Analysis of sequence at junctions between elements in *Drosophila* arrays shows that these elements do not use the precise sequence pairing that telomerase uses to align each repeat.[27] Such alignment is necessary to obtain a consistent sequence when the template is only a few nucleotides in length. On the other hand, the *Drosophila* templates range from six to thirteen kb so imprecise alignment at junctions would have little impact, especially since each junction starts with a variable run of As.

Our working model of the *Drosophila* telomere (Fig. 4) draws from retrovirus and yeast biology, as well as our own results.[28] Transcription of elements in telomere arrays produces RNA which is transported to the cytoplasm where the coding regions are translated to yield the structural protein, Gag and, except for *HeT-A*, the enzymatic protein, Pol. These proteins associate with their own RNA and move into the nucleus where the Gag protein is targeted to telomere regions. We

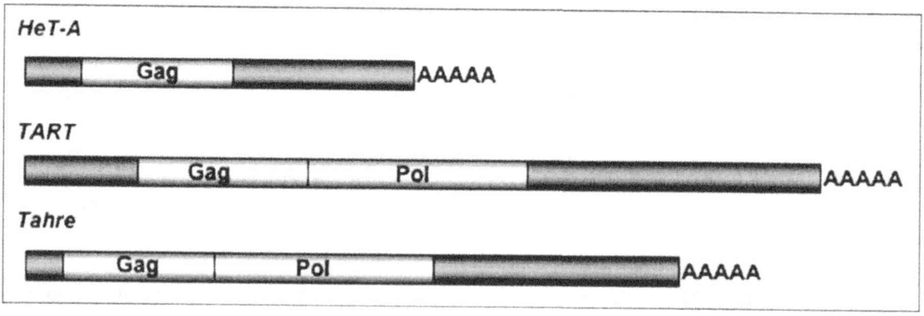

Figure 2. The three *D. melanogaster* telomere retrotransposons drawn as their putative RNA transposition intermediates. Coding regions, Gag and Pol, are labeled. Gray regions indicate 5′ and 3′ untranslated regions. AAAA indicates the 3′ poly(A) tail on each RNA. It is the source of the (dA/T)n that joins each DNA copy to the chromosome when the element transposes. Sizes are only approximate because individual elements can differ in length of both coding and noncoding regions. *HeT-A* elements are~6 kb. The 5′ end of TART has not been completely defined but subfamilies appear to be 10-13 kb. *Tahre* is~10.5 kb.

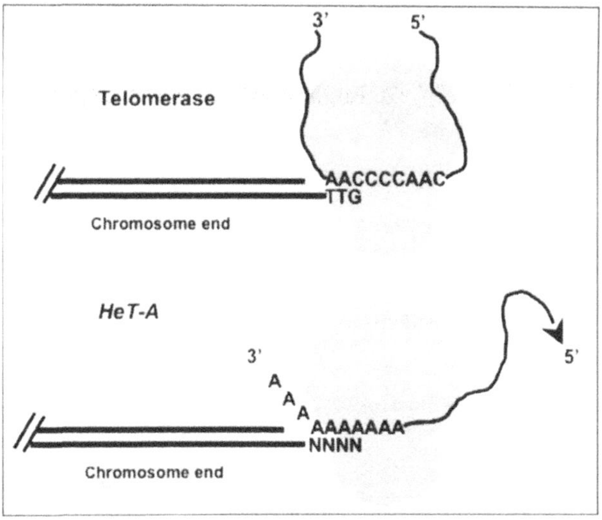

Figure 3. Telomere element retrotransposition resembles telomere extension by telomerase. In both cases the catalytic subunit (gray) with its RNA template (black wavy line) associates with the end of the chromosome. Telomerase aligns the first nucleotides of the sequence to be copied with their complement in the chromosome before copying sequence on to the end of that complement, thus assuring precise replication for each addition. *HeT-A* and other telomere elements, do not require complementary sequence for alignment so the sequence to which the initial Ts are added is shown as NNNN. Retrotransposon additions copy variable amounts of the poly(A) tail at each transposition.

suggest that this targeting is directed by an end-associated protein, perhaps analogous to Est1 or Cdc13 in budding yeast.

Drosophila Telomere Retrotransposons Have Special Features

Retrotransposon telomeres were discovered in *D. melanogaster* and have been most extensively characterized in this species. We will confine discussion in this section to *D. melanogaster* elements and discuss other species later.

There are three telomere-specific retrotransposons in *D. melanogaster*, *HeT-A*, *TART* and *Tahre* (Fig. 2). Sequences of their gag and pol coding regions group these elements in the jockey clade of non-LTR retrotransposons.[29,30] This clade also contains several of the parasitic elements abundant in nontelomeric parts of the *D. melanogaster* genome (e.g., *jockey, Doc and F*). The primary structural feature that distinguishes the telomeric elements from other transposable elements is a very long 3′ UTR (untranslated region). Their relatives in the jockey clade, like transposable elements in general, have very little sequence that does not code for something needed for transposition.

The many available sequences of *HeT-A* and *TART* show that each element has multiple forms that differ both by nucleotide differences and insertions/deletions and yet appear to be fully functional. There are not yet enough *Tahre* sequences to determine whether the *Tahre* sequence is equally variable.

HeT-A is unusual because it does not encode its own reverse transcriptase. Nevertheless *HeT-A* is the most abundant of the elements[31] and the one most often found transposing to heal broken ends.[27,32] It must obtain the necessary enzyme activity from another source—either another element or a nuclear reverse transcriptase. *HeT-A* expression is developmentally regulated.[33,34] It is not expressed in polyploid cells and therefore there is little expression

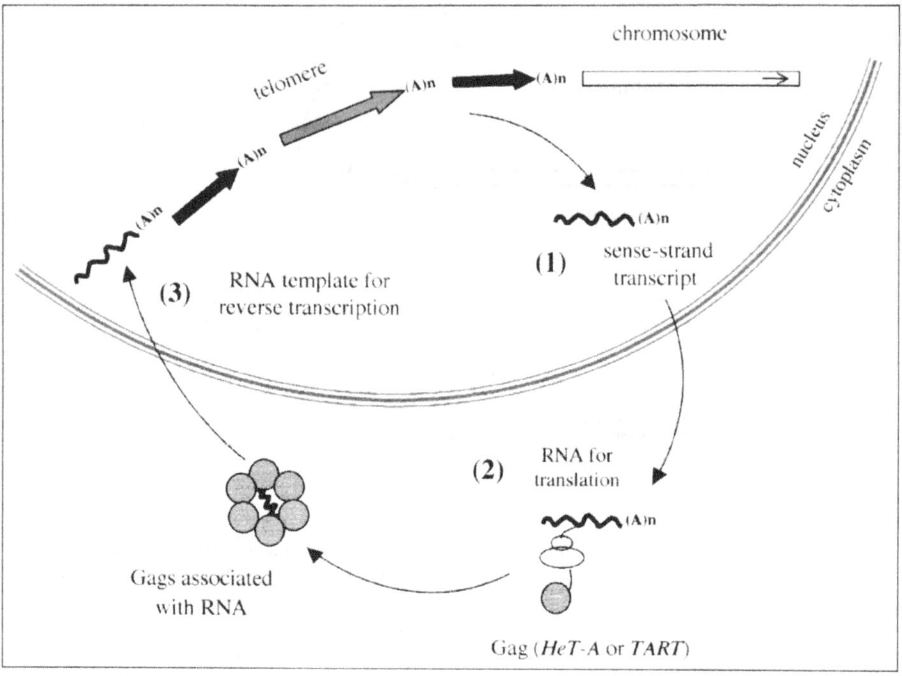

Figure 4. Model for maintenance of chromosome ends by telomeric retrotransposons. Retrotransposons yield sense-strand transcripts that serve as both mRNAs and transposition intermediates. This diagram shows our current model for the path of these RNAs from transcription until they are reverse-transcribed to add another repeat onto the telomere array. Gray arrows represent *HeT-A* (dark) and *TART* (light) elements attached to the end of the chromosome. A poly(A) sense strand RNA is transcribed from a member of the array (step 1). For the telomeric retrotransposons there is evidence suggesting that this RNA must be translated (step 2) before serving as a template (step 3) for telomere addition. This suggestion is now supported by the finding that translation products (Gags) of these RNAs appear capable of delivering the transposition template specifically to its target at the telomere. Gray circles in the diagram represent Gags of either *HeT-A* or *TART*. Analogy with retroviruses suggests that reverse transcriptase is also included in the Gag-RNA complex; however, there is no evidence on this point. Reproduced from J Cell Biol 2002; 159:397-402 by copyright permission of the Rockefeller University Press.[28]

during most of larval growth. Also, *HeT-A* Gag is efficiently targeted to telomeres in interphase diploid nuclei, but not in polyploid cells.[28,35] This targeting appears to be important for telomere-specific transposition.

TART encodes both Gag and Pol and is always less abundant than *HeT-A*. Both elements are present in every *Drosophila* stock, cell line and species we have studied. Measurements across different stocks and cell types[31] show a strong correlation in the relative abundance of the two elements, even where the total number of elements differs significantly (Fig. 5). Unlike *HeT-A*, *TART* produces both sense and antisense transcripts. Antisense transcripts are much more abundant and display little, if any, developmental regulation.[33,36] *TART* Gag, like *HeT-A* Gag, is efficiently transported to interphase nuclei; however, it does not associate with telomeres by itself. When *HeT-A* Gag is present the two proteins colocalize to telomeres.[28] This suggests that the two elements collaborate, with *HeT-A* providing telomere specificity and *TART* providing reverse transcriptase. Such collaboration could explain why the two elements are found together in every stock or cell line studied.

Tahre (*T*elomere-*A*ssociated and *HeT-A*-*R*elated *E*lement) was discovered recently[37] and has been less studied than the others. However much of its sequence is so strongly related to *HeT-A* that some conclusions can be drawn from experiments using *HeT-A* probes. The 5' and 3' UTR and *gag* coding regions are very similar to *HeT-A* while the *pol* coding region resembles, but is less closely related to, *TART*. *Tahre* is very rare. One complete and three truncated copies were reported from a study of BACs made for the *D. melanogaster* Genome Project. The expression pattern has not been reported but a BLAST search of the database of cDNA clones found no sequences except those that were also found with *HeT-A* sequence. If abundance in the telomere array is determined by the ability to transpose, it is surprising that an element combining *HeT-A*'s ability to target telomeres with *TART*'s reverse transcriptase is not more abundant.

The powerful genetic tools available only for *Drosophila* have made it possible to produce chromosomes with broken ends that evade checkpoints and can then be retained in the genome by strong selection. These experiments have shown that broken ends can be healed by transposition of telomere retrotransposons. More recently, it has been shown that the rate of healing is affected by specific genes and that these genes are acting through components of the RNAi machinery.[38,39]

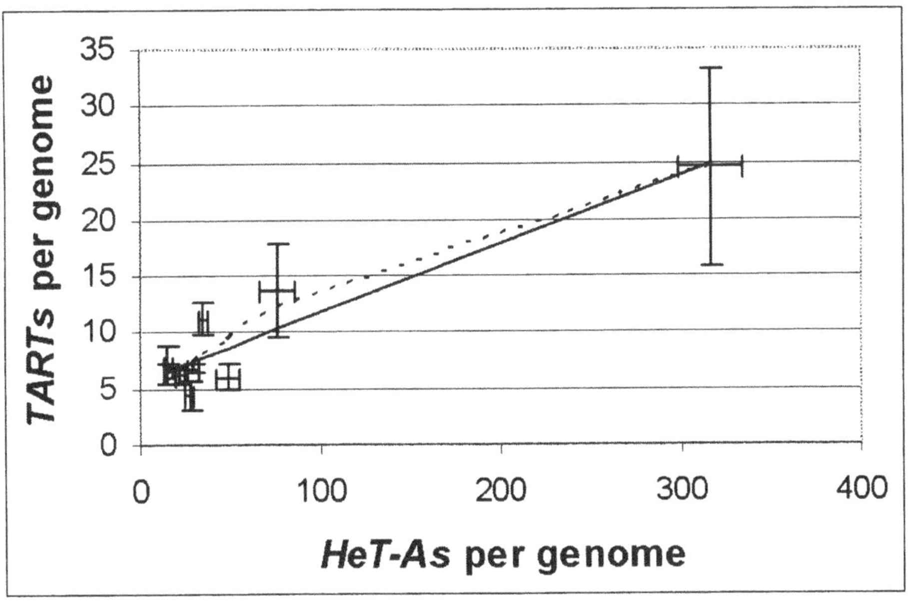

Figure 5. The number of *HeT-A* and *TART* elements per genome are correlated in *D. melanogaster* stocks, cells and tissue types. This figure shows data for a cultured cell line (S2), for diploid cells and for polytene salivary gland cells from four stocks (Oregon R, 2057, Su(var) 205[4] and G3). When analyzed, the data show that the number of the two elements present are linearly correlated at better than the 95% confidence level when polytene salivary gland measurements are compared, when diploid cell measurements are compared, or when all data is pooled and compared.[31] The solid black line is the best linear fit to the data, the dotted line is the best quadratic fit. If the data point in the extreme right-hand upper corner is omitted from the analysis, the best linear fit for the remaining data is indistinguishable from the quadratic fit at the level of detail in this plot.

Telomere Retrotransposons Are Almost Completely Segregated from Other Transposable Elements in the Genome

In spite of their similarity to other retroelements in the *D. melanogaster* genome, the telomere retrotransposons differ markedly from those elements in transposition targeting. As a result, there is little mixing of the two types of elements. The euchromatic regions of the *D. melanogaster* genome have been sequenced completely. No sequence with significant similarity to *HeT-A*, *TART*, or *Tahre* is found in these regions except for the *pol* coding region of *BS*, a non-LTR element found at several sites in euchromatin, which has a small region with similarity to 90 bp of *TART pol*.[31] Thus there is no evidence that telomere elements can transpose into these euchromatic regions, although other retrotransposons are found at many sites in this part of the genome.

The exclusion of telomere elements from euchromatin may be explained by their specific targeting to ends. (Telomere elements do transpose onto euchromatin if a chromosomal break causes that euchromatin to be at the end of the chromosome.[27]) Even if targeting is not perfect, internal insertion might also be forbidden for other reasons. For example, telomere elements may unable to form a junction at the 5' end after reverse transcription and the resulting loss of the part of the chromosome distal to an internal insertion would likely be lethal.

As noted above, telomere arrays and telomere-associated sequences present formidable challenges for correct assembly of sequence. However, the *Drosophila* Heterochromatin Genome Project now has assembled sequence extending into the telomeres on the right end of chromosome 4 (4R) and the left end of the X chromosome (XL). These assemblies (Fig. 6) contain 75,946 bp of telomere transposons on 4R and 19,199 bp of these sequences on XL.[31] It might be supposed that these long telomere arrays would be safe landing sites for parasitic elements because they contain no vital genes to be disrupted. This does not seem to be the case. In both XL and 4R, the interior of the chromosome, peppered with nontelomeric elements, is separated from the distal array of uninterrupted telomere elements by a short transition region comprised of mixed fragments of telomere and nontelomere elements. On XL the transition region is approximately 300 bp and the assembled portion of the distal array of uninterrupted *HeT-A* elements is ~19 kb. On 4R the

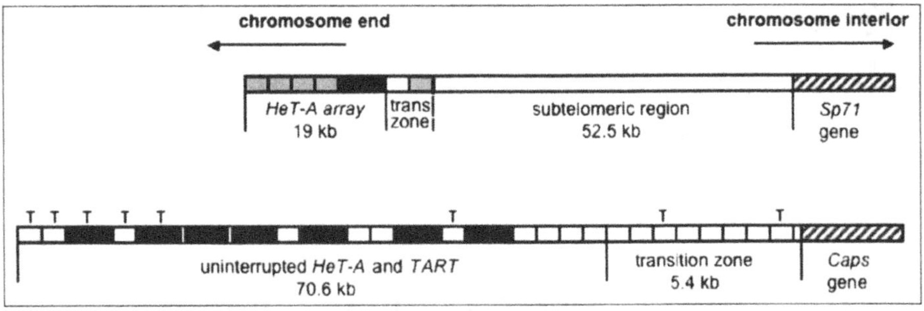

Figure 6. The assembled regions of telomere arrays from two *D. melanogaster* telomeres. The figure, not drawn to scale, illustrates sequence organization from the most distal assembled sequence (left end) to the most distal gene (right end). The more terminal sequences have not been assembled. Upper diagram: array from the left end of the X chromosome. Lower diagram: array from the right end of chromosome 4. Black boxes: full-length telomeric elements. Gray boxes: partial telomeric elements. Boxes marked T are *TART* elements, all other boxes are *HeT-A* elements. All elements have 3′ end toward chromosome interior. All partial elements are truncated at the 5′ end except for one in the transition zone. Differences in size of partial elements not shown. White regions: other transposable elements or regions rich in these elements. Striped regions: the most distal gene on each chromosome. trans zone: the tiny (320 nt) transition zone on the XL telomere.

transition zone is 5.4 kb and the assembled *HeT-A/TART* array is 70.6 kb. The transition zones show that the distal separation is not due to incompatibility of the DNA sequences. Thus the lack of mixing in the distal arrays is probably due to a specific chromatin structure in the distal regions which has been partially invaded by nontelomeric elements at the proximal end.

The distinction between distal telomere arrays and telomere-associated sequences affects not only elements that are long-time residents in the *D. melanogaster* genome but may have at least a partial effect on the recent invader, the *P-element*. *P-elements* were discovered because they disrupt chromosomes when they invade a naïve host and have, therefore, become a powerful tool for geneticists wishing to manipulate chromosomes. Attempts to insert *P-elements* in telomere regions have found hotspots for insertion in telomere-associated sequences[40,41] but there is only one report of inserts in the telomere array.[42] The collection of ~20,000 random *P-element* inserts produced by the *Drosophila* genome project was screened for elements flanked by telomere retrotransposon sequence.[42] Seven inserts were identified that mapped to telomere arrays. Most were inserted in a short region very near the 3' end of *TART;* a smaller hotspot for insertion was seem near the 3' end of *HeT-A*. Thus the *P-element* may be revealing the beginning of a footprint of the telomere chromatin structure. It is interesting that the only retrotransposon that has been found within the telomere array, a *roo* element, was found near one of the *P-element* inserts, suggesting that the *P-element* may have affected chromatin structure, allowing entry of *roo*.

The apparent exclusion of non-telomere elements from the distal array is also seen in telomeres of the silkworm, *Bombyx mori*, which has extremely long TTAGG arrays made by telomerase. *B. mori* has two families of retrotransposons, TRAS and SART, which insert at specific nucleotides in the TTAGG sequence. TRAS and SART are abundant in proximal parts of the TTAGG array but are not found in the distal six to eight kb.[43] As with the *Drosophila* telomere, terminal insertions in *B. mori* seem to be prevented by something other than lack of DNA insertion sites, but nothing is known about its chromatin structure. Other insect species have yet to be studied.

Very Long 3' UTR Sequences Seem to Have a Role in Forming Heterochromatin Structure

One of the distinguishing features of telomere retrotransposons is their very long 3' UTR (Fig. 2). We have suggested that this sequence plays a role in forming telomeric chromatin[44] which, as Muller observed, is heterochromatic. The 3' sequence is overrepresented in telomere arrays because some of the elements are truncated at the 5' end,[31] either because of incomplete reverse transcription or because of erosion during the time when they form the extreme end.

This 3' sequence is also found in another class of *D. melanogaster* heterochromatin, the Y chromosome. Like other heterochromatin, the Y presents significant problems for sequence assembly; however sequence scaffolds of seven Y chromosome genes have been assembled.[45] Four of these contain fragments of *HeT-A* or *TART*.[31] These fragments are distinguished from sequences in telomere arrays in two important ways. First, they have been inserted into the interior of the chromosome, rather than added to the end. Second, the fragments contain only sequence from the 3' UTR and some do not contain the extreme 3' sequences thought necessary for reverse transcription.[24] This suggests that the sequences have been transposed to the Y by some other mechanism. Much of the Y sequence has not been assembled. There is evidence that there is more 3' sequence in unassembled parts of the Y[46-48] and perhaps other unassembled heterochromatic regions of the genome.

These observations of the segregation of 3' UTR sequence into telomeres and other heterochromatic regions shows that these sequences have a special relation to this type of chromatin, possibly because they are involved in forming its structure.

Telomere Retrotransposons Have a Symbiotic Relationship with *Drosophila* Cells

Our hypothesis that telomeric transposons are targeted to telomeres by Gag proteins (Fig. 4) initially had two bases. First, although gag genes of individual *HeT-A* and *TART* elements differ by both indels and nucleotide changes, all these coding regions are open, suggesting that each element must be successfully translated in order to transpose.[31] Second, retroviral Gags had been shown to escort viral RNA through the transport path specific for their virus.[49] *HeT-A* and *TART* Gags share important motifs with retroviral Gags and we surmised they also share functions. Transient expression in cultured cells of *HeT-A* and *TART* Gags tagged with Green Fluorescent Protein supported this hypothesis; in these cells, which normally express both telomeric transposons, Gags of both elements are efficiently transported into the nucleus where *HeT-A* Gag moves to telomeres and also directs TART Gag to the same targets.[28,50,51]

These studies of Gag localization show that the telomere elements have co-evolved with their hosts an ability to interact beneficially with cellular components. This ability is not seen in their nontelomeric relatives. Direct comparison with Gags from *jockey*, *Doc* and I *factor* showed that these proteins were mostly, if not entirely, constrained to remain in the cytoplasm.[51] We suggest that cytoplasmic retention is a reflection of the cell's efforts to keep nontelomeric elements out of their chromosomes.

We have also expressed *HeT-A* Gag in live flies from a transgene driven by a promoter that is active in all tissues of the fly. Normally, *HeT-A* is active in diploid cells and, in these cells, transgenic Gag forms dots as it does in cultured cells. Polyploid cells do not express *HeT-A* and, when transgenic *HeT-A* Gag is expressed in polyploid tissues, the protein does not enter the nucleus. Instead large amounts of the protein accumulate in cytoplasmic regions that differ from tissue to tissue.[35] These transgenic experiments show that cells actively regulate Gag targeting in a cell-type-specific manner.

Retrotransposon Telomeres Probably Predate the Genus *Drosophila*

Telomere retrotransposon sequences diverge rapidly. For example, the six complete *HeT-A* elements found in the assembled 4R and XL sequence from *D. melanogaster* have between 68% and 99% nucleotide identity when pairwise comparisons are made. Even the coding sequence has only 80% to 100% nucleotide identity (76% to 100% amino acid identity), depending on the elements compared.[31] This divergence makes it difficult to search genomes of other species on the basis of sequence homology. Nevertheless we have found both *HeT-A* and *TART* homologues in every *Drosophila* species we have studied,[52-55] including *D. virilis*, a species originally reported to depend on recombination to maintain its telomeres.[56]

Our search for *D. virilis* telomere elements was initiated by using the most conserved part of the *D. melanogaster TART* pol gene in low stringency hybridization experiments to search for a fragment that could then be used to probe a lambdaphage library of *D. virilis* DNA.[55] We found a cross-hybridizing fragment in *D. americana*, a species closely related to *D. virilis*. With this fragment, two phage in the *D. virilis* library were found and sequenced. Both had tandem copies of *TART*, showing that our strategy had been successful. The 3' end of one *TART* was joined to 5' sequence of an unidentified element which had been truncated by cloning. This 5' sequence was used to reprobe the lambda phage library. Two new phage were selected and sequenced.[54] Both contained tandem copies of *HeT-A*, revealing that the truncated element in the *TART* clone was *HeT-A*. The tandem array of elements in one of these phage also contained a novel element, *Uvir*.

Uvir looks like a non-LTR retrotransposon that has 5' and 3' UTRs very similar to those of *HeT-A* but lacks a Gag coding region; instead it has a coding region that most closely resembles the Pol coding region of *jockey*. Because it encodes Pol, but not Gag, *Uvir* represents a new kind of non-LTR element and it is not clear that it is a successful one. Searches through the *D. virilis* genome database show only a few partial copies.[54] However, it is important to note that even the *D. melanogaster* genome, the first and most thoroughly sequenced *Drosophila* genome, is now the focus of a large Heterochromatin Genome Project to sequence the repeated parts of the genome. The project is finding new sequence and correcting assembly of other sequence in *D. melanogaster*. Much less is known about other *Drosophila* genomes.

Hybridization to total *D. virilis* DNA shows that there are very few copies of the *Uvir pol* gene. We have suggested that the *Uvir* ORF might come from a cellular reverse transcriptase.[54] If so, that cellular gene might be the ancestral source of enzyme for *HeT-A*. Indeed, it might still be functioning in *HeT-A* transposition.

Although the *D. virilis* elements, *HeT-A*^{vir} and *TART*^{vir}, differ markedly from their homologues in other species (Fig. 7), their sequences also group in the telomere element group of the jockey clade of non-LTR elements. The cloned sequences are found in mixed tandem arrays and hybridize only to telomere regions of *D. virilis* polytene chromosomes. Thus there is strong evidence that these *D. virilis* elements are true homologues.

In addition, *HeT-A*^{vir} has the same structural features that distinguish *HeT-A*^{mel} from other non-LTR elements; it has unusually long 3' UTRs with an irregular pattern of A-rich repeats and does not encode reverse transcriptase. *TART*^{vir} differs more markedly from *TART*^{mel}, having a significantly shorter 3' UTR and a large domain of unknown function (the X domain) at the C-terminal end of its Pol coding region. A similar C-terminal domain is seen in *D. americana* but not in other species. *TART*^{vir}, like *TART* in other species, yields both sense and antisense transcripts. The conservation of unusual features in spite of marked sequence change suggests that these features are important for telomere function.

One of the remarkable properties of *HeT-A*^{mel} is the specific localization of its Gag protein to telomeres in interphase cells.[28] It seems likely that this localization is important for targeting transposition to chromosome ends. *HeT-A*^{vir} Gag shows similar telomere localization in spite of the large difference in the amino acid sequences of the two proteins. These species-specific differences in amino acid sequence might be driven by need to coevolve with the various cellular components that Gag must interact with as it moves to telomeres, but this does not seem to be the case, *HeT-A*^{vir} Gag forms Het dots when it is expressed in *D. melanogaster* cells and *HeT-A*^{mel} forms Het dots in *D. virilis* cells. Thus both proteins interact appropriately with cellular targeting proteins in the other species.[57]

Gag sequences of *TART*^{vir} and *HeT-A*^{vir} are about equally diverged from their *D. melanogaster* homologs but *TART*^{vir} Gag targeting differs from that of *TART*^{mel} Gag.[57] In both *D. melanogaster* and *D. virilis* cells, *TART*^{mel} Gag enters the nucleus and interacts with *HeT-A*^{mel} Gag to be directed to telomere regions. These localization experiments were done with tagged proteins transiently expressed in cultured cells. In similar experiments, *TART*^{vir} Gag entered the nucleus only if the last ~200 amino acids had been deleted. We believe that the behavior of *TART*^{vir} Gag is a reflection of the experimental design, rather than its normal behavior, but, in either case, *TART*^{vir} Gag differs more from *TART*^{mel} Gag than *HeT-A*^{vir} Gag differs from *HeT-A*^{mel} Gag.

Given that *D. melanogaster* and *D. virilis* are about as widely separated as any members of the *Drosophila* genus (~60 My),[58] it is reasonable to assume that retrotransposon telomeres antedate the genus and will eventually be found in other Diptera.

Drosophila Telomeres Resemble Other Telomeres Both Structurally and Functionally

Transposable elements are generally considered to be parasitic DNA. *HeT-A*, *TART* and *Tahre* are the first elements that appear to be entirely beneficial to the cell. At first glance, *Drosophila* telomeres seem very different from those produced by telomerase but in fact the two telomeres are basically very similar. Both are extended by reverse transcription of RNA templates that produces long arrays of tandem repeats. In fact, telomerase appears to be closely related to the reverse transcriptase of non-LTR retrotransposons.[59] Both kinds of telomeres can be extended by recombination-based mechanisms but, for both, this mechanism appears to be primarily a back-up mechanism.[11-14,60-62]

The three elements that are the *Drosophila* repeats are much longer than repeats produced by telomerase but the total length of the telomere array is similar to telomere arrays in other multicellular eukaryotes. The lengths of these arrays fluctuate around a set point and, in organisms as different as yeast and man, that set point can be changed by genetic background and environment.[5,63,64] There are technical difficulties in accurately measuring *Drosophila* telomere length; however, several genes have been shown to affect the length set point or to affect the rate of transposition to a broken

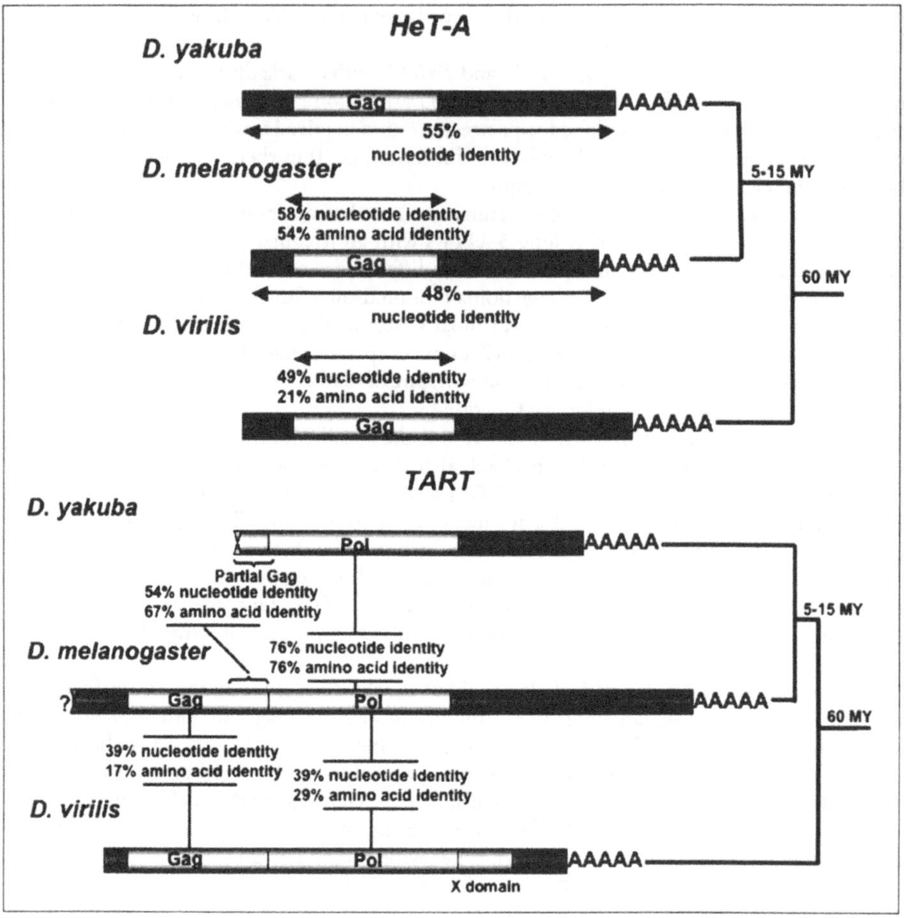

Figure 7. Comparisons of *HeT-A* and *TART* elements from three *Drosophila* species. Elements are drawn approximately to scale but individual elements vary in length of both coding and noncoding regions. Because individual elements differ in sequence, % identity differs depending on elements compared. Numbers shown are typical. Dark gray: 5′ and 3′ untranslated regions. Light gray: Gag and Pol coding regions. Scale on right indicates separation of species in millions of years. For *HeT-A* (top): Full length arrows between species indicate % nucleotide identity between elements in that pair of species. Note that, although *D. virilis* is much more distant from *D. melanogaster* than is *D. yakuba*, the *D. virilis* element shows significant nucleotide identity, showing strong conservation of sequence in the 5′ and 3′ untranslated regions. Note also that in both comparisons the nucleotide sequence of the Gag coding region is more conserved than is the amino acid sequence. For TART (bottom): Only the coding regions are compared and the X domain found only in *D. virilis* Pol is not included. The available sequences of *D. yakuba TART* are 5′-truncated so only a partial gag sequence is shown. The untranslated regions are too different for any meaningful alignment of any two species. ? indicates that the 5′ end of $TART^{mel}$ has not been completely defined.

chromosome end.[31,32,38,61,62] Thus, length regulation is seen with both types of telomeres. It may be that RNA-templated extension is the predominant mechanism for telomere maintenance because it can be easily regulated and produce rapid change in length.

An increasing number of proteins are considered to be telomere-associated either because they have been found on telomeres or because mutation of the protein causes chromosome end stickiness. Many proteins that are telomere-associated in other organisms are also telomere-associated in *Drosophila*. The list includes proteins involved in DNA damage response and repair, such as ATM, RAD50, MRE11, Ku70 and Ku80 and chromatin structure, such as HP1 (see ref. 65 for review).

It is sometimes incorrectly said that *Drosophila* telomeres are unusual because they do not need special telomere sequences; this is based on a misinterpretation of some experiments that have been done with *Drosophila*. The straightforward experiment of simply inducing a chromosome break in a mitotic cell results in activation of a checkpoint and eventual cell death for all organisms studied, including *Drosophila*.[66] In contrast to this simple experiment, *Drosophila* experiments that allow recovery of broken chromosomes without a cap of telomere DNA are multigenerational experiments that involve checkpoint evasion followed by strong selection to maintain the broken chromosome in subsequent generations. These experiments utilize genetic tools not available in other organisms and, in fact, offer clues to telomere behavior that may also pertain to other organisms.

Drosophila geneticists have found two ways to produce broken chromosomes that evade checkpoints. Breaks can be induced in the ovaries of *mu2/mu2* females[67] or they can be induced by *P-element* transposition.[68] In either case, once through the checkpoint, the broken end will not activate a checkpoint in subsequent cell cycles. Broken chromosomes that slip through one checkpoint have received scant attention in other organisms; however, there is one study in budding yeast[69] that suggests that checkpoint-evaders will not be stopped at that checkpoint in subsequent divisions in organisms. In these yeast experiments, after being held by a check point for a period of time, some broken chromosomes managed to complete the cell cycle. Although those that completed the cell cycle did not activate the checkpoint in later cell divisions, they showed a relatively high rate of loss in subsequent divisions in organisms other than *Drosophila*. Sandell and Zakian point out that this experiment demonstrates two critical functions of telomeres in yeast: telomeres distinguish ends from breaks and also prevent chromosome loss. These two functions are separated in time; end identification is needed only before the first checkpoint is passed while the tendency for broken chromosomes to be lost continues through subsequent cell generations.

Similarly, in the *Drosophila* experiments, end identification is needed only for passing the first checkpoint because broken ends induced in a *mu2/mu2* background can be maintained in a wild type background after the first checkpoint.[67] The loss of broken chromosomes in these *Drosophila* experiments was prevented by other genetic tools that select for the broken chromosome. For example, a broken X chromosome can be retained by a mating scheme that causes the broken X to be passed only from father to son. Because there is only one X in males, this broken chromosome is essential for maintenance of the stock and all surviving males will carry the broken chromosome. In these experiments, broken chromosomes shorten by an average of seventy nucleotides per fly generation, a rate consistent with loss of the terminal RNA primer in DNA replication.[68,70,77]

The experiments just described dissect some aspects of telomere function, giving clues that probably apply to other organisms. Telomeres distinguish ends from breaks, but only until the first checkpoint is passed. Lack of telomere sequences does not necessarily make a "sticky end" because surviving broken chromosomes do not form end-to-end attachments. In fact the presence or lack of telomere DNA may not be relevant to "sticky ends" because both *Drosophila* and human chromosomes can form end-to-end junctions that contain significant amounts of telomere DNA.[71,72] Without a telomere, chromosomes shorten but this happens so slowly that it could be many generations before a vital gene activity is lost, even if the end is not healed by transposition of retrotransposons.

It should be noted that these *Drosophila* which carry a broken chromosome do not test several important aspects of telomere function. These flies are living in relatively non-stressful conditions and, importantly, the broken chromosome has no competition from a telomere-containing homolog until a healed chromosome begins to take over the line. Thus only the most drastic effects on fitness will be detected. In addition, all of the other chromosomes have normal telomeres so the loss of one telomere may have little effect on phenotypes that depend on the total amount of cellular telomere sequence or on the organization of all chromosomes in the nucleus.

Evolution of Retrotransposon Telomeres

Understanding the origin of *Drosophila* telomeres would be a significant step toward understanding the evolution of both telomeres and transposable elements. There are several possibilities: (1) Telomerase could be the ancestral mechanism and the *Drosophila* telomeres could have evolved from telomerase, (2) *Drosophila* telomeres could be remainders of the ancestral mechanism, (3) *Drosophila* telomeres could be derived from transposable elements that had no relation to telomeres until they were co-opted to substitute for a lost telomerase. None of these explanations can be eliminated with confidence. However we will offer one possible scenario that is consistent with what is now known and also satisfies the requirements of Occam's razor in that only one reassortment of existing genes is required for the minimally needed functionality; other changes would be the natural result of long-term coevolution with the *Drosophila* cell.

We have suggested that telomerase is the ancestral mechanism and that telomeric retrotransposons are derived from telomerase.[73] We speculate that somewhere in the lineage leading to *Drosophila*, the gene for one of the proteins required to deliver telomerase to its target fused to the gene for the telomerase RNA template. This would be analogous to having a translocation fuse the 3′ end of the *Cdc13* gene to the *Tlc1* gene in yeast. Transcripts of this compound gene would still be translated to yield a protein designed to take the template RNA to the telomere where it would be reverse transcribed by the catalytic subunit of telomerase. The compound gene, comprised of a coding region (derived from the telomerase-related protein) and a long 3′ UTR (derived from the telomerase RNA template), could be the ancestor of *HeT-A*.

This hypothesis provides a relatively easy transition between telomerase and telomeric retrotransposons. The two mechanisms might even coexist for some time in diploid cells because both the retrotransposon-encoded protein and its RNA should still be able to interact with cellular components necessary for telomerase function. Our hypothesis also suggests that the gene for the telomerase catalytic subunit might persist in the *Drosophila* genome. The *D. melanogaster* genome has no sequence with all the hallmarks of telomerase but these might not have been conserved after the RNA template changed so drastically.

The invariant partnership between *HeT-A* and *TART* suggests that *TART* subsequently might have been coopted to take over the enzymatic function from the telomerase catalytic subunit. *TART* differs from *HeT-A* in its promoters, the organization of its untranslated regions and its pattern of transcription. In fact, the predominant similarity between the two elements is in their Gag proteins, which target them to telomeres, this suggests that *TART* may have been a preexisting retrotransposon that acquired Gag coding from *HeT-A* and therefore become targeted to telomeres. The partnership with *TART* might be favored because it offers *HeT-A* more sources of enzyme activity than the single copy telomerase catalytic subunit. Recent evidence that transposition and expression of *HeT-A* and *TART* are sensitive to disruptions in the RNAi pathway suggests another possible advantage of a *HeT-A*/*TART* partnership.[38,39] *TART* appears to be the principal target of this newly recognized regulation mechanism with *HeT-A* possibly controlled by *TART*. As mentioned above, the number of *HeT-A* and *TART* elements per genome varies in a correlated manner between different stocks and cell lines.[31]

It has recently been reported that LINE-1 elements lacking endonuclease activity can transpose in an orientation-specific manner onto telomere ends in Chinese Hamster cells that have dysfunctional telomeres caused by loss of DNA-PKcs.[74] This study and earlier work showing that these endonuclease-independent elements can integrate into internal DNA lesions provide support for the authors' suggestion that non-LTR retrotransposons served to repair DNA lesions before these elements acquired endonuclease activity.[75] It would be tempting to suggest that the telomere retrotransposons owe their end-specificity to lack of endonuclease activity. However, the pol gene sequences of *TART*, *Tahre* and *Uvir* all have well-conserved endonuclease coding sequences, suggesting that these sequences are important even for transposition to chromosome ends.

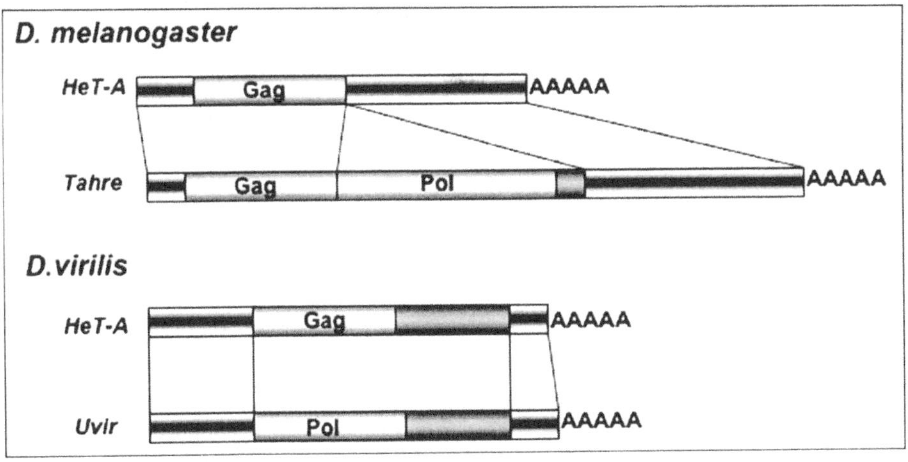

Figure 8. *HeT-A* sequences are strongly conserved in *Tahre* and *Uvir*. Elements are drawn approximately to scale. In *D. melanogaster, Tahre* sequence is highly similar to *HeT-A*^{mel} in the gag gene (light gray) and most of the untranslated regions (horizontal black stripe). In *D. virilis, Uvir* is highly similar to *HeT-A*^{vir} in the 5′ UTR and the last ~600 nt of the 3′UTR (horizontal black stripe). Dark gray regions indicate 3′UTR sequences specific to *Tahre* or *Uvir*.

It has been suggested that *Tahre* is an ancestral element from which *HeT-A* was derived by loss of the Pol coding.[37] However, *HeT-A* is much more abundant than *Tahre* and thus apparently more successful. It is not obvious why loss of an essential function should make the progeny more successful than the parent; although such an outcome might be evolutionarily favored, if cellular well-being requires closely regulated rapid changes in telomere length. On the other hand, *Tahre* could have arisen from *HeT-A* by acquiring a *pol* gene, just as retroviral oncogenes have been acquired from the cellular genome. If so, the scarcity of *Tahre* suggests that combining the two activities is not beneficial in this environment.

Both *Tahre* in *D. melanogaster* and *Uvir* in *D. virilis* have 5' and 3' UTR sequences very closely related to the *HeT-A* elements of their respective species (Fig. 8). Although it is not possible to say whether coding region differences in these elements are the result of gain or loss, taken together, *HeT-A*, *Tahre* and *Uvir* show that sequence changes within *HeT-A*-related UTR sequences may be relatively frequent on an evolutionary time scale. Because these three of the four elements found in telomere arrays have related 3' and 5' UTR sequences, these sequences must have a special role in telomeres. It will be important to look for more elements so that we can use sequence analysis to deduce the history of their components.

Conclusion

It is intriguing that retrotransposons have so completely adapted to an essential cellular role. Although the ways in which *HeT-A* and *TART* have coevolved to perform these functions are interesting in their own right and important to our understanding what is essential to the role of the telomere, we find the more general evolutionary implications most fascinating.

It should be noted that the scenario for deriving retrotransposons from the telomerase machinery, described above, could also occur in organisms that do not lose telomerase. As noted above, if the organism is diploid, modification of one copy of telomerase components leaves the other copy in the genome functional. The modified copy of the telomerase components might be then be lost from the population after the newly generated retrotransposon has occupied other sites in the genome. (Of course, telomerase components may not be the only cellular genes that can give rise to retrotransposable elements.) Thus there may have been multiple sources of retrotransposons in different organisms. The variant telomeres of *Drosophila* raise many questions about the evolution of telomeres and of transposable elements, two topics about which we know little.

Acknowledgements

Work in the authors' laboratory is supported by National Institutes of Health Grant 50315.

References

1. Muller HJ. The remaking of chromosomes. Collecting Net 1938; 13:181-195.
2. Heitz E. Uber α- und β-Heterochromatin Sowie Konstanz und Bau der Chromomeren bei Drosophila. Biol Zentralbl 1934; 54:588-609.
3. McClintock B. The fusion of broken ends of sister half-chromatids following chromatid breakage at meiotic anaphases. Mo Agric Exp Res Stn Res Bull 1938; 290:1-4.
4. Blackburn EH. Telomerases. Annu Rev Biochem 1992; 61:113-129.
5. Greider CW. Telomere length regulation. Annu Rev Biochem 1996; 65:337-365.
6. Pryde FE, Gorham HC, Louis EJ. Chromosome ends: all the same under their caps. Curr Opin Genet Dev 1997; 7(6):822-828.
7. Kornberg A, Baker TA. DNA Replication. 2 ed. New York: WH Freeman; 1992.
8. Blackburn EH. Switching and signaling at the telomere. Cell 2001; 106(6):661-673.
9. Collins K. Mammalian telomeres and telomerase. Curr Opin Cell Biol 2000; 12(3):378-383.
10. Greider CW. Telomerase activity, cell proliferation and cancer. Proc Natl Acad Sci USA 1998; 95(1):90-92.
11. Lundblad V, Blackburn EH. An alternative pathway for yeast telomere maintenance rescues est1- senescence. Cell 1993; 73(2):347-360.
12. Teng SC, Zakian VA. Telomere-telomere recombination is an efficient bypass pathway for telomere maintenance in Saccharomyces cerevisiae. Mol Cell Biol 1999; 19(12):8083-8093.
13. Henson JD, Neumann AA, Yeager TR et al. Alternative lengthening of telomeres in mammalian cells. Oncogene 2002; 21(4):598-610.
14. Lundblad V. Telomere maintenance without telomerase. Oncogene 2002; 21(4):522-531.
15. Ford LP, Zou Y, Pongracz K et al. Telomerase can inhibit the recombination-based pathway of telomere maintenance in human cells. J Biol Chem 2001; 276(34):32198-32203.
16. Perrem K, Colgin LM, Neumann AA et al. Coexistence of alternative lengthening of telomeres and telomerase in hTERT-transfected GM847 cells. Mol Cell Biol 2001; 21(12):3862-3875.
17. Frydrychova R, Grossmann P, Trubac P et al. Phylogenetic distribution of TTAGG telomeric repeats in insects. Genome 2004; 47(1):163-178.
18. Vitkova M, Kral J, Traut W et al. The evolutionary origin of insect telomeric repeats, (TTAGG)n. Chromosome Res 2005; 13(2):145-156.
19. Osanai M, Kojima KK, Futahashi R et al. Identification and characterization of the telomerase reverse transcriptase of Bombyx mori (silkworm) and Tribolium castaneum (flour beetle). Gene 2006; 376(2):281-289.
20. Cohn M, Edstrom JE. Telomere-associated repeats in Chironomus form discrete subfamilies generated by gene conversion. J Mol Evol 1992; 35(2):114-122.
21. Rosen M, Kamnert I, Edstrom JE. Extrachromosomal RNA-DNA complex containing long telomeric repeats in chironomids. Insect Mol Biol 2002; 11(2):167-174.
22. Roth CW, Kobeski F, Walter MF et al. Chromosome end elongation by recombination in the mosquito Anopheles gambiae. Mol Cell Biol 1997; 17(9):5176-5183.
23. Voytas DF, Boeke JD. Ty1 and Ty5 of Sacharomyces cerevisiae. In: Craig NL, Craigie R, Gellert M, Lambowitz AM eds. Mobile DNA II. Washington, D.C.: American Society for Microbiology; 2002:631-662.
24. Luan DD, Korman MH, Jakubczak JL et al. Reverse transcription of R2Bm RNA is primed by a nick at the chromosomal target site: a mechanism for non-LTR retrotransposition. Cell 1993; 72(4):595-605.
25. Christensen SM, Eickbush TH. R2 target-primed reverse transcription: ordered cleavage and polymerization steps by protein subunits asymmetrically bound to the target DNA. Mol Cell Biol 2005; 25(15):6617-6628.
26. Christensen SM, Ye J, Eickbush TH. RNA from the 5' end of the R2 retrotransposon controls R2 protein binding to and cleavage of its DNA target site. Proc Natl Acad Sci USA 2006; 103(47):17602-17607.
27. Biessmann H, Mason JM, Ferry K et al. Addition of telomere-associated HeT DNA sequences "heals" broken chromosome ends in Drosophila. Cell 1990; 61(4):663-673.
28. Rashkova S, Karam SE, Kellum R et al. Gag proteins of the two Drosophila telomeric retrotransposons are targeted to chromosome ends. J Cell Biol 2002; 159(3):397-402.

29. Malik HS, Burke WD, Eickbush TH. The age and evolution of non-LTR retrotransposable elements. Mol Biol Evol 1999; 16(6):793-805.
30. Casacuberta E, Pardue ML. HeT-A and TART, two Drosophila retrotransposons with a bona fide role in chromosome structure for more than 60 million years. Cytogenet Genome Res 2005; 110(1-4):152-159.
31. George JA, DeBaryshe PG, Traverse KL et al. Genomic organization of the Drosophila telomere retrotransposable elements. Genome Res 2006; 16(10):1231-1240.
32. Savitsky M, Kravchuk O, Melnikova L et al. Heterochromatin protein 1 is involved in control of telomere elongation in Drosophila melanogaster. Mol Cell Biol 2002; 22(9):3204-3218.
33. George JA, Pardue ML. The promoter of the heterochromatic Drosophila telomeric retrotransposon, HeT-A, is active when moved into euchromatic locations. Genetics 2003; 163(2):625-635.
34. Walter MF, Biessmann H. Expression of the telomeric retrotransposon HeT-A in Drosophila melanogaster is correlated with cell proliferation. Dev Genes Evol 2004; 214(5):211-219.
35. Pardue ML, Rashkova S, Casacuberta E et al. Two retrotransposons maintain telomeres in Drosophila. Chromosome Res 2005; 13(5):443-453.
36. Danilevskaya ON, Traverse KL, Hogan NC et al. The two Drosophila telomeric transposable elements have very different patterns of transcription. Mol Cell Biol 1999; 19(1):873-881.
37. Abad JP, De Pablos B, Osoegawa K et al. TAHRE, a novel telomeric retrotransposon from Drosophila melanogaster, reveals the origin of Drosophila telomeres. Mol Biol Evol 2004; 21(9):1620-1624.
38. Savitsky M, Kwon D, Georgiev P et al. Telomere elongation is under the control of the RNAi-based mechanism in the Drosophila germline. Genes Dev 2006; 20(3):345-354.
39. Casacuberta E, Pardue ML. RNA interference has a role in regulating Drosophila telomeres. Genome Biol 2006; 7(5):220.
40. Karpen GH, Spradling AC. Analysis of subtelomeric heterochromatin in the Drosophila minichromosome Dp1187 by single P element insertional mutagenesis. Genetics 1992; 132(3):737-753.
41. Cryderman DE, Morris EJ, Biessmann H et al. Silencing at Drosophila telomeres: nuclear organization and chromatin structure play critical roles. EMBO J 1999; 18:3724-3735.
42. Biessmann H, Prasad S, Semeshin VF et al. Two distinct domains in Drosophila melanogaster telomeres. Genetics 2005; 171(4):1767-1777.
43. Fujiwara H, Osanai M, Matsumoto T et al. Telomere-specific non-LTR retrotransposons and telomere maintenance in the silkworm, Bombyx mori. Chromosome Res 2005; 13(5):455-467.
44. Danilevskaya ON, Lowenhaupt K, Pardue ML. Conserved subfamilies of the Drosophila HeT-A telomere-specific retrotransposon. Genetics 1998; 148(1):233-242.
45. Carvalho AB, Dobo BA, Vibranovski MD et al. Identification of five new genes on the Y chromosome of Drosophila melanogaster. Proc Natl Acad Sci USA 2001; 98(23):13225-13230.
46. Danilevskaya O, Lofsky A, Kurenova EV et al. The Y chromosome of Drosophila melanogaster contains a distinctive subclass of Het-A-related repeats. Genetics 1993; 134(2):531-543.
47. Danilevskaya ON, Kurenova EV, Pavlova MN et al. He-T family DNA sequences in the Y chromosome of Drosophila melanogaster share homology with the X-linked stellate genes. Chromosoma 1991; 100(2):118-124.
48. Losada A, Agudo M, Abad JP et al. HeT-A telomere-specific retrotransposons in the centric heterochromatin of Drosophila melanogaster chromosome 3. Mol Gen Genet 1999; 262(4-5):618-622.
49. Swanstorm R, Wills JW. Synthesis, assembly and processing of viral proteins. In: Coffin JM, Hughes SH, Varmus HE, eds. Retroviruses. Cold Spring Harbor: Cold Spring Harbor Laboratory; 1997:263-334.
50. Rashkova S, Athanasiadis A, Pardue ML. Intracellular targeting of Gag proteins of the Drosophila telomeric retrotransposons. J Virol 2003; 77(11):6376-6384.
51. Rashkova S, Karam SE, Pardue ML. Element-specific localization of Drosophila retrotransposon Gag proteins occurs in both nucleus and cytoplasm. Proc Natl Acad Sci USA 2002; 99(6):3621-3626.
52. Danilevskaya ON, Tan C, Wong J et al. Unusual features of the Drosophila melanogaster telomere transposable element HeT-A are conserved in Drosophila yakuba telomere elements. Proc Natl Acad Sci USA 1998; 95(7):3770-3775.
53. Casacuberta E, Pardue ML. Coevolution of the telomeric retrotransposons across Drosophila species. Genetics 2002; 161(3):1113-1124.
54. Casacuberta E, Pardue ML. HeT-A elements in Drosophila virilis: retrotransposon telomeres are conserved across the Drosophila genus. Proc Natl Acad Sci USA 2003; 100(24):14091-14096.
55. Casacuberta E, Pardue ML. Transposon telomeres are widely distributed in the Drosophila genus: TART elements in the virilis group. Proc Natl Acad Sci USA 2003; 100(6):3363-3368.
56. Biessmann H, Zurovcova M, Yao JG et al. A telomeric satellite in Drosophila virilis and its sibling species. Chromosoma 2000; 109(6):372-380.
57. Casacuberta E, Marín FA, Pardue M-L. Intracellular targeting of telomeric retrotransposon Gag proteins of distantly related Drosophila species. Proc Natl Acad Sci USA 2007; 104(20):8391-8396.

58. Beverley SM, Wilson AC. Molecular evolution in Drosophila and the higher Diptera II. A time scale for fly evolution. J Mol Evol 1984; 21(1):1-13.
59. Nakamura TM, Morin GB, Chapman KB et al. Telomerase catalytic subunit homologs from fission yeast and human. Science 1997; 277(5328):955-959.
60. Kahn T, Savitsky M, Georgiev P. Attachment of HeT-A sequences to chromosomal termini in Drosophila melanogaster may occur by different mechanisms. Mol Cell Biol 2000; 20(20):7634-7642.
61. Melnikova L, Georgiev P. Enhancer of terminal gene conversion, a new mutation in Drosophila melanogaster that induces telomere elongation by gene conversion. Genetics 2002; 162(3):1301-1312.
62. Siriaco GM, Cenci G, Haoudi A et al. Telomere elongation (Tel), a new mutation in Drosophila melanogaster that produces long telomeres. Genetics 2002; 160(1):235-245.
63. Askree SH, Yehuda T, Smolikov S et al. A genome-wide screen for Saccharomyces cerevisiae deletion mutants that affect telomere length. Proc Natl Acad Sci USA 2004; 101(23):8658-8663.
64. Smogorzewska A, de Lange T. Regulation of telomerase by telomeric proteins. Annu Rev Biochem 2004; 73:177-208.
65. Cenci G, Ciapponi L, Gatti M. The mechanism of telomere protection: a comparison between Drosophila and humans. Chromosoma 2005; 114(3):135-145.
66. Ahmad K, Golic KG. Telomere loss in somatic cells of Drosophila causes cell cycle arrest and apoptosis. Genetics 1999; 151(3):1041-1051.
67. Mason JM, Strobel E, Green MM. mu-2: mutator gene in Drosophila that potentiates the induction of terminal deficiencies. Proc Natl Acad Sci USA 1984; 81(19):6090-6094.
68. Levis RW. Viable deletions of a telomere from a Drosophila chromosome. Cell 1989; 58(4):791-801.
69. Sandell LL, Zakian VA. Loss of a yeast telomere: arrest, recovery and chromosome loss. Cell 1993; 75(4):729-739.
70. Mikhailovsky S, Belenkaya T, Georgiev P. Broken chromosomal ends can be elongated by conversion in Drosophila melanogaster. Chromosoma 1999; 108(2):114-120.
71. Oikemus SR, Queiroz-Machado J, Lai K et al. Epigenetic telomere protection by Drosophila DNA damage response pathways. PLoS Genet 2006; 2(5):e71.
72. Bi X, Wei SC, Rong YS. Telomere protection without a telomerase; the role of ATM and Mre11 in Drosophila telomere maintenance. Curr Biol 2004; 14(15):1348-1353.
73. Pardue ML, Danilevskaya ON, Traverse KL et al. Evolutionary links between telomeres and transposable elements. Genetica 1997; 100(1-3):73-84.
74. Morrish TA, Garcia-Perez JL, Stamato TD et al. Endonuclease-independent LINE-1 retrotransposition at mammalian telomeres. Nature 2007; 446(7132):208-212.
75. Morrish TA, Gilbert N, Myers JS et al. DNA repair mediated by endonuclease-independent LINE-1 retrotransposition. Nat Genet 2002; 31(2):159-165.
76. Sahara K, Marec F, Traut W. TTAGG telomeric repeats in chromosomes of some insects and other arthropods. Chromosome Res 1999; 7(6):449-460.
77. Biessmann H, Carter SB, Mason JM. Chromosome ends in Drosophila without telomeric DNA sequences. Proc Natl Acad Sci USA 1990; 1758-1761.

Chapter 3

Alternative Lengthening of Telomeres in Mammalian Cells

Anthony J. Cesare and Roger R. Reddel*

Abstract

For human cells to achieve immortalization they must bypass multiple proliferative checkpoints and acquire a telomere maintenance mechanism to counteract the natural telomere attrition that results from the end-replication problem. A number of human tumors and cells immortalized in culture maintain their telomeres by a telomerase independent mechanism termed Alternative Lengthening of Telomeres (ALT). The available data indicate that ALT involves homologous recombination-mediated DNA replication and requires the activity of the MRE11/RAD50/NBS1 recombination complex. Increased levels of various types of telomere recombination events in ALT cells suggest that the cellular mechanisms which normally regulate recombination at mammalian telomeres have been lost. We review here the current literature regarding ALT and telomere biology and discuss possible mechanisms that have evolved in mammalian cells (primarily human) to inhibit deregulated homologous recombination at the telomeres and thus prevent telomere elongation and cellular immortalization.

Introduction

The chromosome ends (telomeres) of mammalian cells contain tandemly arrayed hexanucleotide repeats with the sequence 5'-TTAGGG-3'.[1] This telomeric DNA is mostly double-stranded, but it terminates in a single-stranded 3' overhang.[2] In human somatic cells, each telomere is 4-12 kb long and the single-stranded overhang contains 100-200 nucleotides (Fig. 1A). Telomeres need to be distinguished from double strand breaks (DSBs), to avoid being fused to each other by normal DNA repair mechanisms. This is achieved in part by the proteins that bind to telomeric DNA, forming a "cap" structure (Fig. 1B)(ref. 3 for review) Additionally, mammalian telomeres form a higher order structure by sequestering the 3' overhang in cis within the duplex telomeric DNA, resulting in a telomere loop (t-loop) that likely contributes to the capping mechanism.[4]

Due to the end-replication problem,[5,6] the ends of linear chromosomes shorten with each round of DNA replication.[7] In human somatic cells, the progressive telomere shortening that occurs with continued proliferation eventually results in the triggering of a replicative checkpoint. Telomere shortening and the structural changes that it presumably causes, leads to a DNA-damage checkpoint response at the telomere and induction of a permanent p53- and Rb-dependent growth arrest (i.e., replicative senescence).[8-10] Because this limits the proliferative capacity of somatic cells, including those that have accumulated oncogenic mutations, telomere shortening and replicative senescence are a potent tumor suppressor mechanism.

*Corresponding Author: Roger R. Reddel—Cancer Research Unit, Children's Medical Research Institute, 214 Hawkesbury Road, Westmead, Sydney, New South Wales 2145, Australia. Email: rreddel@cmri.usyd.edu.au

Origin and Evolution of Telomeres, edited by Jozef Nosek and Ľubomír Tomáška.

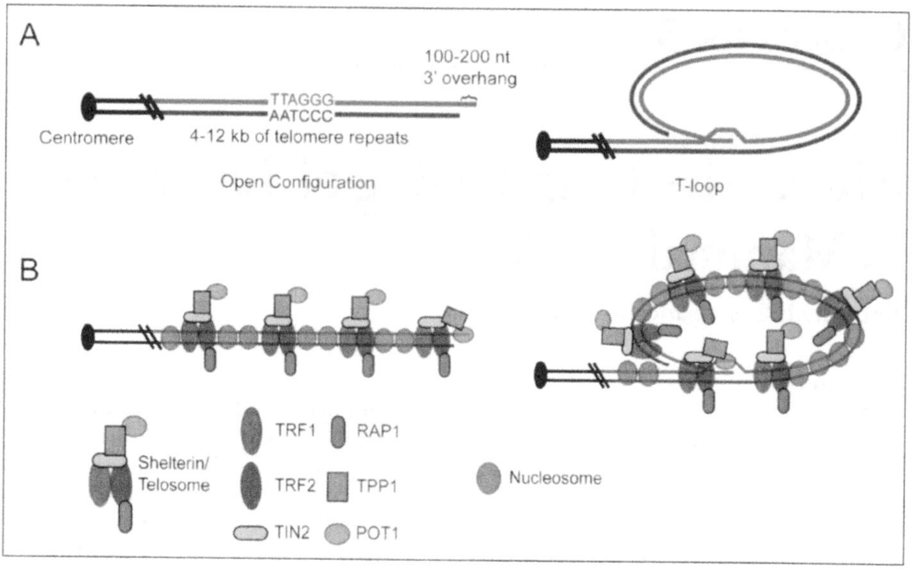

Figure 1. Human telomere components and structure. A) Graphic representation of the telomeric DNA in human cells, which is normally composed of 4-12 kb of G-rich repeats (TTAGGG in red, AATCCC in blue), culminating in a 100-200 nt 3′ overhang. Open and t-loop configurations are shown. B) Graphic representation of telomeric DNA with associated proteins. The six subunit "shelterin" or "telosome" complex coats the length of the duplex telomeric DNA via the direct interaction of TRF1 and TRF2 with telomeric DNA. TIN2 interacts with both TRF1 and TRF2 but not with telomeric DNA. TPP1 (formerly PTOP, PIP1 or TINT1) bridges TIN2 with POT1. POT1 interacts specifically with single-stranded G-rich telomeric DNA, presumably at the chromosome end when the telomere is in open configuration, or at the single-stranded region within the t-loop junction. Telomeres are also assembled into chromatin.

If senescence pathways are absent, due for example to loss of p53 and Rb function, cells will continue to divide until the telomeres become almost completely eroded, leading to crisis, a period characterized by rampant chromosome end-to-end fusions and cell death.[11] Formation of tumors is, in most cases, dependent on the evolution of cells that escape from the barriers that senescence and crisis present to unlimited proliferation. Cells that achieve this are referred to as "immortalized" and in all cases this requires the activation of a mechanism for preventing telomere shortening. In most cases this is accomplished by upregulating the activity of telomerase,[12,13] a ribonucleoprotein enzyme that adds new telomeric repeats to chromosome termini. Telomerase has an important role in cells of the germ line and in normal somatic biology, especially in those tissue compartments that depend upon extensive cellular proliferation. Nevertheless, in normal somatic cells telomerase is not expressed at sufficient levels to prevent telomere shortening and telomere length maintenance in many cancers requires dysregulated levels of telomerase. A substantial minority of immortalized cell lines and tumors are telomerase-negative, however, and in these cells telomere length maintenance can be achieved instead by a telomerase-independent mechanism, termed Alternative Lengthening of Telomeres (ALT).[14,15]

ALT may resemble (or represent) the earliest telomere maintenance mechanism (TMM), which preceded the evolution of telomerase-dependent maintenance of chromosomal termini. While the possibility cannot be excluded that a low level of ALT-like activity occurs at normal mammalian telomeres, the telomere phenotype seen in ALT-positive immortalized cells and tumors is not found in normal cells. The current data strongly support ALT being a homologous recombination

(HR)-mediated DNA replication mechanism, which occurs in the context of telomere instability resulting from loss of several controls over telomere function. Here we discuss the literature regarding telomere biology and ALT, with particular attention to the possible mechanisms that have evolved in mammalian cells (especially human) to prevent aberrant telomere maintenance by HR.

Phenotypic Identifiers of ALT Cells

ALT is defined as telomere length maintenance that is not dependent on telomerase activity. It is currently not clear whether there is more than one ALT mechanism in mammalian cells and there is no assay for ALT activity. The existence of ALT was deduced from observations of telomere length maintenance over many hundreds of population doublings (PDs) in the absence of detectable telomerase activity.[15,16] Fortunately, it is not necessary to perform this type of experiment to determine whether a cell line or tumor utilizes ALT, because ALT-positive human cells can now be recognized on the basis of a number of hallmarks. Analysis of telomeric DNA from an ALT cell line by pulsed field gel electrophoresis and Southern blotting indicates that within a population of cells the telomeric DNA ranges from <2 to >50 kb in length, with a mean size that is usually around 20 kb.[15] Telomere length heterogeneity is also obvious at the single cell level when observing metaphases from ALT cells by fluorescent in-situ hybridization (FISH) with telomere specific probes.[17] This confirms that some telomeres are very long and, notably, that within the great majority of individual ALT cells there is a subset of chromosome ends that lack any discernable telomere signal. Telomeres in ALT cells are also in a very dynamic state, exhibiting sudden lengthening and shortening events.[18] A substantial portion of the telomeric repeats in ALT cells is extrachromosomal and may be linear[19,20] or circular[21,22] in form. The extrachromosomal telomeric repeat (ECTR) circles (t-circles) are also heterogeneous in size, ranging from <1 to >50 kb and equivalent structures have not been observed in high abundance in normal cells or in telomerase positive mammalian cell lines.[21,22]

Another hallmark of ALT cell lines and tumors is the presence of specialized promyelocytic leukemia nuclear bodies (PNBs), termed ALT-associated PNBs (APBs).[23] In addition to the usual PNB components, including PML and Sp100, APBs are defined by the presence of telomeric DNA and telomere binding proteins and also contain an assortment of DNA replication, recombination and repair factors (Table 1). Large, easily recognizable APBs are present in only a minority of cycling ALT cells,[23] most likely due to their enrichment during the G2 phase of the cell cycle.[24,25]

Occurrence of ALT

The typical ALT phenotype has only been found in abnormal situations, including immortalized human cell lines, human tumors and tumors or cell lines derived from telomerase null mice, suggesting that this an anomalous telomere phenotype.[17,26] Up to 10% of all human cancers and a greater proportion of cells immortalized in culture, utilize the ALT TMM.[17,27] Immortalization via activation of ALT appears to occur readily in cells of some Li-Fraumeni syndrome individuals (p53 +/mut) and in fibroblasts immortalized using the SV40 Large T antigen.[28] ALT is not often detected in carcinomas (tumors of epithelial origin), but, for reasons that are currently unknown, ALT occurs commonly in sarcomas (tumors of mesenchymal origin) and there are some types of sarcomas where more than 50% of tumors are ALT-positive.[29]

Abundant Telomere Recombination in ALT Cells

The rapid dynamics of telomere length polymorphisms in ALT cells suggested that the TMM involves HR.[18] HR-dependent DNA replication of telomeres (Fig. 2B) within ALT cells was demonstrated by following a neomycin resistance marker inserted within the telomere repeats, or immediately proximal to the telomere (i.e., in a subtelomeric location), in the GM847 (ALT) and HT1080 (telomerase positive) cell lines.[30] After many PDs, the telomeric neomycin marker was copied to different telomeres within ALT cells, but no movement was observed for the sub-telomeric marker over the same period. No movement of the telomeric marker occurred in the telomerase

Table 1. Known protein constituents of ALT-associated promyelocytic leukemia nuclear bodies

Protein	Function	Reference
PML	PNB constituent	23
SP100	PNB constituent	49
TRF1	telomere binding protein	23
TRF2	telomere binding protein	23
RAP1	telomere protein	92
TIN2	telomere protein	62
RAD51	homologous recombination	23
RAD52	homologous recombination	23
RAD51D	HR, telomere capping	65
MRE11/RAD50/NBS1	DNA damage response, repair, HR	25,61
Bloom helicase (BLM)	RecQ helicase	93,94
Werner helicase (WRN)	RecQ helicase	95
RAD9-RAD1-HUS1	DNA damage response and repair	47
RAD17	DNA damage response and repair	47
BRCA1	DNA damage response and repair	92
RIF1	DNA damage response	83
Replication Protein A (RPA)	single strand DNA binding protein	23
ERCC1	DNA repair	96
XPF	DNA repair	96
hnRNP A2	ssDNA/RNA molecular adaptor	97

positive HT1080 cell line, suggesting that substantial ongoing copying of telomeric DNA to other telomeres occurs exclusively in ALT cells.

ALT telomeres also undergo abundant postreplicative exchanges, compared to non-ALT cells, as assayed by telomere specific chromosome-orientation FISH (CO-FISH).[31,32] While these exchanges are commonly referred to as telomere-sister chromatid exchanges (T-SCEs; Fig. 2C), it is possible that they may also arise from recombination with nonsister telomeres or ECTR elements. It has been proposed that unequal exchanges between telomeres could lead to telomere length changes.[33]

Abundant t-circles also suggest that there is an elevated rate of intra-telomeric recombination-related events in ALT cells. It seems that t-circles arise from t-loops: the size of the t-circles in the GM847 ALT cell line closely correlates with the size of the loop portion of t-loops, as measured by electron microscopy.[21] T-loop formation and stability are thought to require TRF2 function.[4,34] Expression in mammalian cells of a truncated allele of TRF2 lacking the basic domain ($TRF2^{\Delta B}$) results in telomere rapid deletions (TRD) and formation of t-circles, suggesting that the TRF2 basic domain protects against improper resolution of t-loop junctions (referred to here as t-loop junction resolution; t-loop JR)(Fig. 2A).[22] Consistent with this interpretation, the TRF2 basic domain interacts in vitro with four-way DNA junctions, regardless of whether the sequence consists of telomeric repeats.[35] Induction of t-circles by $TRF2^{\Delta B}$ is dependent on NBS1 and XRCC3 and RNAi knockdown of NBS1 or XRCC3 in several ALT cell lines diminishes t-circle abundance.[37] Recent experiments show that deletion of POT1A in the mouse also results in NBS1-dependent formation of t-circles.[36] Therefore, the rapid telomere shortening events[18] and abundant t-circles seen in ALT cells suggest that, in these cells, improper resolution of the HR-intermediate structures represented by t-loop junctions occurs at an increased rate. Furthermore, it is likely that these t-loop JR events are mediated by NBS1 and XRCC3 and are repressed in non-ALT cells by proteins such as TRF2 and POT1.

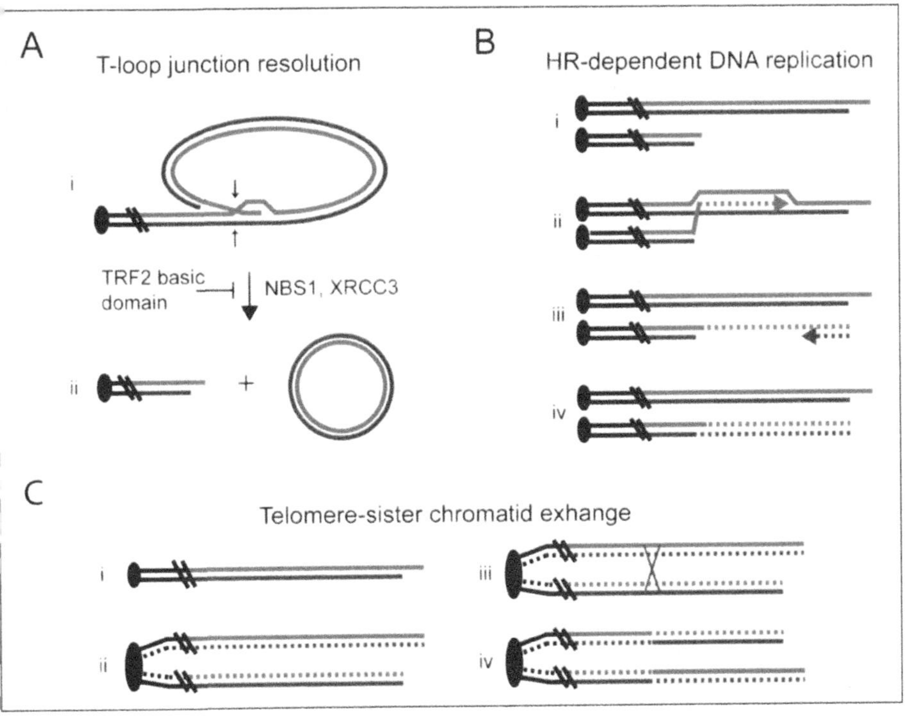

Figure 2. ALT associated telomere recombination. A) T-loop junction resolution. (i) Resolution of the t-loop junction in a NBS1 and XRCC3 dependent manner at sites denoted by arrows results in (ii) a free t-circle and a shortened telomere. Branch migration at the t-loop junction may be necessary prior to resolution (not shown). B) HR-dependent DNA replication. (ii) Invasion of a 3′ overhang within the duplex telomeric DNA of an adjacent telomere is followed by extension via DNA polymerases (red dotted line). (iii) The C-rich strand is filled in (blue dotted line) resulting in (iv) telomere elongation. C) Telomere-sister chromatid exchange. Telomeres are (ii) replicated in a semi-conservative fashion (dashed lines represent newly synthesized DNA). Following replication, homologous recombination between sister chromatid telomeres (iii) can lead to T-SCE (iv).

There is no increase in genomic HR in ALT cells as compared to non-ALT controls,[38,39] suggesting a telomere-specific dysfunction rather than a general increase in recombination in these cells. ALT telomeres thus appear to be susceptible to three distinct types of recombination events: (1) HR-dependent DNA-replication telomere copying, (2) postreplicative exchanges and (3) HR-mediated t-loop junction resolution.

Possible ALT Mechanisms

Although HR-mediated copying of one telomere by another is the simplest explanation for the spread of a DNA tag from one telomere to others,[30] other types of elongation events may also occur as observed in the telomerase null Type II survivors from the budding yeast species *Saccharomyces cerevisiae* and *Kluyveromyces lactis*.[40-42] It was suggested that in these species rolling circle replication on t-circles served as the initial lengthening event followed by HR spreading of the newly elongated telomere to other chromosome ends ("roll and spread" model). Evidence supporting this hypothesis was obtained by transfecting exogenous t-circles into telomerase null *K. lactis* cells.[43,44] The result was extension of the chromosomal telomeres with repetitive units of exogenous t-circle DNA and spreading of this sequence to other chromosome termini. Moreover, it was shown that

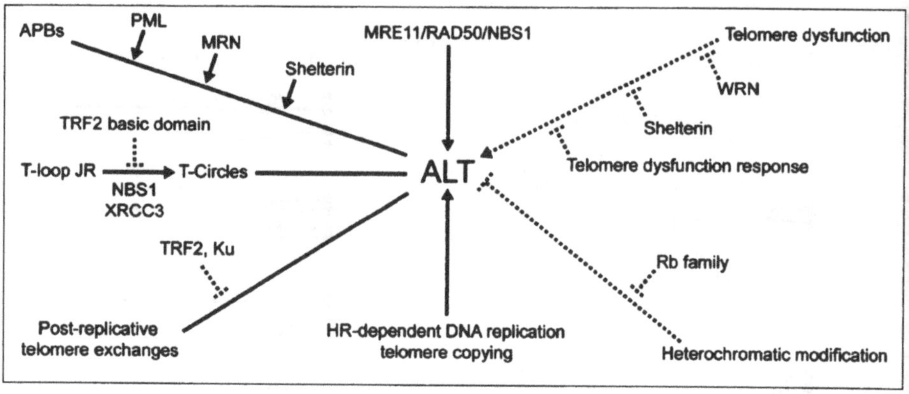

Figure 3. Factors that are known or thought to contribute to the ALT phenotype. Known associations with the ALT phenotype are signified by solid lines and more speculative associations are shown by dotted lines. APBs, t-circles and T-SCEs are known to associate with ALT, but it is not yet known if they are required for ALT to occur.

a single elongated telomere is sufficient to drive extension of the other shortened telomeres in *K. lactis* cells.[45] Physical evidence of roll and spread was also documented in the mitochondria of *Candida parapsilosis* which contain a linear genome capped by telomeres maintained in a telomerase independent manner.[46] Because all ALT cell lines examined so far contain t-circles (refs. 21,22 and C. Fasching and R. Reddel, unpublished data), it seems feasible that roll and spread could also occur in mammalian ALT cells. Additional possibilities for telomere elongation mechanisms in ALT cells include DNA replication primed from the terminal hydroxyl group of the 3' overhang within the t-loop, or via HR with ECTR DNA molecules. Presumably, once the generation of long tracts of telomeric DNA by one or more of the above means results in a threshold quantity of telomere sequence being attained, the reservoir of telomeric DNA that this constitutes will permit ongoing telomere maintenance by HR.

Although it is often assumed that APBs serve as sites of telomere extension in ALT cells, this remains to be fully validated. Evidence that is consistent with this notion includes the observation that BrdU incorporation within APBs is caffeine sensitive suggesting ATM- or ATR-dependent DNA synthesis, possibly in response to a DNA damage signal.[47] The periodic association and dissociation of chromosomal telomeres and APBs seen in live cell imaging experiments suggests APBs may colocalize with telomeres for extension.[48] Morever, when ALT is inhibited (see below) the percentage of APB positive cells decreases (ref. 49 and Z. Zhong and R. Reddel, unpublished data).

Genes Involved in ALT

It is becoming apparent that the relationship between DNA repair proteins and human telomeres is complex (ref. 50 for review). In this light and considering the phenotypic characteristics of ALT cells, almost any protein involved in telomere function, HR, DNA damage response and repair, DNA replication, or APBs could be involved in ALT. In Type II *S. cerevisiae* telomerase-null survivors, which have telomeres that phenotypically resemble those of ALT-positive human cells, the HR protein Rad52 and epistasis group members Rad50 and Rad59, as well as the RecQ Sgs1 helicase are required for telomere maintenance.[41,51-53] There are no data regarding human RAD52 involvement in ALT, although this seems likely given its function in HR. While the Sgs1 protein is the only RecQ helicase in *S. cerevisiae*, human cells contain several orthologs, with the Werner syndrome helicase (WRN) drawing considerable attention in the telomere field. Mutation in the WRN helicase leads to a premature aging phenotype (ref. 54 for review) that is recapitulated in mouse models only in the context of telomere dysfunction in late generation telomerase null

(TERC -/-) mice.[55,56] WRN telomere functions appear to include unwinding of G-quartets created in the lagging strand during DNA synthesis, that if left unresolved lead to sister chromatid loss and genomic instability due to chromosome fusions.[57,58] ALT tumors form in late generation (G5) TERC -/- WRN -/- mice following loss of p53 function.[59] Thus, WRN is not essential for ALT and this conclusion is supported by the observation that the W-v human Werner syndrome cell line is a typical ALT line.[60] This does not exclude the possibility that other RecQ family members are required for ALT.

Recent experiments have, however, implicated the complex containing human MRE11, RAD50 and NBS1 (MRN) proteins as being required for ALT activity. Over-expression of the PNB component, Sp100, led to the sequestration of MRN in Sp100 microbodies via interaction with NBS1.[49] Long term expression of Sp100 resulted in ALT suppression for >80 PD in the IIICF/c ALT cell line as characterized by a steady decrease in telomere lengths consistent with natural telomere attrition and reduction of APBs. In a follow up study, individual components of the MRN complex were knocked down by long term expression of shRNAs in the IIICF/c ALT cell line (Z. Zhong and R. Reddel, unpublished data). Knockdown of NBS1 resulted in suppression of ALT, as characterized in the above study for >70 PD, although variable results were observed in individual clones. Similar ALT suppression was seen in clones following knockdown of RAD50 and to a lesser extent following knockdown of MRE11. However, knockdown of RAD50 or MRE11 also resulted in a reduction of NBS1, or NBS1 and RAD50 levels, respectively. Thus, it is difficult to draw conclusions about the contributions of the components of the MRN complex to the ALT mechanism. Interestingly, the MRN complex associates with normal telomeres, suggesting that there must be a mechanism to control its function and thus to prevent ALT-like activity.[61]

Another recent study elucidated a series of genes required for APB formation.[62] Following the observation that methionine restriction enhances the abundance of APB positive cells, a screen for APB genes was carried out by transfecting siRNAs prior to methionine restriction. In this study, the telomere proteins TRF1, TRF2, RAP1 and TIN2, PML and all three components of the MRN complex were shown to be required for APB formation. Therefore, these proteins may be required for ALT, while the DNA response and repair protein 53BP1 was shown to be dispensable.

Telomere Capping and ALT Inhibition

A series of recent reports suggest functional telomeres are recognized by the DNA damage machinery during G2 and the action of HR is necessary to cap the chromosome ends before entering mitosis.[63,64] Furthermore, the RAD51D recombination protein interacts with telomeric DNA and its deletion leads to a telomere uncapping phenotype.[65] The MRN recombination complex, which is essential for ALT,[49] also functions at normal human telomeres.[61] Therefore, normal cells need to achieve a fine balance where the beneficial capping-associated HR at the telomere is permitted while the HR-mediated telomere lengthening associated with ALT is inhibited.

Telomeres in budding yeast appear to become much more recombinogenic in the absence of proper capping. In Type II *S. cerevisiae* survivors, recombinational telomere elongation occurs predominantly on the shortest telomeres[41] and telomeres in *K. lactis* only become recombinogenic after extreme shortening following telomerase inhibition.[66] An interpretation of these data is that as telomeres become very short, they bind far fewer telomere associated proteins, become uncapped and lose their ability to repress telomere HR. Consistent with this, inhibition of the function of the capping proteins Cdc13 in *S. cerevisiae* and Stn1 in *K. lactis*, or the telomere protein Rif2 in *S. cerevisiae*, leads to induction of Type II-like telomere recombination.[41,67,68] Uncapping of telomeres in *K. lactis* due to mutations in the telomeric DNA that diminish binding of the telomere protein Rap1 also result in increased telomere recombination, t-loop JR and t-circle generation. (refs. 42,69,70 and A. Cesare, M. McEachern and J. Griffith, unpublished).

Telomere capping in mammalian cells is a major function of TRF2. Following TRF2 disruption, mammalian cells exhibit a telomeric DNA checkpoint response[71] coinciding with p53/Rb-dependent senescence[9] or ATM/p53-dependent apoptosis.[8] Decreased TRF2 function in the absence of p53 results in escape from cellular arrest and extensive chromosome fusions in a non-homologous end joining (NHEJ) dependent manner.[72,73] T-loop formation and stability are associated with TRF2 function and sequestration of the 3' overhang with a t-loop is proposed to hide the chromosome end from NHEJ.[4,34] POT1 in humans and POT1A and B in mice, also appear to have a role in telomere capping, although decreased levels of any of these proteins do not result in the same drastic uncapping phenotypes as inhibition of TRF2.[36,74-76]

It seems likely that proper telomere capping in mammalian cells plays a role in suppressing the telomere length maintenance associated with ALT. As described above, improper resolution of t-loop structure in mammals is prevented by the basic domain of TRF2,[22] and in mice by POT1A.[36] TRF2 also functions with the NHEJ protein Ku70 to inhibit T-SCE at mouse telomeres.[77] When chromosome fusions are prevented in mouse embryo fibroblasts (MEFs) by deletion of DNA ligase 4, deletion of both TRF2 and Ku70, but not either of these genes alone, causes a remarkable increase in the abundance of T-SCEs, similar to what is observed in ALT cells.[77] The increase in T-SCEs is prevented by $TRF2^{\Delta B}$ expression indicating that repression of t-loop JR and T-SCE are separate functions. POT1A and POT1B deletion in mice also results in an increase in T-SCE, although to much more modest levels than seen in ALT cells.[36,76] Thus, proteins associated with telomere capping inhibit two of the three types of telomere recombination events associated with ALT cells. It has not yet been determined if normal capping also inhibits lengthening of ALT telomeres by HR-dependent DNA replication.

The connection between telomeric uncapping and abnormal HR events suggests that ALT cells might have capping dysfunction. An important component of mammalian telomere capping is believed to be the formation of t-loops that sequester the 3' overhang from the NHEJ machinery in the G1 phase of the cell cycle, when this form of DSB repair is most active. T-loops are formed in the GM847 ALT cell line[21] and rampant chromosome fusions are not observed in ALT cells consistent with proper capping in G1. During S phase, it has been proposed that DNA replication opens up the t-loops, resulting in chromosome ends being recognized by the DNA damage response machinery in G2.[63] HR proteins are then proposed to function in capping, possibly by re-establishing the t-loop.[64] T-loop JR and T-SCE are both post-replicative events,[22,78] so if they are due to an ALT-associated capping dysfunction, it is likely that this occurs in the G2 phase of the cell cycle, when HR is most active. The observation that APBs are associated with G2 is consistent with ALT activity occurring during this period.[24]

ALT activity results from the loss of an inhibitory function that is present in normal somatic cells and also in telomerase-positive cells. This was demonstrated by fusing GM847 ALT cells with HFF5 normal fibroblasts, or with HT1080 or T24 telomerase-positive cells and observing suppression of ALT activity.[79] This raises the possibility that both the putative ALT capping dysfunction and the derepression of ALT result simply from the loss of a specific cap component. All of the known telomere associated proteins are present in ALT cells, however and mechanisms of DNA metabolism show no obvious defects (although this has not been analyzed in detail). Long term TRF2 overexpression in the SUSM-1 ALT cell line did inhibit some phenotypic features of ALT cells in one study, supporting the concept that ALT cells have a telomere capping dysfunction, although a prolonged and complete ALT inhibition was not observed (L. Colgin and R. Reddel, unpublished data). Rather than loss of a single telomere cap component, it seems more likely that higher order control over the intricate functions of DNA repair and telomere capping are altered in ALT cells.

The persistence of very short telomeres in ALT cells may also contribute to aberrant telomere recombination in these cells, but does not appear to be essential. The short telomeres in ALT cells may be more prone to initiating recombination, especially HR-dependent DNA replication, as seen in budding yeast.[66] Consistent with this, the shortest telomeres are preferentially, but not exclusively, elongated in ALT cells.[18] Expression of exogenous telomerase elongates the shortest telomeres in ALT cells, but does not usually repress the phenotypic characteristics of ALT,[80-82] including t-circles[21] and post replicative exchanges,[32] suggesting that ALT activity is not dependent on the presence of very short telomeres.

Telomere Structural Dysfunction Response

At uncapped telomeres in mammalian cells, the DNA damage response proteins γ-H2AX, 53BP1, Rad17-ser645 and ATM-ser1981 accumulate in telomere dysfunction induced foci (TIFs), suggesting the chromosome end is now recognized as a DSB.[10,71,83] This may result in p53- and Rb-dependent senescence and occurs in response to inhibition of TRF2 function,[71] t-loop JR induction by TRF2ΔB,[22] or in aging cells with shortened telomeres.[10] Persistent t-loop JR and the short telomeres in ALT cells would thus be expected to induce a similar checkpoint response. Although TIFs are observed in ALT cells, no arrest occurs.(refs. 47,83 and A. Cesare and R. Reddel unpublished). This is possibly due to a lack of signal transduction as the majority of ALT cell lines and tumors are p53-deficient. Of the human ALT cell lines that have been examined to date, only one (U-2 OS) is known to retain functional wild-type p53 and a recent study of glioblastoma multiforme showed that, of 18 tumors using the ALT pathway, 14 (78%) were p53-deficient, while 26 of the 33 telomerase positive tumors (79%) had wild-type p53.[84] In mouse cells, p53 loss has been shown previously to be permissive for telomere dysfunction,[85] and telomere dysfunction induced by WRN deletion leads to ALT tumors following loss of p53 function.[59]

Telomeric Epigenetic Modification

Mammalian telomeres are assembled into constitutive heterochromatin as defined by specific modifications to the basic tails of the histones H3 and H4.[86,87] In MEFs with a double knock out (RB and RBL1), or triple knock out (RB, RBL1 and RBL2) of the Rb family proteins, the telomeres display increasing length and heterogeneity with progressive passages, with a concomitant decrease in the constitutive heterochromatin marker histone H4 tri-methyl lysine 20.[86,88] In a similar experiment, passage of embryonic stem (ES) cells or MEFs deleted for the Suv39h1 and Suv39h2 histone methyltransferases also resulted in increased telomere length and heterogeneity, coinciding with a significant decrease in the heterochromatin markers di- and tri-methylated histone H3 lysine 9 and their interacting partners, Cbx3 and Cbx5, at MEF or ES cell telomeres.[87] Similar telomeric DNA phenotypes were observed following deletion of DNA methyltransferases, DNMT1 or DMNT3a and DMNT3b, which resulted in decreased sub-telomeric DNA methylation (TTAGGG telomeres lack the CpG methylation site) in ES cells.[89] Additionally, mutation of the DMNT family members was reported to increase post-replicative telomeric exchanges and APBs, suggestive of ALT.[89] Finally, the epigenetic changes associated with euchromatin were observed with telomere shortening in MEFs from late generation TERC -/- mice and these changes were accompanied by reported increases in T-SCEs and in the proportion of APB positive cells.[89,90]

These data suggest telomere epigenetic modifications may also regulate telomeric HR and thus inhibit ALT. In the above studies investigating the knockout of epigenetic regulatory genes, the cells used retained telomerase activity, therefore the observed telomere length increase could conceivably be due to increased access of telomerase to chromosome ends. Furthermore, no evidence was obtained for recombination-mediated telomere extension in these cells. Nevertheless, it is an attractive concept that changes in chromatin state may occur at the telomere and that a euchromatic state may result in the telomere being more open to HR. Since many ALT cell lines have functional deficiencies in Rb family proteins (e.g., due to expression of SV40 large T antigen which binds to each of these proteins), they may also have similar epigenetic alterations at the telomeres. Observation of ALT t-loops by electron microscopy indicated that the loop portion of the t-loop were similar in size to loops seen in telomerase positive or mortal cells, suggesting that a significant portion of the telomere may remain outside the loop in ALT cells.[21] Therefore exclusion of much of the telomere from the loop, together with a more open state, may leave this portion of the telomere more prone to recombination or invasion by an adjacent telomere 3' overhang, leading to a HR-mediated DNA replication event.

What Is ALT and Why Does It Exist?

It has been postulated previously that a telomerase-independent TMM preceded telomerase in the course of eukaryotic evolution,[91] though it remains to be determined if such a mechanism functions at telomeres in normal mammalian cells. It is an intriguing possibility that human cells, especially those that are not normally rapidly dividing, may use such a mechanism to repair an accidentally truncated telomere without having to express telomerase and thereby risk immortalization. However, if one or more such mechanisms exist, they would need to be tightly controlled to avoid the deleterious consequences of telomere recombination and the possibility of cellular immortalization. In this view, the ALT phenotype as observed in immortalized cell lines and cancers results from loss of the normal mechanisms for controlling this telomere repair mechanism and consequent telomere dysfunction. Presumably, loss of the normal control mechanisms is selected for in the context of tumor suppressor pathway deficiencies that allow continued propagation of cells with shortened telomeres that are then under selection pressure to acquire a TMM. In some cases, the TMM is provided by dysregulated telomerase activity and in others by dysregulated HR-mediated telomere lengthening, i.e., ALT activity. Repression of ALT activity in normal mammalian cells appears to be a function that is conserved in eukaryotes, suggesting that telomeres have evolved intricate systems to utilize HR for beneficial purposes (e.g., capping) while inhibiting HR-mediated events that have deleterious consequences. These processes of HR-related mechanisms at the telomere and their systems of control represent the outcome of a long period of development, which has occurred in the context of the co-development of telomere binding proteins and control systems for telomerase, from the earliest eukaryotes with linear chromosomes and repetitive telomere sequences through to mammalian cells.

Acknowledgements

Members of the CMRI Cancer Research Unit, Clare Fasching, Axel Neumann, Ze-Huai Zhong and Lorel Colgin are thanked for critical review of the manuscript. A.J.C. is supported by a Sir Keith Murdoch Fellowship from the American Australian Association and the USA National Science Foundation International Research Fellowship Program. Research in the authors' laboratory is supported by a Program Grant from the Cancer Council New South Wales and project grants from the National Health and Medical Research Council Australia.

References

1. Moyzis RK, Buckingham JM, Cram LS et al. A highly conserved repetitive DNA sequence, $(TTAGGG)_n$, present at the telomeres of human chromosomes. Proc Natl Acad Sci USA 1988; 85:6622-6626.
2. Makarov VL, Hirose Y, Langmore JP. Long G tails at both ends of human chromosomes suggest a C strand degradation mechanism for telomere shortening. Cell 1997; 88:657-666.
3. de Lange T. Shelterin: the protein complex that shapes and safeguards human telomeres. Genes Dev 2005; 19:2100-2110.
4. Griffith JD, Comeau L, Rosenfield S et al. Mammalian telomeres end in a large duplex loop. Cell 1999; 97:503-514.
5. Watson JD. Origin of concatemeric T7 DNA. Nat New Biol 1972; 239:197-201.
6. Olovnikov AM. [Principle of marginotomy in template synthesis of polynucleotides]. Doklady Akademii Nauk SSR 1971; 201:1496-1499.
7. Harley CB, Futcher AB, Greider CW. Telomeres shorten during ageing of human fibroblasts. Nature 1990; 345:458-460.
8. Karlseder J, Broccoli D, Dai Y et al. p53- and ATM-dependent apoptosis induced by telomeres lacking TRF2. Science 1999; 283:1321-1325.
9. Smogorzewska A, de Lange T. Different telomere damage signaling pathways in human and mouse cells. EMBO J 2002; 21:4338-4348.
10. d'Adda di Fagagna F, Reaper PM, Clay-Farrace L et al. A DNA damage checkpoint response in telomere-initiated senescence. Nature 2003; 426(6963):194-198.
11. Counter CM, Avilion AA, LeFeuvre CE et al. Telomere shortening associated with chromosome instability is arrested in immortal cells which express telomerase activity. EMBO J 1992; 11:1921-1929.
12. Shay JW, Bacchetti S. A survey of telomerase activity in human cancer. Eur J Cancer 1997; 33:787-791.
13. Kim NW, Piatyszek MA, Prowse KR et al. Specific association of human telomerase activity with immortal cells and cancer. Science 1994; 266:2011-2015.
14. Bryan TM, Englezou A, Dalla-Pozza L et al. Evidence for an alternative mechanism for maintaining telomere length in human tumors and tumor-derived cell lines. Nat Med 1997; 3:1271-1274.

15. Bryan TM, Englezou A, Gupta J et al. Telomere elongation in immortal human cells without detectable telomerase activity. EMBO J 1995; 14:4240-4248.
16. Rogan EM, Bryan TM, Hukku B et al. Alterations in p53 and p16^{INK4} expression and telomere length during spontaneous immortalization of Li-Fraumeni syndrome fibroblasts. Mol Cell Biol 1995; 15:4745-4753.
17. Henson JD, Neumann AA, Yeager TR et al. Alternative lengthening of telomeres in mammalian cells. Oncogene 2002; 21(4):598-610.
18. Murnane JP, Sabatier L, Marder BA et al. Telomere dynamics in an immortal human cell line. EMBO J 1994; 13:4953-4962.
19. Tokutake Y, Matsumoto T, Watanabe T et al. Extra-chromosome telomere repeat DNA in telomerase-negative immortalized cell lines. Biochem Biophys Res Commun 1998; 247:765-772.
20. Ogino H, Nakabayashi K, Suzuki M et al. Release of telomeric DNA from chromosomes in immortal human cells lacking telomerase activity. Biochem Biophys Res Commun 1998; 248:223-227.
21. Cesare AJ, Griffith JD. Telomeric DNA in ALT cells is characterized by free telomeric circles and heterogeneous T-loops. Mol Cell Biol 2004; 24(22):9948-9957.
22. Wang RC, Smogorzewska A, de Lange T. Homologous recombination generates T-loop-sized deletions at human telomeres. Cell 2004; 119(3):355-368.
23. Yeager TR, Neumann AA, Englezou A et al. Telomerase-negative immortalized human cells contain a novel type of promyelocytic leukemia (PML) body. Cancer Res 1999; 59:4175-4179.
24. Grobelny JV, Godwin AK, Broccoli D. ALT-associated PML bodies are present in viable cells and are enriched in cells in the G_2/M phase of the cell cycle. J Cell Sci 2000; 113(Pt 24):4577-4585.
25. Wu G, Lee WH, Chen PL. NBS1 and TRF1 colocalize at promyelocytic leukemia bodies during late S/G2 phrases in immortalized telomerase-negative cells. Implication of NBS1 in alternative lengthening of telomeres. J Biol Chem 2000; 275:30618-30622.
26. Chang S, Khoo CM, Naylor ML et al. Telomere-based crisis: functional differences between telomerase activation and ALT in tumor progression. Genes Dev 2003; 17:88-100.
27. Reddel RR, Bryan TM, Colgin LM et al. Alternative lengthening of telomeres in human cells. Radiat Res 2001; 155:194-200.
28. Neumann AA, Reddel RR. Opinion: Telomere maintenance and cancer—look, no telomerase. Nat Rev Cancer 2002; 2:879-884.
29. Henson JD, Hannay JA, McCarthy SW et al. A robust assay for alternative lengthening of telomeres (ALT) in tumors demonstrates the significance of ALT in sarcomas and astrocytomas. Clin Cancer Res 2005; 11:217-225.
30. Dunham MA, Neumann AA, Fasching CL et al. Telomere maintenance by recombination in human cells. Nat Genet 2000; 26:447-450.
31. Bechter OE, Zou Y, Walker W et al. Telomeric recombination in mismatch repair deficient human colon cancer cells after telomerase inhibition. Cancer Res 2004; 64(10):3444-3451.
32. Londono-Vallejo JA, Der-Sarkissian H, Cazes L et al. Alternative lengthening of telomeres is characterized by high rates of telomeric exchange. Cancer Res 2004; 64(7):2324-2327.
33. Bailey SM, Brenneman MA, Goodwin EH. Frequent recombination in telomeric DNA may extend the proliferative life of telomerase-negative cells. Nucleic Acids Res 2004; 32(12):3743-3751.
34. Stansel RM, de Lange T, Griffith JD. T-loop assembly in vitro involves binding of TRF2 near the 3' telomeric overhang. EMBO J 2001; 20(19):5532-5540.
35. Fouche N, Cesare AJ, Willcox S et al. The basic domain of TRF2 directs binding to DNA junctions irrespective of the presence of TTAGGG repeats. J Biol Chem 2006; 281:37486-37495.
36. Wu L, Multani AS, He H et al. Pot1 deficiency initiates DNA damage checkpoint activation and aberrant homologous recombination at telomeres. Cell 2006; 126:49-62.
37. Compton SA, Choi JH, Cesare AJ et al. Xrcc3 and Nbs1 are required for the production of extrachromosomal telomeric circles in human alternative lengthening of telomere cells. Cancer Res 2007; 67:1513-1519.
38. Bechter OE, Zou Y, Shay JW et al. Homologous recombination in human telomerase-positive and ALT cells occurs with the same frequency. EMBO Rep 2003; 4:1138-1143.
39. Bechter OE, Shay JW, Wright WE. The frequency of homologous recombination in human ALT cells. Cell Cycle 2004; 3(5):547-549.
40. Teng SC, Zakian VA. Telomere-telomere recombination is an efficient bypass pathway for telomere maintenance in Saccharomyces cerevisiae. Mol Cell Biol 1999; 19:8083-8093.
41. Teng SC, Chang J, McCowan B et al. Telomerase-independent lengthening of yeast telomeres occurs by an abrupt Rad50p-dependent, Rif-inhibited recombinational process. Mol Cell 2000; 6:947-952.
42. McEachern MJ, Blackburn EH. Cap-prevented recombination between terminal telomeric repeat arrays (telomere CPR) maintains telomeres in Kluyveromyces lactis lacking telomerase. Genes Dev 1996; 10:1822-1834.

43. Natarajan S, McEachern MJ. Recombinational telomere elongation promoted by DNA circles. Mol Cell Biol 2002; 22(13):4512-4521.
44. Natarajan S, Groff-Vindman C, McEachern MJ. Factors influencing the recombinational expansion and spread of telomeric tandem arrays in Kluyveromyces lactis. Eukaryot Cell 2003; 2(5):1115-1127.
45. Topcu Z, Nickles K, Davis C et al. Abrupt disruption of capping and a single source for recombinationally elongated telomeres in Kluyveromyces lactis. Proc Natl Acad Sci USA 2005; 102:3348-3353.
46. Nosek J, Rycovska A, Makhov AM et al. Amplification of telomeric arrays via rolling-circle mechanism. J Biol Chem 2005; 280:10840-10845.
47. Nabetani A, Yokoyama O, Ishikawa F. Localization of hRad9, hHus1, hRad1 and hRad17 and caffeine-sensitive DNA replication at ALT (alternative lengthening of telomeres)-associated promyelocytic leukemia body. J Biol Chem 2004; 279:25849-25857.
48. Molenaar C, Wiesmeijer K, Verwoerd NP et al. Visualizing telomere dynamics in living mammalian cells using PNA probes. EMBO J 2003; 22(24):6631-6641.
49. Jiang WQ, Zhong ZH, Henson JD et al. Suppression of alternative lengthening of telomeres by Sp100-mediated sequestration of MRE11/RAD50/NBS1 complex. Mol Cell Biol 2005; 25:2708-2721.
50. d'Adda di Fagagna F, Teo SH, Jackson SP. Functional links between telomeres and proteins of the DNA-damage response. Genes Dev 2004; 18(15):1781-1799.
51. Chen Q, Ijpma A, Greider CW. Two survivor pathways that allow growth in the absence of telomerase are generated by distinct telomere recombination events. Mol Cell Biol 2001; 21:1819-1827.
52. Cohen H, Sinclair DA. Recombination-mediated lengthening of terminal telomeric repeats requires the Sgs1 DNA helicase. Proc Natl Acad Sci USA 2001; 98(6):3174-3179.
53. Le S, Moore JK, Haber JE et al. RAD50 and RAD51 define two pathways that collaborate to maintain telomeres in the absence of telomerase. Genetics 1999; 152:143-152.
54. Martin GM, Oshima J. Lessons from human progeroid syndromes. Nature 2000; 408(6809):263-266.
55. Chang S, Multani AS, Cabrera NG et al. Essential role of limiting telomeres in the pathogenesis of Werner syndrome. Nat Genet 2004; 36:877-882.
56. Du X, Shen J, Kugan N et al. Telomere shortening exposes functions for the mouse werner and bloom syndrome genes. Mol Cell Biol 2004; 24(19):8437-8446.
57. Crabbe L, Verdun RE, Haggblom CI et al. Defective telomere lagging strand synthesis in cells lacking WRN helicase activity. Science 2004; 306(5703):1951-1953.
58. Crabbe L, Jauch A, Naeger CM et al. Telomere dysfunction as a cause of genomic instability in Werner syndrome. Proc Natl Acad Sci USA 2007; 104:2205-2210.
59. Laud PR, Multani AS, Bailey SM et al. Elevated telomere-telomere recombination in WRN-deficient, telomere dysfunctional cells promotes escape from senescence and engagement of the ALT pathway. Genes Dev 2005; 19:2560-2570.
60. Fasching CL, Bower K, Reddel RR. Telomerase-independent telomere length maintenance in the absence of ALT-associated PML bodies. Cancer Res 2005; 65:2722-2729.
61. Zhu XD, Kuster B, Mann M et al. Cell-cycle-regulated association of RAD50/MRE11/NBS1 with TRF2 and human telomeres. Nat Genet 2000; 25:347-352.
62. Jiang WQ, Zhong ZH, Henson JD et al. Identification of candidate alternative lengthening of telomeres genes by methionine restriction and RNA interference. Oncogene 2007; In Press.
63. Verdun RE, Crabbe L, Haggblom C et al. Functional human telomeres are recognized as DNA damage in G2 of the cell cycle. Mol Cell 2005; 20:551-561.
64. Verdun RE, Karlseder J. The DNA damage machinery and homologous recombination pathway act consecutively to protect human telomeres. Cell 2006; 127:709-720.
65. Tarsounas M, Munoz P, Claas A et al. Telomere maintenance requires the RAD51D recombination/repair protein. Cell 2004; 117(3):337-347.
66. McEachern MJ, Iyer S. Short telomeres in yeast are highly recombinogenic. Mol Cell 2001; 7(4):695-704.
67. Grandin N, Damon C, Charbonneau M. Cdc13 prevents telomere uncapping and Rad50-dependent homologous recombination. EMBO J 2001; 20(21):6127-6139.
68. Iyer S, Chadha AD, McEachern MJ. A mutation in the STN1 gene triggers an alternative lengthening of telomere-like runaway recombinational telomere elongation and rapid deletion in yeast. Mol Cell Biol 2005; 25:8064-8073.
69. Groff-Vindman C, Cesare AJ, Natarajan S et al. Recombination at long mutant telomeres produces tiny single- and double-stranded telomeric circles. Mol Cell Biol 2005; 25(11):4406-4412.
70. Underwood DH, Carroll C, McEachern MJ. Genetic dissection of the Kluyveromyces lactis telomere and evidence for telomere capping defects in TER1 mutants with long telomeres. Eukaryot Cell 2004; 3(2):369-384.

71. Takai H, Smogorzewska A, de Lange T. DNA damage foci at dysfunctional telomeres. Curr Biol 2003; 13(17):1549-1556.
72. Celli GB, de Lange T. DNA processing is not required for ATM-mediated telomere damage response after TRF2 deletion. Nat Cell Biol 2005; 7:712-718.
73. Smogorzewska A, Karlseder J, Holtgreve-Grez H et al. DNA ligase IV-dependent NHEJ of deprotected mammalian telomeres in G1 and G2. Curr Biol 2002; 12(19):1635-1644.
74. Hockemeyer D, Sfeir AJ, Shay JW et al. POT1 protects telomeres from a transient DNA damage response and determines how human chromosomes end. EMBO J 2005; 24:2667-2678.
75. Hockemeyer D, Daniels JP, Takai H et al. Recent expansion of the telomeric complex in rodents: two distinct POT1 proteins protect mouse telomeres. Cell 2006; 126:63-77.
76. He H, Multani AS, Cosme-Blanco W et al. POT1b protects telomeres from end-to-end chromosomal fusions and aberrant homologous recombination. EMBO J 2006; 25:5180-5190.
77. Celli GB, Denchi EL, de Lange T. Ku70 stimulates fusion of dysfunctional telomeres yet protects chromosome ends from homologous recombination. Nat Cell Biol 2006; 8:885-890.
78. Bailey SM, Goodwin EH, Cornforth MN. Strand-specific fluorescence in situ hybridization: the CO-FISH family. Cytogenet Genome Res 2004; 107(1-2):14-17.
79. Perrem K, Bryan TM, Englezou A et al. Repression of an alternative mechanism for lengthening of telomeres in somatic cell hybrids. Oncogene 1999; 18:3383-3390.
80. Perrem K, Colgin LM, Neumann AA et al. Coexistence of alternative lengthening of telomeres and telomerase in hTERT-transfected GM847 cells. Mol Cell Biol 2001; 21(12):3862-3875.
81. Grobelny JV, Kulp-McEliece M, Broccoli D. Effects of reconstitution of telomerase activity on telomere maintenance by the alternative lengthening of telomeres (ALT) pathway. Hum Mol Genet 2001; 10:1953-1961.
82. Cerone MA, Londono-Vallejo JA, Bacchetti S. Telomere maintenance by telomerase and by recombination can coexist in human cells. Hum Mol Genet 2001; 10:1945-1952.
83. Silverman J, Takai H, Buonomo SB et al. Human Rif1, ortholog of a yeast telomeric protein, is regulated by ATM and 53BP1 and functions in the S-phase checkpoint. Genes Dev 2004; 18(17):2108-2119.
84. Costa A, Daidone MG, Daprai L et al. Telomere maintenance mechanisms in liposarcomas: association with histologic subtypes and disease progression. Cancer Res 2006; 66:8918-8924.
85. Chin L, Artandi SE, Shen Q et al. p53 deficiency rescues the adverse effects of telomere loss and cooperates with telomere dysfunction to accelerate carcinogenesis. Cell 1999; 97:527-538.
86. Gonzalo S, Garcia-Cao M, Fraga MF et al. Role of the RB1 family in stabilizing histone methylation at constitutive heterochromatin. Nat Cell Biol 2005; 7:420-428.
87. Garcia-Cao M, O'Sullivan R, Peters AH et al. Epigenetic regulation of telomere length in mammalian cells by the Suv39h1 and Suv39h2 histone methyltransferases. Nat Genet 2004; 36(1):94-99.
88. Garcia-Cao M, Gonzalo S, Dean D et al. A role for the Rb family of proteins in controlling telomere length. Nat Genet 2002; 32:415-419.
89. Gonzalo S, Jaco I, Fraga MF et al. DNA methyltransferases control telomere length and telomere recombination in mammalian cells. Nat Cell Biol 2006; 8:416-424.
90. Benetti R, Garcia-Cao M, Blasco MA. Telomere length regulates the epigenetic status of mammalian telomeres and subtelomeres. Nat Genet 2007; 39:243-250.
91. de Lange T. Opinion: T-loops and the origin of telomeres. Nat Rev Mol Cell Biol 2004; 5(4):323-329.
92. Wu G, Jiang X, Lee WH et al. Assembly of functional ALT-associated promyelocytic leukemia bodies requires Nijmegen breakage syndrome 1. Cancer Res 2003; 63(10):2589-2595.
93. Yankiwski V, Marciniak RA, Guarente L et al. Nuclear structure in normal and Bloom syndrome cells. Proc Natl Acad Sci USA 2000; 97:5214-5219.
94. Stavropoulos DJ, Bradshaw PS, Li X et al. The Bloom syndrome helicase BLM interacts with TRF2 in ALT cells and promotes telomeric DNA synthesis. Hum Mol Genet 2002; 11:3135-3144.
95. Johnson FB, Marciniak RA, McVey M et al. The Saccharomyces cerevisiae WRN homolog Sgs1p participates in telomere maintenance in cells lacking telomerase. EMBO J 2001; 20(4):905-913.
96. Zhu XD, Niedernhofer L, Kuster B et al. ERCC1/XPF removes the 3' overhang from uncapped telomeres and represses formation of telomeric DNA-containing double minute chromosomes. Mol Cell 2003; 12(6):1489-1498.
97. Moran-Jones K, Wayman L, Kennedy DD et al. hnRNP A2, a potential ssDNA/RNA molecular adapter at the telomere. Nucleic Acids Res 2005; 33(2):486-496.

Chapter 4

T-Loops, T-Circles and Slippery Forks

Sarah A. Compton, Anthony J. Cesare, Nicole Fouche, Sezgin Ozgur and Jack D. Griffith*

Abstract

All species with linear chromosomes require telomeres, whose role is to stabilize chromosome ends and prevent undesirable recombination-mediated or DNA repair-mediated events involving these DNA ends. The telomeres of most higher eukaryotic species are composed of very long tracts of a short repeated DNA sequence that is G-rich on one strand. These tracts are variable in length, ranging from approximately 3 kb in *Arabidopsis*, 15 to 50 kb in some rodents, to 100 kb or longer in some plants such as garden peas and tobacco.[1-5] Telomeric DNA interacts with histones and other chromatin proteins to form chromatin, which in turn forms a higher order looped structure called a t-loop.[6] Under some circumstances, t-loops may be converted to or generate extrachromosomal t-circles; for example, t-circles are associated with the Alternative Lengthening of Telomeres (ALT) pathway, which maintains telomere length by a telomerase-independent recombination-dependent mechanism.[7,8] Recent studies show that formation of t-circles in human ALT cells is dependent on several recombination proteins.[9]

Telomeric DNA faces unusual impediments to replication; in particular, the replication fork has a tendency to stall in tracts of short DNA repeats. To facilitate replication of telomeric repeats, the replication fork may interact with telomere-specific factors, such as TRF2, which may prevent replication fork slippage. While telomeric DNA has several unique properties and is compacted differently from euchromatic DNA, telomeric DNA may share some traits and behaviors with other tracts of short repeats such as the triplet repeats associated with Huntington disease, Fragile X syndrome and Myotonic Dystrophy. Thus, studies of telomeric DNA may yield insight into mechanisms involved in triplet repeat expansion.

This chapter reviews recent insights into unique structural elements of telomeres including t-loops and t-circles and discusses possible relationships between telomere biology and human triplet diseases.

Introduction: Unusual Physical Properties of Telomeric DNA

The long arrays of 6-7 nucleotide repeats constitute a high concentration of binding sites for telomere-specific DNA binding proteins. This has consequences for the manner in which TRF1 interacts with long TTAGGG repeat tracts in telomeric DNA. In unpublished studies we have noted that while the off-rate of TRF1 from each telomeric repeat is relatively fast, once the protein has left its binding site, it immediately encounters multiple potential new binding sites and thus is likely to be "recaptured" and remain bound in the proximity of the original binding site. For example, purified TRF1 protein binds avidly to circular plasmid DNA carrying a several hundred bp tract of TTAGGG repeats. Furthermore, TRF1 comigrates with the plasmid during gel filtration, suggesting that it is tightly bound and has a relatively slow off-rate. However, TRF1 can be

*Corresponding Author: Jack D.Griffith—Lineberger Comprehensive Cancer Center, University of North Carolina, Mason Farm Road CB7095, Chapel Hill, NC 27599-7295 USA
Email: jdg@med.unc.edu

Origin and Evolution of Telomeres, edited by Jozef Nosek and Ľubomír Tomáška.

readily displaced from the circular plasmid DNA by excess linear DNA that also carries a tract of telomeric repeats. Thus, the off-rate of TRF1 is sufficiently fast to allow release of the protein and rebinding to a competing telomeric repeat in the local DNA environment. Thus, an array of low affinity binding sites can generate a high affinity binding locale for TRF1. This scenario is likely to influence telomere biology and the types of DNA transactions that occur in and near telomeric DNA.

The free energy of nucleosome formation in TTAGGG repeats is somewhat higher than the free energy of nucleosome formation in mixed sequence DNA, resulting in lower stability of nucleosomes in telomeric DNA than in bulk chromatin.[10] This lower stability likely reflects unique structural characteristics of nucleosomes formed on telomeric DNA: even if such structural perturbations are very small, they would be magnified as much as 25-fold over the length of a 145-160 bp nucleosomal core fragment. The combination of a higher free energy of nucleosome formation and the lack of a free energy barrier could increase the ability of nucleosomes to slide in telomeric DNA. This could result in arrays of nucleosome cores lacking linkers and histone H1. This model is supported by evidence from de Lange and colleagues, showing that the octamer to octamer repeat in telomeric DNA is approximately 160 bp, while it is 200 bp in bulk chromatin.[11] However, the model has not yet been tested and confirmed by performing in vitro chromatin reconstitution experiments.

The T-Loop Model

Classic studies of RecA and U vs X recombinases show that linear DNA with a 3' protruding single stranded (ss) DNA tail efficiently forms a "D-loop" with homologous double stranded (ds) DNA, such that the ssDNA tail displaces one strand of the homologous dsDNA.[12] These observations led to the proposal that a related structure, the t-loop, could form in telomeric DNA.[6] The t-loop model proposes that t-loops have a high propensity to form in mammalian telomeres, because telomeres carry multiple homologous dsDNA target sites for the terminal 3' ssDNA tail on mammalian chromosomes. If both the ssDNA tail and the homologous dsDNA target are on the same molecule as they are in the telomere, the self-insertion reaction proceeds rapidly and with high efficiency. The most compelling aspect of this model is that it is an extremely simple mechanism by which chromosome ends could be protected from recombination- or DNA repair-mediated transactions. Both in vitro and in vivo evidence supports the validity of the t-loop model in human, murine, chicken, pea, trypanosome and yeast mitochondria.[2,6,13-15] In vitro studies implicate TRF2, a telomere-specific binding protein, in the formation of t-loops in mammalian systems and an analogous role for Taz1, a TRF2 homolog in *Schizosaccharomyces pombe*.[6,16,17]

Furthermore, recombination proteins such as RecA or Rad51 in combination with their single stranded binding proteins SSB or RPA also form t-loop like structures in vitro. In these studies the single stranded telomeric overhang invades the homologous telomere tract on the same telomere forming a looped structure identical to that observed with TRF2.[16] Like the t-loops formed by TRF2 the looped structure can be stabilized by psoralen crosslinking and can visualized by EM after the removal of the recombination proteins (Fig. 1). These findings are consistent with the idea that recombination proteins actively promote the formation of t-loops on DNA templates resembling telomere structures in vitro. Therefore it is entirely possible that recombination proteins assist in t-loop formation in vivo. However, it remains unclear which specific telomeric or recombination protein complexes actively generate t-loops in vivo.

Many questions remain unanswered concerning t-loop structure and dynamics. Griffith et al (1999) noted that the loop portion of t-loops tended to be relatively longer in mouse liver cells than in HeLa cells, when measured relative to total t-loop size (loop relative to loop plus linear tail).[6] The size of t-loops in pea cells was highly variable, from a few kb to >80 kb.[2] In contrast, in minichromosomes from trypanosomes, TTAGGG telomeric repeat tracts were approximately 5 kb by TRF analysis, but the loop of the t-loop tended to be <1 kb.[14]

More studies are required to understand what determines the size of the loop portion of t-loops and this will undoubtedly require better understanding of how t-loops are formed. In the

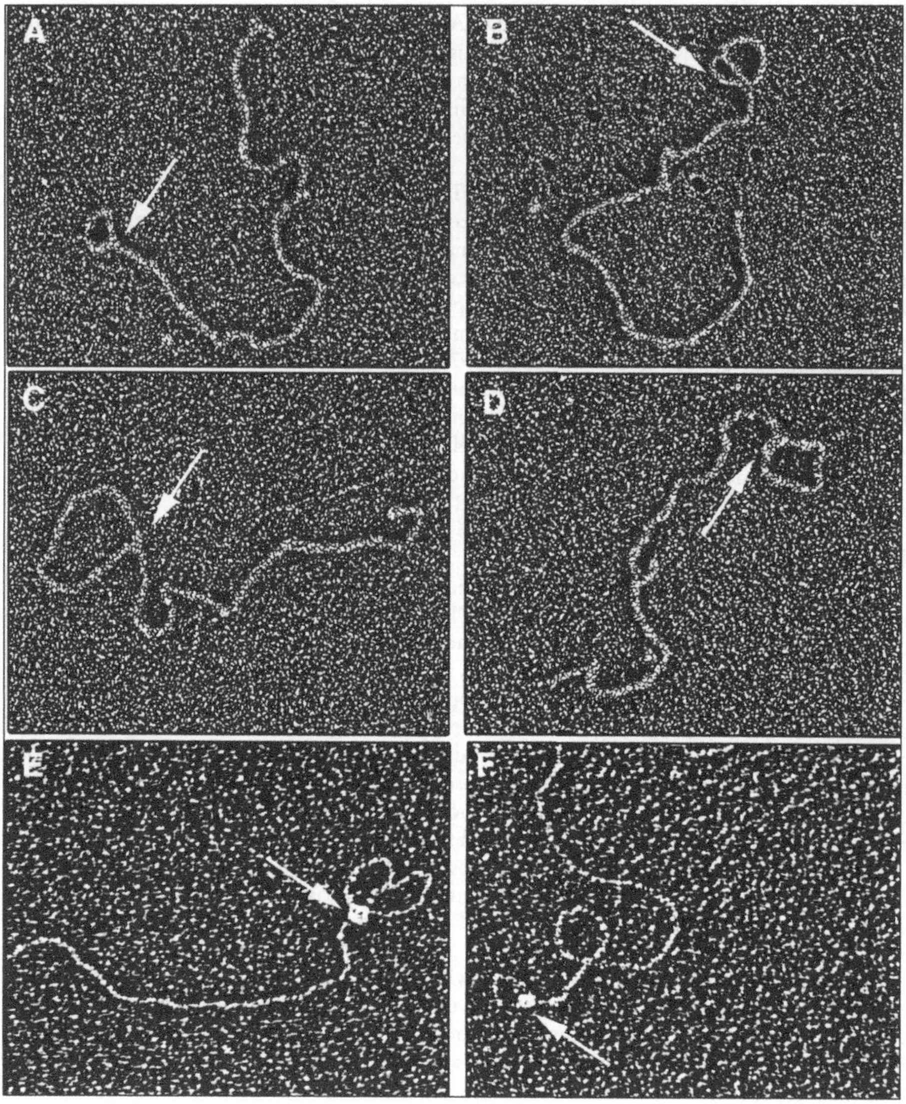

Figure 1. The formation of looped structures on model telomeres by RecA and SSB. RecA and SSB were incubated with telomere templates for 30 min at 37°C. The reactions were either psoralen crosslinked and proteins removed by treatment with SDS and proteinase K (A-D) or fixed in 0.6% glutaraldehyde (E-F) and passed over a 2 ml column of 2% agarose beads equilibrated with 10 mM Tri-HCl (pH 7.6) and 0.1 mM EDTA. The DNA rich fractions were mounted on glow charged carbon foil grids for 3 min in the presence of spermidine. Carbon grids were washed in water and dehydrated in a series of ethanol washes, air-dried and rotary shadowcast with tungsten. Samples were visualized on a Tecnai 12 transmission electron microscope at 40 kV. Images were captured at 52,000X magnification using a Gatan slow-scan charge-coupled device camera and supporting software. Arrows indicate the looped structure with (E-F) or without (A-D) RecA bound to the junction.

model proposed for *Schizosaccharomyces pombe* Taz1, Taz1 resembles a donut and the telomeric DNA threads through the "donut" hole.[17] T-loops form as Taz1 moves along the DNA, so that the size of the loop increases as the protein progresses. One possibility is that Taz1 scans the DNA sequence and stops when it encounters nontelomeric DNA sequences. One consequence of such a mechanism would be that loop size is determined by the first encounter from the terminus with a lesion or non-TTAGGG repeat (such as TTTGGG). An alternative model would be that the telomere terminus bends backwards until it contacts and interacts with an internal segment of the telomere. In nonchromatinized DNA, the loop size would follow a Shore and Baldwin distribution, which predicts relatively small loops of approximately 500 bp.[18] However, with the DNA stiffened and compacted 7 fold by nucleosome formation, the size of the smallest possible loop would be much larger. The fact that extremely large loops have been observed in t-loops from pea cells (80 kb) suggests that loop size is determined by an active mechanism, because a passive mechanism would tend to favor a smaller loop size.[2]

The minimal size of the loop portion of the t-loop is also not known. While DNA circles as small as 100 bp have been observed in vitro in the absence of any bound protein, the characteristics of nucleosome assembly may be relevant to this question and may indicate a much larger minimal size for t-loop formation. For example, circular 5 kb SV40 DNA readily assembles into 21 nucleosomes and forms a compact structure in the core of the SV40 virus particle.[19] Inspection of these and smaller DNAs assembled into nucleosomes suggests that a chain of roughly 12 nucleosomes or less might encounter significant difficulty in forming a loop and this size translates to 2 to 3 kb. This size constraint is consistent with the observation that telomere erosion and cellular senescence is associated with telomeres shorter than 2 to 3 kb in human cells. Some models propose that a minimal size for t-loop formation could be a mechanism for signaling cellular senescence. Thus, in the chromatinized state, a terminal telomeric ssDNA end would fold back and form a stable t-loop; however, through progressive telomere erosion, shortened telomeric DNA assembled into chromatin would no longer be able to form a stable t-loop, the telomere would appear "uncapped," and additional degradation and cellular senescence would become likely. This model requires further testing and validation; for example, nucleosome reconstitution studies should be carried out with telomeric DNA of variable length and the flexibility of telomeric DNA should be determined in the presence and absence of histones, TRF1 and TRF2.

Recombinational Telomere Elongation and T-Circles

Repetitive telomeric DNA and a homologous ssDNA 3' protruding end are structurally essential for t-loops to form. However, these elements also make the telomere a prime target for homologous recombination. The t-loop is a structural mechanism that sequesters the ends of linear chromosomes, prevents recognition of the chromosome end as a dsDNA break and inhibits exonucleolytic degradation; but the t-loop itself poses a conundrum. In particular, t-loops resemble homologous recombination intermediates and branch migration of a t-loop could generate a Holliday junction. This raises the question, how does telomeric DNA and/or its associated proteins prevent inappropriate recombination at the telomere or t-loop? This is especially puzzling given the large number of DNA repair proteins that associate with telomeres.

Studies in the yeast, *Saccharomyces cerevisiae*, provided the first direct evidence that recombination-dependent mechanisms play a role in maintaining telomeres in telomerase-deficient cells.[20] In yeast that lack telomerase, telomeres undergo gradual shortening, leading to widespread cell death. However, a small fraction of telomerase-null yeast survive and these cells maintain telomere length by two different mechanisms: type I survivors have short telomeres with amplified subtelomeric regions and type II survivors have long telomeres that are maintained by a Rad51-dependent recombination pathway.[20] Telomerase null *Kluyveromyces lactis* resemble type II telomerase null yeast survivors and also maintain telomere length by a recombination-dependent mechanism.[21] These cells contain small extrachromosomal circular DNA molecules with telomeric repeats (ECTR), which have been called "t-circles."[22] T-circles, which are likely generated during telomeric recombination events, are thought to be templates for rolling circle replication. This has led to a proposed "roll and spread" mechanism of telomere elongation, in which rolling circle replication elongates one or a few telomeres and these elongated telomeres are subsequently used as templates in trans for additional telomere lengthening.[20,23] T-circles have been observed in numerous species including *Kluyveromyces lactis, Candida salmanticensis, Pichia philodendri* and *Candida parapsilosis.*[24,25] In some cells, t-circles may also play a role in maintaining mitochondrial telomeres.[23,26] Together these data support the conclusion that yeast telomeres can be maintained by a homologous recombination-dependent mechanism, which is characterized by the presence of t-circles. Indeed the first reports of extrachromosomal telomeric circles and suggestion of their role in telomere maintenance came from the studies of *Candida parapsilosis* mitochondrial telomeres.[25]

A similar mechanism of telomere maintenance has been described in human cells that lack telomerase.[27,28] In human cells, this mechanism has been called alternative lengthening of telomeres (ALT) and it is characterized by heterogeneous telomere lengths, increased telomere sister chromatid exchange (T-SCE) and t-circles.[29] Evidence for recombination between telomeres in human ALT cells includes the observation that heterologous DNA inserted into one telomere appeared on multiple telomeres after several rounds of replication.[30] ALT cells also contain unusual promyelocytic leukemia (PML) bodies (or APBs), which contain telomere binding proteins, telomeric DNA and several recombination proteins including the MRN complex, Rad51, Rad52, RPA and Werner and Bloom helicases; this result suggests that these enzymes may participate in the ALT pathway.[31]

Proteins Involved in T-Circle Formation

Telomeric repeat binding factor 2 (TRF2) is a DNA binding protein with high specificity and high affinity for telomeric dsDNA.[32,33] In human cells that express a truncated TRF2 lacking the unique basic domain ($TRF2^{\Delta B}$), telomeres undergo stochastic shortening, a DNA damage response is activated and cellular senescence is induced.[7] Interestingly, the extent of telomere shortening correlates with the size of the looped portion of the t-loop and the size of t-circles, suggesting that t-loops may have been converted to t-circles by homologous recombination. In support of this model, the formation of t-circles is dependent on Nbs1 and Xrcc3. Xrcc3 is a human Rad51 paralog associated with a Holliday junction resolvase activity.[7,34] These data suggest that TRF2 may prevent recombination-dependent reactions involving t-loops. Interestingly, t-circles are not detected in mouse embryonic fibroblasts lacking TRF2 or cells expressing a dominant negative TRF2 ($TRF2^{\Delta B\Delta M}$), suggesting that additional mechanisms may exist to prevent recombination in telomeric DNA.[35,36]

Nbs1 dependent t-circles have been detected in human cells lacking telomerase or expressing $TRF2^{\Delta B}$ and in murine cells lacking POT1a.[7,9,37] Nbs1 is a component of the Mre11-Rad50-Nbs1 (MRN) complex however, the exact role that Nbs1 plays in generating t-circles remains unclear. The MRN complex is a key player in DSB repair and plays integral roles in homologous recombination, nonhomologous end joining (NHEJ) and the DNA damage checkpoint.[38] Interestingly, the MRN complex interacts with TRF2-Rap1, suggesting that it is directly involved in telomere maintenance and while Mre11 and Rad50 bind to telomeres throughout the cell cycle; Nbs1 binds to telomeres specifically in S-phase.[39] Production of t-circles in human cells lacking telomerase or expressing $TRF2^{\Delta B}$ also requires the recombination protein Xrcc3.[7,9]

T-circles have been observed in cells that lack telomerase or other proteins that bind to and protect the telomere. However, some cancer cells that proliferate in the absence of telomerase do not have t-circles.[40] T-circles and rolling circle replication using t-circles, may be required for telomere maintenance in ALT cells, when rapid elongation of degraded telomeres may be essential to cellular survival. However, t-circles may only be required transiently to repair seriously degraded chromosomes, after which a recombination-dependent mechanism of telomere maintenance may become dominant.[9]

The Role of Recombination Proteins at Mammalian Telomeres

Xrcc3 is one of five rad51 paralogs in mammalian cells.[41] The others include Rad51B, Rad51C, Rad51D and Xrcc2. These Rad51 paralogs form two distinct complexes in vivo; one complex includes Rad51B-Rad51C-Rad51D-Xrcc2 (BCDX2) and the other complex includes Rad51C-Xrcc3 (CX3).[41] The Rad51 paralogs play roles in early and late stages of homologous recombination and bind to Holliday junctions. Therefore, Rad51 paralogs may recognize and bind to the t-loop and play roles in homologous recombination-mediated reactions at telomeres. In fact, an interaction between Rad51D and telomeric DNA was observed by chromatin immunoprecipitation experiments and it also colocalizes with TRF2 at telomeres.[42] Furthermore, a loss of Rad51D results in telomere dysfunction.[42] Rad54 may also play a role in telomere maintenance, since in Rad54-defective mice, shorter telomeres and chromosomal aberrations (i.e., end-to-end fusions) involving telomeres have been observed.[43] Interactions between other Rad51 paralogs and telomeres have not yet been detected, but additional experiments are required before the existence of such interactions is discounted.[44]

Replication and Replication Restart at Telomeres

Telomeres are replicated by the normal cellular replication machinery.[45-47] However, because of the repetitive nature of telomeric DNA, it presents an inherently difficult template for DNA synthesis. G-quartets on the leading and lagging strands may inhibit replication in telomeric DNA and telomeres may accumulate more oxidative DNA lesions than the rest of the genome, because of its enrichment in runs of guanine residues.[48-51] T-loops must also be unfolded during replication and high affinity telomere-binding proteins must be displaced from their high affinity binding sites.

We recently demonstrated that stalled replication forks have a higher propensity to regress at telomeric repeats; such fork regression generates stable four-stranded "chickenfoot structures" that need to be resolved before replication resumes.[52] The kinetics of telomere replication is the same as for bulk DNA; therefore, telomere-specific factors may play a role in facilitating telomere replication.[53] Indeed, *Schizosaccharomyces pombe* Taz1 is required for efficient telomere replication, as are Bloom's (BLM) and Werner's (WRN) syndrome helicases in human cells.[54-61] These RecQ helicases have the ability to unwind G quartets and D-loops and to promote branch migration of four-stranded junctions (i.e., chickenfoot structures) in telomeric DNA templates.[62-69] Telomere-specific helicase activity may also be required, because at least four helicases have been reported to play a role in normal telomere function in *Saccharomyces cerevisiae*. In mice, the putative DNA helicase, Rtel, is also a putative regulator of telomere length.[70,71]

The replication fork could stall or collapse, when encountering DNA damage in telomeric repeats.[48,72,73] Stalled replication forks are thought to restart by a recombination-dependent process that resolves chickenfoot structures. Interestingly, TRF2 recognizes and binds with high affinity to chickenfoot structures in vitro.[52] Given the natural tendency of repetitive DNA to form chickenfoot structures, it seems plausible that telomere replication would have evolved to take advantage of this propensity, selecting for replication restart via homologous recombination over a damage-bypass mechanism.[74]

Consistent with this model, recombination proteins including Rad51D and Rad54 are required for proper telomere maintenance and BCDX2, which includes Rad51B, C and D, promotes formation of chickenfoot structures on synthetic replication forks and binds to and facilitates cleavage of Holliday junctions in vitro (Compton and Griffith unpublished data).[42,43] In yeast, Rad51 may be involved in an active process for restarting stalled replication forks, which may involve TRF2, Holliday junction resolvases and other telomere-specific DNA binding proteins; these proteins may direct fork regression and subsequent cleavage of the G-strands by Holliday junction resolvases, releasing the lagging strand telomere as a free end.

As previously discussed, the MRN complex is associated with telomeres in human cells.[39] However, Nbs1 only binds telomeres during S phase, suggesting that MRN may play a role in telomere replication.[39] Mre11 is capable of tethering DNA fragments, such as the free end of the lagging strand telomere and it was recently suggested that the essential role of MRN in vertebrates might be to promote restart of collapsed replication forks.[75-78]

Subsequent steps in recombination-mediated replication restart would presumably involve strand resection of the lagging C strand 5' terminus and strand invasion of the resultant 3' G overhang into the leading strand duplex, with strand migration or lagging strand elongation of the complementary strand of the D-loop, to form a Holliday junction structure. The final step would then be identical to the first, involving Holliday junction resolution by an activity that cleaves the G strands. Evidence for this mechanism has come from studies of long tracts of CTG repeats associated with Myotonic Dystrophy; these tracts undergo deletion at a high rate in plasmids in *Escherichia coli*.[79]

Because the above-mentioned steps resemble t-loop formation and t-loop homologous recombination, these processes would no doubt require several proteins already implicated in telomere end capping or t-loop homologous recombination. Thus the 5' C strand exonuclease would likely be the same protein involved in C strand resection during telomere end-processing and would likely be regulated by Pot1 binding to the overhang. Strand invasion during homologous recombination is typically catalyzed by the combined actions of Rad51, Rad52 and RPA and defects in two of these proteins, Rad51 and RPA, cause defects in telomere maintenance.[80-82] WRN colocalizes with RPA during replication arrest and prevents aberrant recombination at stalled replication forks.[63] Since Pot1 cooperates with WRN to efficiently unwind telomeric substrates in vitro, both proteins may also be required for formation t-loops.[83] Pot1 may also be required to stabilize the t-loops by binding to the displaced single stranded G-rich DNA. Moreover, WRN may be required for branch migration of the D-loop or lagging strand synthesis required to establish a bona fide recombination substrate between the two replicating strands. Consistent with this idea, deletion of Pot1 in mice leads to the preferential loss of lagging strand telomeres and human cells lacking WRN also exhibit deletion of telomeres replicated by lagging strand synthesis.[37,56] Telomeric G- and C strand synthesis are coordinately regulated in *Euplotes crassus*, suggesting that leading strand replication may not continue if lagging strand replication stalls.[46] Thus, a possible explanation for the above-mentioned phenotypes is that collapsed replication forks that are cleaved fail to restart. Indeed, if this occurred close to the sites of replication initiation, the lagging strand telomeres would be removed. In this case, lack of replication of the lagging strand telomeres causes telomere loss.

TRF2 is likely the key factor regulating telomere replication. In human cells, TRF2 colocalizes and physically interacts with WRN and it binds to and stimulates the activities of both WRN and BLM helicases in vitro.[60,84] Also, the dominant negative mutant of TRF2 ($TRF2^{\Delta B \Delta M}$) binds to Pot1 and prevents it from localizing to telomeres, suggesting that TRF2 binding to telomeric DNA may be required for Pot1 recruitment to recombination intermediates with minimal ssDNA.[85] However, it is the separation of function mutant of TRF2 ($TRF2^{\Delta B}$) that may reveal the most about TRF2 function during telomere replication. When $TRF2^{\Delta B}$ is overexpressed, it induces catastrophic deletions of telomeric DNA, in which both the C-rich and the G-rich strands are shortened.[7] However, as previously mentioned, $TRF2^{\Delta B}$-dependent deletion of leading strand telomeres requires Nbs1 and Xrcc3, suggesting a role for telomere-targeted homologous recombination in the effects of $TRF2^{\Delta B}$ on telomere stability.[7]

Although $TRF2^{\Delta B}$ binds poorly to chickenfoot structures, it still binds duplex telomeric repeats and can presumably recruit Pot1, WRN and possibly other proteins required for efficient telomere replication.[52] Thus, the loss of leading strand telomeres may be limited to problems encountered during telomere end capping. However, it is also possible that TRF2 is needed to direct the preferential cleavage of the G rich strands of chickenfoot intermediates. This would account for the postreplicative loss of leading strand telomeres in $TRF2^{\Delta B}$ cells. Evidence suggests that the C rich strands may indeed be cleaved by default during Holliday junction resolution. For example, in humans, the branch migration-associated resolvase (of which Rad51C and Xrcc3 may be components) cleaves Holliday junctions preferentially between two cytosine residues.[34,86]

Human Triplet Disease Expansion: Possible Parallels with ALT?

In the ALT pathway for maintaining human telomeric DNA, it is thought that telomeres can be elongated by recombination with extrachromosomal telomeric DNA (i.e., t-circles). Are there parallels for this in other systems? The classic explanation for triplet DNA expansion has been that the replication fork slips during DNA synthesis and the slippage generates a ssDNA loop, which is then converted to a duplex insertion.[87-89] This model remains attractive for small increases in the number of repeats. However, repeat tracts associated with triplet diseases can undergo rapid increases in size, expanding over a generation to 10,000 bp for the CCTG repeat in DM-2 and to 25 kb for repeat tracts in SCA-10.[90,91] It is difficult to imagine how such expansions could occur via a simple polymerase slippage mechanism.

An alternative model has been proposed by Fouche et al (2006).[74] This model suggests that replication forks that encounter repeat blocks slip backwards to generate chickenfoot structures. Resolution of this structure could involve cleavage in the stem of the four-way junction and release of a linear dsDNA carrying multiple triplet repeats. If this dsDNA has ssDNA protruding ends, or is processed by exonucleases to generate such ends, it could rapidly circularize generating a triplet repeat-containing circular dsDNA with or without a ssDNA gap. This molecule would be an ideal template for rolling circle DNA replication and could generate long tracts of triplet repeats, which could recombine via homologous recombination with target triplet repeats in the chromosome.

This model may provide an explanation for the rapid expansion of triplet repeats in the human genome. It may also provide an answer for the perplexing fact that irrespective of the triplet disease, there is a precise barrier to expansion for tracts shorter than 35 triplet repeats. In specific, triplet repeat tracts are highly stable and refractory to expansion if they contain fewer than 35 repeats. In contrast, the triplet tracts are highly "expansion-prone" and can undergo rapid expansion if they contain more than 35 repeats.

The explanation based on our model is that there are physical constraints that restrict formation of circular repeat containing DNAs below a certain size. One constraint is the requirement that the DNA termini anneal efficiently. Notably, the efficiency of annealing varies with exact length of terminal ssDNA ends (i.e., in steps of 3 base pairs for a triplet repeat). A second constraint is the stiffness of the DNA. Classic studies of Shore and Baldwin show that 100 bp is a minimal size for circularization of dsDNA.[18] Finally, there will be a rotational constraint. Here when the two DNA ends that are to pair approach each other in space, the 3' and 5' termini need to be aligned close to each other in space as contrasted to being 180 degrees out of phase in which a 3' terminus of one end would be juxtaposed to the 3' terminus of the other DNA end. Close rotational juxtaposition would be required for efficient ligation of the DNA ends. As the length of the DNA changes, the two ends will rotate in space and then come in and out of juxtaposition or phase. This phasing problem becomes critical as the length of the DNA decreases and the stiffness of the DNA increases. For a DNA close to 100 bp in length, proper phasing may be the dominant factor limiting circularization; in contrast, for a DNA molecule 200 bp or greater, higher flexibility may make it possible to overcome imperfect phasing.

These three factors will combine to limit the size of possible circles formed by triplet repeat containing DNAs and may inhibit formation of triplet repeat circles, when repeat tracts contain fewer than 35 repeats (<105 bp). Preferred lengths for circularization of triplet repeat tracts, based only on base pairing, can be predicted and would include 91, 94, 97, 99, 102, 105, 108 bp. Based on 10.5 bp/turn for DNA, one can also predict which of these lengths will have optimal phasing. This would further limit the possible repeat lengths to values of 84, 105 and 126 bp the first of which is likely to be excluded due to DNA stiffness considerations. Hence 105 bp, or 35 repeats, may be a critical minimal repeat length for circularization of a tract of triplet repeats, based on physical considerations. For a four-base repeat, the smallest DNA that would circularize would be 116 bp, for a five-base repeat it would be 105 bp and for a six-base (telomeric) repeat, the optimal minimal length would be 126 bp.

Evolution of TRF2 and Telomere Related Proteins

The discovery that TRF2 has a second "face" that had not been recognized previously raises interesting evolutionary questions related to its function and telomeres. Long nucleotide repeats occur in nearly all genomes and likely predate the evolution of the TTAGGG/TTTAGGG repeats at chromosome ends. In most cases, such as the expanded triplet repeats in the evolutionarily recent human genome, these long repeat blocks do not convey any advantage to the cell and indeed can wreck havoc in cells where they occur. Indeed the high probability that these repeats would lead to fork stalling and generation of four-stranded structures may help explain why long triplet bocks are highly unstable and are not more prevalent in various genomes. With the advent of the TTAGGG/TTTAGGG repeat at plant and animal telomeres, the highly slippery nature of the telomeric DNA and propensity of the replication to stall and generate a four-stranded chickenfoot structure had to be addressed at a molecular level. Otherwise these repeat blocks would have been highly disruptive to the genome as the chickenfoot structures induced unwanted recombination and deletions at the telomere. It is of great interest therefore to ask whether TRF2 first arose as a factor whose role was to monitor the generation of four-stranded structures and halt their formation and that over time this protein has acquired additional telomere specific roles, such as aiding in the generation and stability of t-loops and in interactions with other possibly more modern telomere specific factors including Pot1, TPP1 and Rap1. Our recent finding that p53 had a highly homologous role in recognizing regressed forks raises the parallel question as to whether this function may have been the earliest role of p53 and that this was followed later in time with its role as a transcriptional activator. A further related question is the evolutionary relation of TRF1 and TRF2. Hopefully in the future identification of plant homologs of these proteins may aid in understanding some of these evolutionary questions.

It will be interesting to determine experimentally whether extrachromosomal circular dsDNA containing disease-associated repeats are detected in cells of patients with unstable repeat tracts. However, the finding of Compton et al (2007), indicating that t-circles are not detected in all human ALT cells, may indicate that such molecules exist transiently.[9] Thus, it may only be possible to detect extrachromosomal circular DNA containing disease–associated repeats in cells actively undergoing repeat expansion and it may not be possible to detect them in cells with stably expanded repeat tracts.

References

1. Shakirov EV, Shippen DE. Length regulation and dynamics of individual telomere tracts in wild-type Arabidopsis. Plant Cell 2004; 16:1959-67.
2. Cesare AJ, Quinney N, Willcox S et al. Telomere looping in P. sativum (common garden pea). Plant J 2003; 36:271-9.
3. Yang SW, Jin E, Chung IK et al. Cell cycle-dependent regulation of telomerase activity by auxin, abscisic acid and protein phosphorylation in tobacco BY-2 suspension culture cells. Plant J 2002; 29:617-26.
4. Kipling D, Cooke HJ. Hypervariable ultra-long telomeres in mice. Nature 1990; 347: 400-2.
5. Starling JA, Maule J, Hastie ND et al. Extensive telomere repeat arrays in mouse are hypervariable. Nucleic Acids Res 1990; 18:6881-8.
6. Griffith JD, Comeau L, Rosenfield S et al. Mammalian telomeres end in a large duplex loop. Cell 1999; 97:503-14.
7. Wang RC, Smogorzewska A, de Lange T. Homologous recombination generates T-loop-sized deletions at human telomeres. Cell 2004; 119:355-68.
8. Cesare AJ, Griffith JD. Telomeric DNA in ALT cells is characterized by free telomeric circles and heterogeneous t-loops. Mol Cell Biol 2004; 24: 9948-57.
9. Compton SA, Choi JH, Cesare AJ et al. Xrcc3 and Nbs1 are required for the production of extrachromosomal telomeric circles in human alternative lengthening of telomere cells. Cancer Res 2007; 67:1513-9.
10. Filesi I, Cacchione S, De Santis P et al. The main role of the sequence-dependent DNA elasticity in determining the free energy of nucleosome formation on telomeric DNAs. Biophys Chem 2000; 83:223-37.
11. Tommerup H, Dousmanis A, de Lange T. Unusual chromatin in human telomeres. Mol Cell Biol 1994; 14:5777-85.
12. Griffith JD, Harris LD. DNA strand exchanges. CRC Crit Rev Biochem 1988; 23(Suppl 1):S43-86.

13. Nikitina T, Woodcock CL. Closed chromatin loops at the ends of chromosomes. J Cell Biol 2004; 166:161-5.
14. Munoz-Jordan JL, Cross GA, de Lange T et al. t-loops at trypanosome telomeres. EMBO J 2001; 20:579-88.
15. Tomaska L, Makhov AM, Griffith JD et al. t-Loops in yeast mitochondria. Mitochondrion 2002; 1:455-9.
16. Stansel RM, de Lange T, Griffith JD. T-loop assembly in vitro involves binding of TRF2 near the 3′ telomeric overhang. EMBO J 2001; 20:5532-40.
17. Tomaska L, Willcox S, Slezakova J et al. Taz1 binding to a fission yeast model telomere: formation of telomeric loops and higher order structures. J Biol Chem 2004; 279:50764-72.
18. Shore D, Langowski J, Baldwin RL. DNA flexibility studied by covalent closure of short fragments into circles. Proc Natl Acad Sci USA 1981; 78:4833-7.
19. Griffith J, Dieckmann M, Berg P. Electron microscope localization of a protein bound near the origin of simian virus 40 DNA replication. J Virol 1975; 15:167-72.
20. McEachern MJ, Haber JE. Break-Induced Replication and Recombinational Telomere Elongation in Yeast. Annu Rev Biochem 2006; 75:111-35.
21. McEachern MJ, Blackburn EH. Cap-prevented recombination between terminal telomeric repeat arrays (telomere CPR) maintains telomeres in Kluyveromyces lactis lacking telomerase. Genes Dev 1996; 10:1822-34.
22. Natarajan S, McEachern MJ. Recombinational telomere elongation promoted by DNA circles. Mol Cell Biol 2002; 22:4512-21.
23. Nosek J, Rycovska A, Makhov AM et al. Amplification of telomeric arrays via rolling-circle mechanism. J Biol Chem 2005; 280:10840-5.
24. Underwood DH, Carroll C, McEachern MJ. Genetic dissection of the Kluyveromyces lactis telomere and evidence for telomere capping defects in TER1 mutants with long telomeres. Eukaryot Cell 2004; 3:369-84.
25. Tomaska L, Nosek J, Makhov AM et al. Extragenomic double-stranded DNA circles in yeast with linear mitochondrial genomes: potential involvement in telomere maintenance. Nucleic Acids Res 2000; 28:4479-87.
26. Nosek J, Dinouel N, Kovac L et al. Linear mitochondrial DNAs from yeasts: telomeres with large tandem repetitions. Mol Gen Genet 1995; 247:61-72.
27. Reddel RR, Bryan TM, Murnane JP. Immortalized cells with no detectable telomerase activity. A review. Biochemistry (Mosc) 1997; 62:1254-62.
28. Bryan TM, Englezou A, Dalla-Pozza L et al. Evidence for an alternative mechanism for maintaining telomere length in human tumors and tumor-derived cell lines. Nat Med 1997; 3:1271-4.
29. Muntoni A, Reddel RR. The first molecular details of ALT in human tumor cells. Hum Mol Genet 2005; 14(Spec No. 2):R191-6.
30. Dunham MA, Neumann AA, Fasching CL et al. Telomere maintenance by recombination in human cells. Nat Genet 2000; 26:447-50.
31. Yeager TR, Neumann AA, Englezou A et al. Telomerase-negative immortalized human cells contain a novel type of promyelocytic leukemia (PML) body. Cancer Res 1999; 59:4175-9.
32. Bilaud T, Brun C, Ancelin K et al. Telomeric localization of TRF2, a novel human telobox protein. Nat Genet 1997; 17:236-9.
33. Broccoli D, Smogorzewska A, Chong L et al. Human telomeres contain two distinct Myb-related proteins, TRF1 and TRF2. Nat Genet 1997; 17:231-5.
34. Liu Y, Masson JY, Shah R et al. RAD51C is required for Holliday junction processing in mammalian cells. Science 2004; 303:243-6.
35. van Steensel B, Smogorzewska A, de Lange T. TRF2 protects human telomeres from end-to-end fusions. Cell 1998; 92:401-13.
36. Celli GB, de Lange T. DNA processing is not required for ATM-mediated telomere damage response after TRF2 deletion. Nat Cell Biol 2005; 7:712-8.
37. Wu L, Multani AS, He H et al. Pot1 deficiency initiates DNA damage checkpoint activation and aberrant homologous recombination at telomeres. Cell 2006; 126:49-62.
38. Zhang Y, Zhou J, Lim CU. The role of NBS1 in DNA double strand break repair, telomere stability and cell cycle checkpoint control. Cell Res 2006; 16:45-54.
39. Zhu XD, Kuster B, Mann M et al. Cell-cycle-regulated association of RAD50/MRE11/NBS1 with TRF2 and human telomeres. Nat Genet 2000; 25: 347-52.
40. Cerone MA, Autexier C, Londono-Vallejo JA et al. A human cell line that maintains telomeres in the absence of telomerase and of key markers of ALT. Oncogene 2005; 24:7893-901.
41. Masson JY, Tarsounas MC, Stasiak AZ et al. Identification and purification of two distinct complexes containing the five RAD51 paralogs. Genes Dev 2001; 15:3296-307.

42. Tarsounas M, Munoz P, Claas A et al. Telomere maintenance requires the RAD51D recombination/repair protein. Cell 2004; 117:337-47.
43. Jaco I, Munoz P, Goytisolo F et al. Role of mammalian Rad54 in telomere length maintenance. Mol Cell Biol 2003; 23:5572-80.
44. Tarsounas M, West SC. Recombination at mammalian telomeres: an alternative mechanism for telomere protection and elongation. Cell Cycle 2005; 4:672-4.
45. Diede SJ, Gottschling DE. Telomerase-mediated telomere addition in vivo requires DNA primase and DNA polymerases alpha and delta. Cell 1999; 99:723-33.
46. Fan X, Price CM. Coordinate regulation of G- and C strand length during new telomere synthesis. Mol Biol Cell 1997; 8:2145-55.
47. Greider CW. Telomere length regulation. Annu Rev Biochem 1996; 65:337-65.
48. von Zglinicki T. Oxidative stress shortens telomeres. Trends Biochem Sci 2002; 27:339-44.
49. Henle ES, Han Z, Tang N et al. Sequence-specific DNA cleavage by Fe2+-mediated fenton reactions has possible biological implications. J Biol Chem 1999; 274:962-71.
50. Wang Y, Patel DJ. Solution structure of the human telomeric repeat d[AG3(T2AG3)3] G-tetraplex. Structure 1993; 1:263-82.
51. Phan AT, Gueron M, Leroy JL. The solution structure and internal motions of a fragment of the cytidine-rich strand of the human telomere. J Mol Biol 2000; 299:123-44.
52. Fouche N, Cesare AJ, Willcox S et al. The basic domain of TRF2 directs binding to DNA junctions irrespective of the presence of TTAGGG repeats. J Biol Chem 2006; 281:37486-95.
53. Karlseder J. Telomeric proteins: clearing the way for the replication fork. Nat Struct Mol Biol 2006; 13:386-7.
54. Ten Hagen KG, Gilbert DM, Willard HF et al. Replication timing of DNA sequences associated with human centromeres and telomeres. Mol Cell Biol 1990; 10:6348-55.
55. Miller KM, Rog O, Cooper JP. Semi-conservative DNA replication through telomeres requires Taz1. Nature 2006; 440:824-8.
56. Crabbe L, Verdun RE, Haggblom CI et al. Defective telomere lagging strand synthesis in cells lacking WRN helicase activity. Science 2004; 306:1951-3.
57. Lillard-Wetherell K, Machwe A, Langland GT et al. Association and regulation of the BLM helicase by the telomere proteins TRF1 and TRF2. Hum Mol Genet 2004; 13:1919-32.
58. Bai Y, Murnane JP. Telomere instability in a human tumor cell line expressing a dominant-negative WRN protein. Hum Genet 2003; 113:337-47.
59. Schulz VP, Zakian VA, Ogburn CE et al. Accelerated loss of telomeric repeats may not explain accelerated replicative decline of Werner syndrome cells. Hum Genet 1996; 97:750-4.
60. Opresko PL, von Kobbe C, Laine JP et al. Telomere-binding protein TRF2 binds to and stimulates the Werner and Bloom syndrome helicases. J Biol Chem 2002; 277:41110-9.
61. Hickson ID. RecQ helicases: caretakers of the genome. Nat Rev Cancer 2003; 3:169-78.
62. Yang Q, Zhang R, Wang XW et al. The processing of Holliday junctions by BLM and WRN helicases is regulated by p53. J Biol Chem 2002; 277:31980-7.
63. Constantinou A, Tarsounas M, Karow JK et al. Werner's syndrome protein (WRN) migrates Holliday junctions and colocalizes with RPA upon replication arrest. EMBO Rep 2000; 1:80-4.
64. Karow JK, Constantinou A, Li JL et al. The Bloom's syndrome gene product promotes branch migration of holliday junctions. Proc Natl Acad Sci USA 2000; 97:6504-8.
65. Opresko PL, Otterlei M, Graakjaer J et al. The Werner syndrome helicase and exonuclease cooperate to resolve telomeric D loops in a manner regulated by TRF1 and TRF2. Mol Cell 2004; 14:763-74.
66. Mohaghegh P, Karow JK, Brosh Jr et al. The Bloom's and Werner's syndrome proteins are DNA structure-specific helicases. Nucleic Acids Res 2001; 29:2843-9.
67. Opresko PL, Cheng WH, von Kobbe C et al. Werner syndrome and the function of the Werner protein; what they can teach us about the molecular aging process. Carcinogenesis 2003; 24:791-802.
68. Sun H, Karow JK, Hickson ID et al. The Bloom's syndrome helicase unwinds G4 DNA. J Biol Chem 1998; 273:27587-92.
69. Huber MD, Lee DC, Maizels N. G4 DNA unwinding by BLM and Sgs1p: substrate specificity and substrate-specific inhibition. Nucleic Acids Res 2002; 30:3954-61.
70. Flanary B. Regulation of murine telomere length via Rtel. Rejuvenation Res 2004; 7:168-70.
71. Ding H, Schertzer M, Wu X et al. Regulation of murine telomere length by Rtel: an essential gene encoding a helicase-like protein. Cell 2004; 117:873-86.
72. von Zglinicki T, Pilger R, Sitte N. Accumulation of single-strand breaks is the major cause of telomere shortening in human fibroblasts. Free Radic Biol Med 2000; 28:64-74.
73. Branzei D, Foiani M. The DNA damage response during DNA replication. Curr Opin Cell Biol 2005; 17:568-75.

74. Fouche N, Ozgur S, Roy D et al. Replication fork regression in repetitive DNAs. Nucleic Acids Res 2006; 34:6044-50.
75. de Jager M, van Noort J, van Gent DC et al. Human Rad50/Mre11 is a flexible complex that can tether DNA ends. Mol Cell 2001; 8:1129-35.
76. Costanzo V, Paull T, Gottesman M et al. Mre11 assembles linear DNA fragments into DNA damage signaling complexes. PLoS Biol 2004; 2:E110.
77. Hopfner KP, Karcher A, Craig L et al. Structural biochemistry and interaction architecture of the DNA double-strand break repair Mre11 nuclease and Rad50-ATPase. Cell 2001; 105:473-85.
78. Trenz K, Smith E, Smith S et al. ATM and ATR promote Mre11 dependent restart of collapsed replication forks and prevent accumulation of DNA breaks. EMBO J 2006; 25:1764-74.
79. Kim SH, Pytlos MJ. Sinden RR. Replication restart: a pathway for (CTG).(CAG) repeat deletion in Escherichia coli. Mutat Res 2006; 595:5-22.
80. Smith J, Zou H, Rothstein R. Characterization of genetic interactions with RFA1: the role of RPA in DNA replication and telomere maintenance. Biochimie 2000; 82:71-8.
81. Schramke V, Luciano P, Brevet V et al. RPA regulates telomerase action by providing Est1p access to chromosome ends. Nat Genet 2004; 36:46-54.
82. Ono Y, Tomita K, Matsuura A et al. A novel allele of fission yeast rad11 that causes defects in DNA repair and telomere length regulation. Nucleic Acids Res 2003; 31:7141-9.
83. Opresko PL, Mason PA, Podell ER et al. POT1 stimulates RecQ helicases WRN and BLM to unwind telomeric DNA substrates. J Biol Chem 2005; 280:32069-80.
84. Machwe A, Xiao L, Orren DK. TRF2 recruits the Werner syndrome (WRN) exonuclease for processing of telomeric DNA. Oncogene 2004; 23:149-56.
85. Yang Q, Zheng YL, Harris CC. POT1 and TRF2 cooperate to maintain telomeric integrity. Mol Cell Biol 2005; 25:1070-80.
86. Constantinou A, Chen XB, McGowan CH et al. Holliday junction resolution in human cells: two junction endonucleases with distinct substrate specificities. EMBO J 2002; 21:5577-85.
87. Wells RD, Parniewski P, Pluciennik A et al. Small slipped register genetic instabilities in Escherichia coli in triplet repeat sequences associated with hereditary neurological diseases. J Biol Chem 1998; 273:19532-41.
88. Tam M, Erin Montgomery S, Kekis M et al. Slipped (CTG).(CAG) repeats of the myotonic dystrophy locus: surface probing with anti-DNA antibodies. J Mol Biol 2003; 332:585-600.
89. Sinden RR, Potaman VN, Oussatcheva EA et al. Triplet repeat DNA structures and human genetic disease: dynamic mutations from dynamic DNA. J Biosci 2002; 27:53-65.
90. Cho DH, Tapscott SJ. Myotonic dystrophy: emerging mechanisms for DM1 and DM2. Biochim Biophys Acta 2007; 1772:195-204.
91. Matsuura T, Yamagata T, Burgess DL et al. Large expansion of the ATTCT pentanucleotide repeat in spinocerebellar ataxia type 10. Nat Genet 2000; 26:191-4.

Chapter 5

Molecular Diversity of Telomeric Sequences

Marita Cohn*

Abstract

Telomeric DNA generally comprises a repetitive, noncoding sequence which terminates in a single-stranded extension, the 3' overhang. Although these species-specific repeats show a large diversity in sequence and length, they comprise some specific characteristic conserved features. Therefore, analyses of the telomeric repeats of various organisms have delivered important insights concerning the origin and evolution of telomere structures. The canonical repeats are synthesized onto the 3' overhangs by the specialized telomerase enzyme. Although the general feature of the telomerase is to produce homogeneous repeats, variant repeats and irregular sequences are found in many species. In this chapter some aspects of the conserving and diverging evolutionary forces on the telomeric sequences are discussed. Proteins specifically binding to the telomeric sequences are of importance for nucleating the cap structure and they regulate telomere homeostasis and telomere end protection. Since a conserved binding specificity has been shown in several homologues of both single-stranded and double-stranded telomere-binding proteins, their binding may be considered a major selective force operating on telomeric DNA sequences.

Introduction

The noncoding sequences present at the ends of the linear eukaryotic chromosomes, the telomeres, are seemingly not giving rise to any protein product. Even though RNA transcripts containing telomeric sequences have been identified in some species, it is yet unknown whether those transcripts fulfill a function in the cell. Thus, the general consensus today is that one of the important functions of telomeres is to form a buffer of dispensable sequences that protect against the loss off sequences occurring in every cell division. For the conventional DNA polymerases, the replication of the ends of linear DNA molecules poses a problem, termed the end-replication problem. In human somatic cells this leads to a shortening of the telomeres with every cell division. When the telomeres become critically short a cellular response is triggered, signaling the cell to exit the cell cycling and to senesce.

However, in addition of providing a maintenance mechanism to counterbalance the attrition of chromosome material, the telomeres serve an important role in the protective sealing of the chromosome ends. The telomeric sequences are bound by specific proteins, forming a cap structure that protects the end from nucleolytic degradation and from being detected by the DNA damage response mechanisms. The importance of a proper cap structure is highlighted by the finding that genomic deletion of a telomere-binding protein causes the telomeres to become sticky and fuse together, leading to major genome rearrangements.[1] Furthermore, such dysfunctional telomeres

*Marita Cohn—Department of Cell and Organism Biology, Lund University, Sölvegatan 35, 223 62 Lund, Sweden. Email: marita.cohn@cob.lu.se

Origin and Evolution of Telomeres, edited by Jozef Nosek and Ľubomír Tomáška.

may effectively be formed by mutating the telomeric DNA sequences targeted by the telomere binding proteins.[2]

The telomeric sequences have been isolated from a large variety of species and they show both diverged and conserved features. In an attempt to categorize all known telomeric DNA sequences, I personally group them into (i) short and simple repeats; found in most eukaryotic species, (ii) short and complex repeats; found in various yeast species and (iii) long and complex repeats (including defined retrotransposons); found in some insect and plant families. In this chapter I will focus on the short and simple repeats, the canonical repeats, which as for today's knowledge seem to be the type that is most widespread among eukaryotes. The emphasis will be on the insights gained from analysis of the telomere repeats in yeasts, where a surprisingly high degree of variability has been demonstrated. The canonical repeats are synthesized by the telomerase enzyme, which elongates the 3' single-stranded overhang present at the chromosome terminus. Telomerase is a reverse transcriptase (RT), copying an RNA template to produce a complementary DNA strand. It is, however, an unusual RT since it synthesizes relatively short repeated sequences and because it carries its own internal RNA template. It was placed in a separate subfamily when the sequence of the catalytic protein subunit of telomerase was compared to other RTs[3] and it has been proposed that telomerase may be a relic from the time of the evolutionary transition from RNA to DNA genomes.[4] Its phylogeny is still enigmatic and expanded analyses of the telomere maintenance mechanisms of various organisms of different phyla will be necessary in order to spread light on the evolution of this enzyme.

Telomeric Repeats Are Species-Specific and May Include Variants

In 1978 the DNA sequences present at the termini of the extrachromosomal ribosomal RNA genes in *Tetrahymena* were reported.[5] They were found to be short tandemly repetitive units reading 5'-TTGGGG-3' along the strand towards the terminus of the molecules. It was found that introduction of a telomerase synthesizing a mutated version of the repeats led to the progressive shortening of the telomeres. Eventually, after several more cell divisions, it led to the senescence of the *Tetrahymena* cells.[6] Thus, the telomeric repeats proved to be necessary for stabilizing the chromosomes.

Nuclear telomeric sequences have now been isolated from a huge variety of eukaryotic species (Table 1). The alignment of a large number of species-specific sequences shows intriguing features in several respects. Firstly, all nuclear chromosome termini isolated to date consist of repeated sequences. Secondly, the specific telomeric sequence, as well as the length of the repeated unit, varies between different species and may differ between relatively closely related species. The majority of the canonical repeats are 5-8 bp long. Moreover, the total length of the array of repeats is species-specifically regulated. Thirdly, as in *Tetrahymena*, the canonical repeats are generally rich in T and G nucleotides in one strand. Because all repeats are arranged in a head to tail direction, a TG/CA strand bias of the nucleotide content is formed, with the TG-rich strand running 5'-3' towards the end. The more complex repeats of certain yeasts do not clearly show this strand bias, because they commonly contain all four GATC nucleotides in both strands. But, as I will describe later, a core motif sharing the same features as the canonical short repeats has been identified within those repeats.[7,8] Fourthly, some species contain perfectly homogeneous repeated units along the telomeric arrays. However, some species have patterns of sporadically distributed variant repeats,[7-9] while others comprise variant repeats mainly in the internal centromere-proximal part of the array. Furthermore, a few species have such irregular telomeric sequences that a specific basic repeat unit cannot be distinguished. Concomitantly, there are both conserved and diverged features of the telomeric sequences.

The TTAGGG repeat sequence present in humans has been isolated also from mouse and has been identified by in situ hybridization in many other vertebrates (Table 1).[10-12] The same sequence is found also in such diverse organisms as slime molds, some filamentous fungi and the flagellated protozoan *Trypanosoma brucei*. Moreover, a very similar sequence is found in insects, the pentamer TTAGG. This would indicate a conserved nature of the telomeric sequences. Seemingly

Table 1. Nuclear telomeric repeats from species of various eukaryotic phyla. The table is intended to give a perspective on the general features, such as the TG-richness and the high diversity combined with intrinsic conserved motifs and is not a complete list of all known telomeric sequences. The list only includes species where the telomeric sequences have been either cloned and sequenced, or the telomeric position has been determined, for example, by sensitivity to Bal31 treatment.

Telomeric Repeats Organism	Sequence	Reference
Vertebrates		
Homo sapiens	TTAGGG	63
Mus spp.	TTAGGG	12
Slime Molds		
Physarum polycephalum	TTAGGG	64
Didymium iridis	TTAGGG	64
Dictyostelium discoideum	AG_{1-8}	65
Protozoa		
Trypanosoma brucei	TTAGGG	66, 67
Trypanosoma cruzi	TTAGGG	68
Leishmania donovani	TTAGGG	69
Giardia lamblia	TAGGG	70
Tetrahymena thermophila	TTGGGG	5
Paramecium primaurelia	TT(T/G)GGG	71
Euplotes crassus	TTTTGGGG	72
Oxytricha nova	TTTTGGGG	73
Sporozoa		
Plasmodium berghei	TT(T/C)AGGG	74
Plants		
Othocallis siberica	TTAGGG	75
Lycopersicon esculentum (tomato)	TT(T/A)AGGG	76
Arabidopsis thaliana	TTTAGGG	77
Algae		
Chlorella vulgaris	TTTAGGG	78
Chlamydomonas reinhardtii	TTTTAGGG	79
Invertebrates		
Ascaris lumbricoides	TTAGGC	80
Caenorhabditis elegans	TTAGGC	81
Parascaris univalens	TTGCA	82
Insects		
Bombyx mori, other insects	TTAGG	83
Apis mellifera (honey bee)	TTAGG	84
Fungi		
Neurospora crassa	TTAGGG	85
Fusarium oxysporum	TTAGGG	86
Histoplasma capsulatum	TTAGGG	87
Cladosporium fulvum	TTAGGG	88
Podospora anserina	TTAGGG	89
Ustilago maydis	TTAGGG	90
Magnaporthe grisea	TTAGGG	91

continued on next page

Table 1. Continued

Telomeric Repeats Organism	Sequence	Reference
Pneumocystis carinii	TTAGGG	92
Aspergillus nidulans	TTAGGG	93
Aspergillus oryzae	TTAGGGTCAACA	94
Cryptococcus neoformans	$TTAG_{3\text{-}5}$	95
Schizosaccharomyces pombe	$TTACAC_{0\text{-}1}\ G_{2\text{-}8}$	19, 96
Saccharomyces cerevisiae	$TG_{2\text{-}3}(TG)_{1\text{-}6}$	97
Saccharomyces exiguus	$TG_{2\text{-}3}(TG)_{1\text{-}6}$	8
Saccharomyces castellii	TCTGGGTG	8
Saccharomyces dairenensis	TCTGGGTG	8
Saccharomyces kluyveri	GACATGCGTACTGTGAGGTCTGGGTG	8
Candida albicans	TCTAACTTCTTGGTGTACGGATG	98
Candida guilliermondii	TACTGGTG	7
Candida glabrata	CTGTGGGGTCTGGGTG	7
Candida maltosa	CAGACTCGCTTGGTGTACGGATG	7
Candida tropicalis	TCACGATCATTGGTGTA(A/C)GGATG	7
Candida pseudotropicalis	TGATTAGTTATGTGGTGTACGGATT	7
Kluyveromyces lactis	TGATTAGGTATGTGGTGTACGGATT	7

contradictory though, within the group of ciliated protozoa the respective telomeric sequences are quite diverged. The repeat length is varying from 6 and 7 to 8 nucleotides. Although diverged, the ciliate telomeric sequences show a profound relatedness to each other and the mutational changes are clearly traceable. Investigation of the ascomycetous budding yeasts shows an astonishingly different picture, with a quite remarkable sequence divergence. The repeat unit length varies from 8 nucleotides in *Saccharomyces castellii* and *Saccharomyces dairenensis* (formerly *S. dairensis*) to 26 nucleotides in *Saccharomyces kluyveri*.[8] When an alignment of the various budding yeast telomeric sequences was performed, a conserved TG-rich core sequence was identified.[7,8] Thus, mutations seem to be more prone to appear in regions outside of this core sequence, indicating a selective pressure acting on the core sequence, which may be due to the preservation of binding sites for the telomere binding proteins Rap1 and Cdc13.[8,13] Significantly, also the human and ciliate telomeric sequences conform to this conserved pattern.

Mechanisms for Generation of Irregular Telomeric Sequences

Telomerase synthesizes the telomeric repeats by copying a template comprising a small part of the internal RNA moiety (TR). The template region constitutes approximately 1.5 telomeric repeats and most telomerases produce perfectly homogeneous tandem repeats by annealing to and copying specific parts of the template region. In a diploid organism, the possibility for an allelic variation is obvious, where the two TR genes may contain a slightly different template sequence in respect to each other. This would allow for the incorporation of variant repeats differing in for instance one single nucleotide, which are randomly distributed along the telomeric arrays. However, to my knowledge, no such TR alleles have yet been demonstrated. An alternative mechanism for the generation of degenerate sequences is the misincorporation of nucleotides by an error-prone telomerase enzyme. This mechanism was determined as responsible for the production of the major class of variant T_3G_3 repeats which are randomly distributed among the main population of T_2G_4 repeats in *Paramecium tetraurelia* telomeres.[9,14,15]

The telomeric sequences of *S. cerevisiae* are highly irregular, sometimes abbreviated $TG_{1\text{-}3}$. However, when analyzing the telomeric sequence array in detail, a pattern consisting of a mix of variant repeats can be deciphered, leading to the more accurate consensus $TG_{2\text{-}3}(TG)_{1\text{-}6}$. In this formula we can read out that a repeated unit consists of either a TGG or a TGGG motif followed by a variable number of TG dinucleotides.[8,16] This feature may readily be explained by the telomerase synthesis mechanism being degenerate, copying various stretches on a 17-nucleotide-long template (5'-CACCACACCCACACACA-3') by using alternative start and stop positions on the template.[16,17] Thus, the *S. cerevisiae* telomerase exemplifies a type of low fidelity telomerase enzyme, which however seems to faithfully copy the template RNA, although different parts of it. A resembling pattern, with variability in the number of TG dinucleotides, is found within the species *S. castellii* and *S. dairenensis*.[8] Though, the majority of the arrays consist of homogeneous TCTGGGTG repeats, 26% and 36% of the respective *S. castellii* and *S. dairenensis* repeats are either lacking the last TG dinucleotide or contain additional 1-3 TG dinucleotides. In analogy with *S. cerevisiae*, this variation could possibly be produced by alternative segments being copied on a single telomerase template.[18] A similar situation may be attributed to *Schizosaccharomyces pombe* telomeres, which are composed of highly irregular sequences that may be accounted for as variants of the basic repeat unit TTACAGG.[19] The variation can be described as $TTACAC_{0\text{-}1}\,G_{2\text{-}8}$. However, major conclusions regarding the possibility for the variants to be template-directed will have to await the isolation and analysis of the TR gene in *S. pombe*.

Observations in several organisms demonstrate the presence of large amounts of variant repeats in the centromere-proximal regions of the telomeric sequence arrays.[20-24] For example, all the telomeres of the *Tetrahymena thermophila* micronucleus contain 0.6-1 kb of T_3G_4 repeats, in an inner homogeneous tract adjacent to the distal tract of T_2G_4 repeats (1.4-2.8 kb).[24] The centromere-proximal array of variant repeats is unusually homogeneous in this case, since in most other species, the corresponding region is composed of a mixture of different variants. The micronucleus of *Tetrahymena thermophila* is providing the germ line of the cell and it may be speculated that the T_3G_4 repeats of the inner telomere tracts constitute an evolutionary remnant of telomeric repeats produced by an ancestral telomerase.[24] Thus, an investigation of the inner telomere repeat tracts of various species may be able to provide clues to the evolutionary paths of telomerase and its sequence products.[25] However, because the inner region is likely to be replicated by DNA polymerase rather than by telomerase, there may be alternative explanations for the presence of the inner variant repeats, such as the amplification and homogenization of a repeat mutation. It is well established that the telomeric and sub-telomeric regions are prone to be involved in recombination events,[26] although there are indications of zones of localized suppression of recombination.[25] Moreover, when telomerase function is compromised, the cells may escape senescence by maintaining their telomeres by recombinational mechanisms, so-called alternative lengthening of telomeres (ALT). Because telomerase will only sustain the turnover of the outermost telomeric repeats, the inner repeat tracts may be collecting mutations that are spread among the telomeres due to recombination. This model would fit the situation found in most eukaryotes and applies particularly to *Arabidopsis thaliana* where it was found that the homogeneous outermost repeats are succeeded by heterogeneous variant repeats becoming increasingly degenerate when moving inward to the subtelomeric regions.[23]

Furthermore, telomeric repeats located in nontelomeric internal chromosome loci have been described in several species and may be considered remnants from chromosome fusion events.[27-31] Specifically, the pericentric regions in several vertebrates hold TTAGGG-like sequences.[32] In *Tetrahymena thermophila* such internal telomeric repeats have been found to be frequently associated with a class of transposon-like elements.[33] Thus, telomeric repeats may spread through genomic rearrangements caused by translocations, recombinations, or transpositional events and may subsequently be replicated and handled as any other repeated sequence within the genome. These dynamic properties of the genome seem to be universal and even the pufferfish (*Fugu rubripes*) carries such repeats as minisatellites, in spite of its small-sized genome.[34] Moreover, telomere-like sequences that arise by chance at interior chromosomal positions may be duplicated, leading to the amplification and creation of a minisatellite that has never been in any contact with the telomeres.

Presumably, those internal locations may then be taking part in telomere recombinations, contributing to the propensity for chromosomal translocations to take place.[28]

Conserved Telomeric Sequence Motifs

The conservational pressure on the specific sequence would not be considered as very high, was the only role for the telomeric sequences to counterweigh the gradual erosion. Any repetitive sequence could fulfill such a role if it was promptly added to the end. But, as a fact, the important protective roles of telomeres are mediated by the proteins included in the nucleoprotein structure capping the end, the telosome. The DNA-binding proteins binding sequence-specifically to the telomeric sequences are important nucleators of the formation of the protective cap structure, subsequently recruiting other telosome proteins. Many of the telomere-binding proteins have been demonstrated to be important regulators of the length of the telomeres. Sequence-specific telomere proteins binding to the double-stranded region have been isolated in various organisms and include TRF1 and TRF2 in humans, Rap1 in budding yeast[35] and Taz1 in *S. pombe* (Horvath, this volume). Even though the overall amino acid similarity is low between these proteins, their respective DNA-binding motifs show similarities to the DNA-binding homeodomain motifs of the c-Myb family of transcriptional activators (Horvath, this volume).[36] Thus, they all show similar characteristics in their binding to the telomeric DNA sequences. TRF1 and TRF2 only contain one Myb-like domain each, while Rap1 contains two structurally similar Myb-like domains connected by a loosely folded linker.[37-39] These proteins bind along the length of the telomere and have been demonstrated to be part of a negative feedback mechanism that regulates telomere length (reviewed in ref. 40). Plant-specific telomere proteins have also been identified, which bind telomere repeat sequences in vitro via Myb-like motifs (Horvath, this volume). Thus, the ubiquitous use of the Myb-like/homeodomain motif signifies it as an essential protein domain for the establishment of the chromosome end capping structure during the course of evolution.

In contrast to many other single-stranded DNA-binding proteins, the proteins binding the single-stranded TG-rich overhangs of telomeres (G-overhangs) display a high sequence specificity.[41,42] Telomeric proteins binding the G-overhangs have been isolated from a wide range of organisms, including TEBP in ciliates, Pot1 in vertebrates, plants and fission yeast and Cdc13 in budding yeasts (Horvath, this volume). Even though their overall primary sequences are quite diverged, these proteins recognize the G-overhangs via a structurally conserved domain, the so-called OB-fold.[43] Cdc13 interacts with its partners Stn1 and Ten1, which also contain predicted OB-folds and together they form an RPA-like complex with telomere protective functions.[44] In *S. cerevisiae*, one of the components of the telomerase holoenzyme, the EST1 protein, exhibits a weak but sequence-specific binding to the G-overhangs and notably has a specific requirement of a free 3' terminus.[45,46]

Sequence-specific DNA-binding proteins, which perform essential cellular functions, would be supposed to put a constraint on the evolutionary freedom of the telomeric DNA repeat sequence. Investigations of the telomeric sequences and proteins in various yeast species have provided indications for this notion. Both the Rap1 and Cdc13 homologues have been identified in the budding yeast *S. castellii* (scasRap1, scasCdc13) and the proteins have been determined to bind the double-stranded telomeric DNA and the G-overhangs, respectively.[13,47] Interestingly, the binding sites for full-length scasRap1 and scasCdc13 were found to overlap completely (Fig. 1).[13] In the 8-mer minimal binding site of scasCdc13, the four nucleotides (GT—GG) were found to be of most importance for the sequence-specific interaction (Fig. 1). Of these, the three G-nucleotides had previously been shown to interact with scasRap1 in a DMS footprint analysis.[47] Moreover, when aligning the telomeric sequences of a number of evolutionary diverged species, these four nucleotides could be distinguished as a highly conserved motif (Fig. 2). Thus, both scasRap1 and scasCdc13 bind to the conserved core sequence within the otherwise highly divergent telomeric sequences of a wide variety of organisms. Significantly, scasRap1 and scasCdc13 have been shown to bind in vitro to the specific telomeric sequences of several of these other species, including the human repeats.[13,47] Intriguingly, the *C. guilliermondii* sequence is less well bound than the *K. lactis*

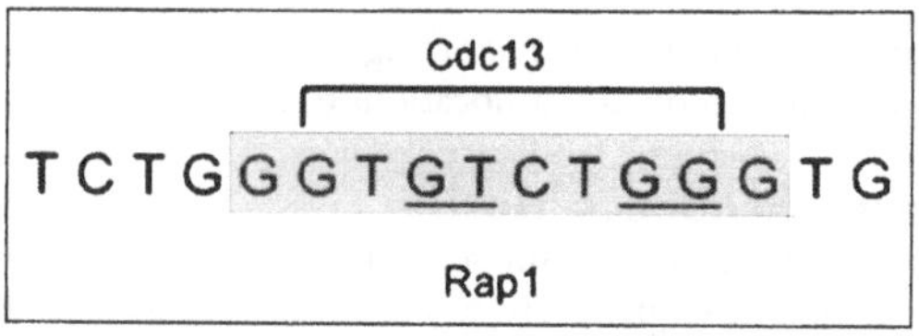

Figure 1. Overlapping binding sites of the respective scasRap1 and scasCdc13 full-length proteins in the telomeric sequence of *S. castellii*. Two telomeric repeats are included. The scasRap1 binding site on the double-stranded DNA is indicated with a grey box. All the G-nucleotides within the box are contacted by the protein, as determined by DMS footprinting.[47] The scasCdc13 minimal binding site on the single-stranded DNA is indicated by the bar above the sequence.[13] The four nucleotides most important for the scasCdc13 sequence-specific binding are underlined.

sequence. Most probably, this lowered affinity is caused by the change of one of the four conserved nucleotides within the *C. guilliermondii* repeat (Fig. 2).

Studies of the *Kluyveromyces lactis* Rap1 showed that in vitro binding of Rap1 was disrupted when the specific positions included in the core motif were mutated.[48] These results corroborate the proposal of the conserved core being used as the binding site by the Rap1 homologues in a wide variety of yeast species.[8] Furthermore, the in vivo length regulation was impaired when the corresponding mutations were incorporated into the telomeres by a mutant telomerase.[2] The severity of the in vivo length regulation defect correlated directly with the degree of loss of Rap1 binding affinity to the mutant repeats in vitro.[48] Notably, a mutation of the first position in the (GT—GG) motif caused an extremely severe phenotype, with an immediate telomere lengthening and reduced growth.[2,49] In view of the above reviewed scasCdc13 results, where this position is particularly important for the scasCdc13 binding, it is very tempting to speculate that the binding of the *K. lactis* Cdc13 homologue is likewise challenged in the mutant *K. lactis* strain. A simultaneous loss of both Rap1 and Cdc13 binding may be responsible for the immediate severe dysregulation of the telomere length (author's own interpretation). It could of course be argued, that Cdc13 binding would be necessary for the production of the elongated telomeres. Hypothetically, however, a combined loss of both Rap1 binding and Cdc13 binding may render the telomeres as accessible for telomerase elongation as within the in vitro telomerase assay where

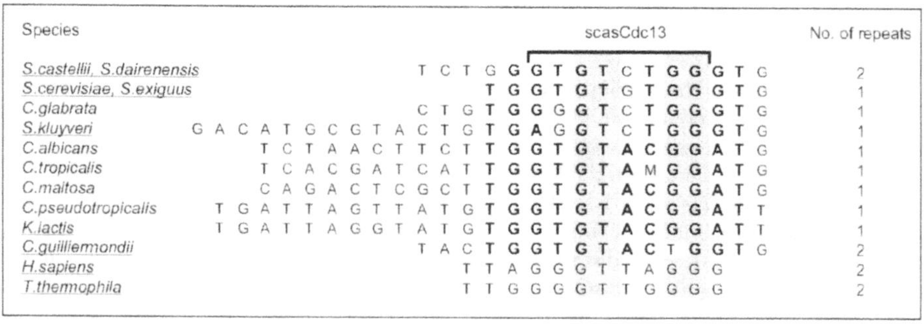

Figure 2. The Rap1 and Cdc13 binding sites are localized within a conserved core of the otherwise highly diverged telomeric sequences of various organisms. Both the scasRap1 and the scasCdc13 homologues have been determined to bind to the telomeric repeats of the species that are underlined.[13,47] The conserved Rap1 binding site is shown in bold letters. The minimal binding site (MBS) of scasCdc13 is indicated by the bar. The four nucleotides most important for sequence-specific binding of scasCdc13 are indicated by the grey boxes. S., C. and K. denote the respective genera *Saccharomyces*, *Candida* and *Kluyveromyces*.

the sole core telomerase is competent for extending the 3' end. Isolation of the *K. lactis* Cdc13 homologue and characterization of its binding specificity will open up for the investigation of the relative contribution of the respective proteins for the obtained phenotype and will also be highly interesting for the evolutionary analyses of binding site specificities.

The single-strand and double-strand DNA-binding proteins may be acting together with the proteins binding to G-quadruplex structures in conserving the G-richness of the telomeric repeats.[50] G-quadruplex binding proteins may similarly be a fundament for the intricate network of transient telomere protein interactions. Although the overall primary sequences of the telomere proteins have not been very well conserved in evolution, their respective DNA-binding domains are showing conservation to a much higher extent, either in their primary sequences or in their three-dimensional structures. Accordingly, the DNA-binding specificities of homologues of telomere-binding proteins have been conserved to a considerable degree, which suggests that their binding is a major selective force operating on telomeric DNA sequences and that a co-evolution is strongly favored.

Molecular Evolution of Telomere Sequences

In an evolutionary perspective, the chromosome end-maintenance system seems to be a dynamic and flexible cellular process. This is illustrated by the variability found, for example, within ascomycetous yeasts and within green plants.[8,51] The *Arabidopsis*-type repeats (TTTAGGG) which are found in various plants, have been replaced by the vertebrate-like repeats (TTAGGG) in a big phylogenetic group of plants belonging to the order of Asparagales.[51] Yet, another group of Asparagales, the *Allium* species, do not comprise any canonical repeats at their telomeres at all, but instead have complex repeat structures where the organization is still not conclusively elucidated. Thus, the telomerase enzyme indeed seems to be prone to mutate. Some of those mutations will be acceptable in terms of telomere maintenance and other cellular functions, while other mutations may be too detrimental and lead to a crisis and the invention of new capping and maintenance solutions. The mutation responsible for the change in the telomeric repeat sequence being synthesized may be located either in the protein (TERT) or in the RNA (TR) component of the telomerase. Obviously, a mutation within the copied region of the TR template will as a direct consequence alter the telomeric sequence. Mutations in the TERT catalytic core component, however, may be assumed to alter the reading frame on the template RNA region, leading to either an expansion or a decrease in the length and/or a permutation of the copied region.

Whether the predominant cause for telomeric sequence evolution lies within the TR or the TERT component is still only speculative. However, the TR genes are so extremely diverged that only the primary sequences of TR genes from very closely related species can be reliably aligned (Brault et al this volume). The reason for the TR genes having such an exceptionally high divergency rate is not yet explained. However, the RNA subunit folds into a highly conserved secondary structure, providing binding sites for structure recognizing proteins, which may cause a relaxed selection pressure.[18,52] This is further corroborated by the fact that large parts of the *S. cerevisiae* telomerase RNA may be deleted and still form a functional telomerase enzyme, elongating the telomeres in a similar way as the wild-type enzyme.[52] Thus, the RNA seems to form a flexible scaffold for bringing the TR-binding proteins into the vicinity of the central RNA core containing the template.[52] In fact, the central core, including the template sequence and the TERT-interaction region, could be considered one of the few major conserved regions.[53-55] It will be necessary to examine more TR template sequences and to correlate them to the actual repeats present at the telomeres, in order to be able to deduce which of the two components has been the major contributor to the molecular evolution of telomeric DNA sequences.

The variant repeats present at internal locations may tell us about the evolution of the telomeric sequences, as they may be considered remnants of previously used telomere sequences. However, due to the high probability of recombinations, caution would need to be taken in the analyses. If several species within a genus would comprise the very same repeat pattern, the ancestral interpretation could be considered reliable. Interestingly, the telomerase RNA template region in

T. thermophila comprises a sequence that is complementary to both the terminal T_2G_4 repeats and the T_3G_4 repeats in the internal regions.[53] It could therefore be hypothesized that, during the evolution, telomerase experienced a mutation leading to a decrease and permutation of its reading frame on the template (author's own interpretation).

Mutations in the telomeric sequences would be predicted to cause severe effects only if they change the nucleotides within the binding motifs of the telomere proteins. The potential harmful consequences of mutations would therefore be considered to differ between species having long and short telomeric repeats. Therefore, for the yeast species harboring the relatively longer and more complex telomere repeats, mutations outside the conserved core sequence would not be predicted to cause any severe harm, since the binding of telomere proteins will not be affected. However, additions and deletions would change the spacing of the telomere proteins, thus potentially affecting the telomere length regulation. For species having the short 5-8 bp repeats, however, two adjacent repeats may be needed to build up the binding site. This is the situation in *S. castellii*, where the Rap1 binding site is formed by two consecutive 8 bp repeats.[47] However, the binding of *S. cerevisiae* Rap1 has been demonstrated to be very flexible and the protein can still bind a sequence where the two binding site motifs are separated by an additional 6 nt.[56] The underlying reason for this flexibility has not yet been determined, but may be explained by the ability of the protein to bend the DNA and thus bring the two binding sites together. An alternative explanation could be that the bipartite subdomain structure of the Rap1 protein confers a dynamic structure to the protein.[56] Because the two Myb-like domains do not form any protein-protein interactions and are connected by an unstructured loop, their relative positioning on the DNA is somewhat flexible.[36] A similar flexibility in the binding is accomplished by the human TRF1 and TRF2 proteins, which bind telomeric DNA as preformed homodimers where the linker regions between the Myb-like DNA-binding domains of the respective monomers have no apparent inherent structure.[36] This flexibility in binding site recognition may be a crucial feature for an evolutionary successful double-strand telomere binding protein. It provides a possibility to cope with mutational changes in repeat sequences, misincorporation of nucleotides by low fidelity telomerases, as well as variant repeats produced by alternative copying frames on the RNA template.[8,9,16,51]

Concluding Remarks

In contrast to the predictions for the molecular evolution of protein-coding genes, it is a complicated task to attempt to define a phylogenetic pathway based on the comparisons of telomeric sequences. The genetic mechanisms may be rather differently applied on these noncoding sequences. One could imagine the molecular evolution of terminal sequences being rather random, moving back and fourth, building a network rather than a tree. However, it is of a high scientific value to elucidate the absolute origin of the telomere sequences. When and how did telomeric sequences appear in evolution? Is telomerase the ancient enzyme which originally tended to the chromosome homeostasis? Alternatively, did transposable elements invade ancient circular chromosomes and provide the basis for the maintenance of linear chromosomes?[57]

In fact, the telomeres of *Drosophila* are composed of two types of nonLTR retrotransposons, HeT-A and TART. But, are the retrotransposable elements and the canonical repeats related at all, or did telomere maintenance mechanisms occur several times in evolution? The retrotransposon mechanism of telomere maintenance in *Drosophila* is basically very similar to the telomerase mechanism.[58] The two retrotransposable elements targeted to the telomeres are however distinct and the conserved differences in these elements have led to the proposal that they have different origins but have converged on their roles at the telomeres.[59] It is speculated that the HeT-A element may be derived from telomerase, while the TART element was present in the genome before it became specifically targeted to telomeres.[59] While telomerase reverse transcribes only a limited region of its RNA, the retrotransposons copy the full RNA sequence. However, the analyses of yeast species have extended our view of the telomerase capacity, demonstrating its ability to copy up to 25 nt RNA templates.[60] It could therefore be speculated that the ancient telomerase used an even longer region. If the telomeric sequences have decreased in length throughout evolution, there has probably been a strong selection for the preservation of the motifs supporting the interaction with the telomeric DNA-binding proteins. One could speculate that the originally long sequences became successively shorter, until only the core domain supporting the binding motifs of the essential DNA-binding proteins remained in some species.

No matter the origin, telomere sequences are today fulfilling essential roles in the genomes. However, seemingly contradictory, they perform the dual role of stabilizing the chromosomes when present at the telomeres, while providing potential new telomeres and thereby contributing to the karyotype evolution when present at interstitial sites.[32] McClintock and Muller made us aware of the telomere questions already in the 1930s. Today, however, our models of telomere evolution are still purely speculative and it will be necessary to collect data from a wider variety of organisms in order to be able to get some conclusive answers to these intriguing evolutionary questions. The telomere structures of linear mitochondrial chromosomes will probably be valuable links when sketching the evolutionary scenario.[57,61] Moreover, the wide spectrum of different telomeric repeats found in various yeast species constitutes a rich source of data for evolutionary studies. Interestingly, an array of degenerate TTAGGG repeats, the sequence of the vertebrate canonical telomere sequence, are found within the X elements in the sub-telomeric locations of the *S. cerevisiae* telomeres (reviewed in ref. 62). The repeats are in the same orientation as the actual yeast telomere. Since a large number of filamentous fungi contain this repeat type at their extreme termini, while unicellular yeasts instead comprise a wide spectrum of different more complex repeats, more thorough studies of sub-telomeric regions and inclusion of additional fungal species, will be specifically informative for deciphering the evolution of telomeric repeats.

References

1. Celli GB, de Lange T. DNA processing is not required for ATM-mediated telomere damage response after TRF2 deletion. Nat Cell Biol 2005; 7(7):712-718.
2. McEachern MJ, Blackburn EH. Runaway telomere elongation caused by telomerase RNA gene mutations. Nature 1995; 376:403-409.
3. Nakamura TM, Cech TR. Reversing time: origin of telomerase. Cell 1998; 92(5):587-90.
4. Blackburn EH. Telomeres and their synthesis. Science 1990; 249(4968):489-90.
5. Blackburn EH, Gall JG. A tandemly repeated sequence at the termini of the extrachromosomal ribosomal RNA genes in Tetrahymena. J Mol Biol 1978; 120(1):33-53.
6. Yu GL, Bradley JD, Attardi LD et al. In vivo alteration of telomere sequences and senescence caused by mutated Tetrahymena telomerase RNAs. Nature 1990; 344(6262):126-32.
7. McEachern MJ, Blackburn EH. A conserved sequence motif within the exceptionally diverse telomeric sequences of budding yeasts. Proc Natl Acad Sci USA 1994; 91:3453-3457.
8. Cohn M, McEachern MJ, Blackburn EH. Telomeric sequence diversity within the genus Saccharomyces. Curr Genet 1998; 33(2):83-91.
9. McCormick-Graham M, Romero DP. A single telomerase RNA is sufficient for the synthesis of variable telomeric repeats in ciliates of the genus Paramecium. Mol Cell Biol 1996; 16(4):1871-1879.
10. Meyne J, Ratliff RL, Moyzis RK. Conservation of the human telomere sequence (TTAGGG)n among vertebrates. Proc Natl Acad Sci USA 1989; 86(18):7049-53.
11. Makarov VL, Lejnine S, Bedoyan J et al. Nucleosomal organization of telomere-specific chromatin in rat. Cell 1993; 73(4):775-87.
12. Kipling D, Cooke HJ. Hypervariable ultra-long telomeres in mice. Nature 1990; 347(6291):400-2.
13. Rhodin J, Astromskas E, Cohn M. Characterization of the DNA Binding Features of Saccharomyces castellii Cdc13p. J Mol Biol 2006; 355(3):335-46.
14. Forney JD, Blackburn EH. Developmentally controlled telomere addition in wild-type and mutant paramecia. Mol Cell Biol 1988; 8:251-258.
15. McCormick-Graham M, Haynes WJ, Romero DP. Variable telomeric repeat synthesis in Paramecium tetraurelia is consistent with misincorporation by telomerase. EMBO J 1997; 16(11):3233-42.
16. Forstemann K, Lingner J. Molecular basis for telomere repeat divergence in budding yeast. Mol Cell Biol 2001; 21(21):7277-86.
17. Prescott J, Blackburn EH. Telomerase RNA mutations in Saccharomyces cerevisiae alter telomerase action and reveal nonprocessivity in vivo and in vitro. Genes Development 1997;11:528-540.
18. Chappell AS, Lundblad V. Structural elements required for association of the Saccharomyces cerevisiae telomerase RNA with the Est2 reverse transcriptase. Mol Cell Biol 2004; 24(17):7720-36.
19. Hiraoka Y, Henderson E, Blackburn EH. Not so peculiar: fission yeast telomere repeats. Trends Biochem Sci 1998; 23(4):126.

20. Allshire RC, Dempster M, Hastie ND. Human telomeres contain at least three types of G-rich repeat distributed nonrandomly. Nucleic Acids Res 1989; 17:4611-4627.
21. Brown WR, MacKinnon PJ, Villasante A et al. Structure and polymorphism of human telomere-associated DNA. Cell 1990; 63(1):119-32.
22. Baird DM, Jeffreys AJ, Royle NJ. Mechanisms underlying telomere repeat turnover, revealed by hypervariable variant repeat distribution patterns in the human Xp/Yp telomere. EMBO J 1995; 14(21):5433-43.
23. Richards EJ, Chao S, Vongs A et al. Characterization of Arabidopsis thaliana telomeres isolated in yeast. Nucleic Acids Res 1992; 20:4039-4046.
24. Kirk KE, Blackburn EH. An unusual sequence arrangement in the telomeres of the germ-line micronucleus in Tetrahymena thermophila. Genes Development 1995; 9:59-71.
25. Baird DM, Coleman J, Rosser ZH et al. High levels of sequence polymorphism and linkage disequilibrium at the telomere of 12q: implications for telomere biology and human evolution. Am J Hum Genet 2000; 66(1):235-50.
26. Schlecht HB, Lichten M, Goldman AS. Compartmentalization of the yeast meiotic nucleus revealed by analysis of ectopic recombination. Genetics 2004; 168(3):1189-203.
27. IJdo JW, Baldini A, Ward DC et al. Origin of human chromosome 2: an ancestral telomere-telomere fusion. Proc Natl Acad Sci USA 1991; 88(20):9051-5.
28. Park VM, Gustashaw KM, Wathen TM. The presence of interstitial telomeric sequences in constitutional chromosome abnormalities. Am J Hum Genet 1992; 50(5):914-23.
29. Hartmann N, Scherthan H. Characterization of ancestral chromosome fusion points in the Indian muntjac deer. Chromosoma 2004; 112(5):213-20.
30. Huang L, Wang J, Nie W et al. Tandem chromosome fusions in karyotypic evolution of Muntiacus: evidence from M feae and M gongshanensis. Chromosome Res 2006; 14(6):637-47.
31. Metcalfe CJ, Eldridge MD, Johnston PG. Mapping the distribution of the telomeric sequence (T2AG3)n in the Macropodoidea (Marsupialia) by fluorescence in situ hybridization. II. The ancestral 2n = 22 macropodid karyotype. Cytogenet Genome Res 2007; 116(3):212-7.
32. Meyne J, Baker RJ, Hobart HH et al. Distribution of nontelomeric sites of the (TTAGGG)n telomeric sequence in vertebrate chromosomes. Chromosoma 1990; 99(1):3-10.
33. Cherry JM, Blackburn EH. The internally located telomeric sequences in the germ-line chromosomes of Tetrahymena are at the ends of transposon-like elements. Cell 1985; 43(3):47-58.
34. Edwards YJ, Elgar G, Clark MS et al. The identification and characterization of microsatellites in the compact genome of the Japanese pufferfish, Fugu rubripes: perspectives in functional and comparative genomic analyses. J Mol Biol 1998; 278(4):843-54.
35. Wahlin J, Cohn M. RAP1 binding and length regulation of yeast telomeres. In: Krupp G, Parwaresch R, eds. Telomeres and Telomerases: Cancer and Biology, Landes Bioscience 2002; 259-281.
36. Rhodes D. The structural Biology of Telomeres, In: de Lange T, Lundblad V, Blackburn EH, eds. Telomeres. 2 ed. Cold Spring Harbor: Cold Spring Harbor Laboratory Press, 2006:317-343.
37. Broccoli D, Smogorzewska A, Chong L et al. Human telomeres contain two distinct Myb-related proteins, TRF1 and TRF2. Nat Genet 1997; 17(2):231-5.
38. Cooper JP, Nimmo ER, Allshire RC et al. Regulation of telomere length and function by a Myb-domain protein in fission yeast. Nature 1997; 385:744-747.
39. König P, Giraldo R, Chapman L et al. The crystal structure of the DNA-binding domain of yeast RAP1 in complex with telomeric DNA. Cell 1996; 85:125-136.
40. Smogorzewska A, de Lange T. Regulation of telomerase by telomeric proteins. Annu Rev Biochem 2004; 73:177-208.
41. Nugent CI, Hughes TR, Lue NF et al. Cdc13p: A single-strand telomeric DNA-binding protein with a dual role in yeast telomere maintenance. Science 1996; 274:249-252.
42. Lin JJ, Zakian VA. The Saccharomyces CDC13 protein is a single-strand TG1-3 telomeric DNA-binding protein in vitro that affects telomere behavior in vivo. Proc Natl Acad Sci USA 1996; 93(24):13760-5.
43. Mitton-Fry RM, Anderson EM, Hughes TR et al. Conserved structure for single-stranded telomeric DNA recognition. Science 2002; 296(5565):145-7.
44. Gao H, Cervantes RB, Mandell EK et al. RPA-like proteins mediate yeast telomere function. Nat Struct Mol Biol 2007; 14(3):208-14.
45. Virta-Pearlman V, Morris DK, Lundblad V. Est1 has the properties of a single-stranded telomere end-binding protein. Genes Development 1996; 10:3094-3104.
46. Zhou J, Hidaka K, Futcher B. The Est1 subunit of yeast telomerase binds the Tlc1 telomerase RNA. Mol Cell Biol 2000; 20(6):1947-55.
47. Wahlin J, Cohn M. Analysis of the RAP1 protein binding to homogeneous telomeric repeats in Saccharomyces castellii. Yeast 2002; 19:241-256.

48. Krauskopf A, Blackburn EH. Control of telomere growth by interactions of RAP1 with the most distal telomeric repeats. Nature 1996; 383:354-357.
49. McEachern MJ, Iyer S, Fulton TB et al. Telomere fusions caused by mutating the terminal region of telomeric DNA. Proc Natl Acad Sci USA 2000; 97(21):11409-14.
50. Hayashi N, Murakami S. STM1, a gene which encodes a guanine quadruplex binding protein, interacts with CDC13 in Saccharomyces cerevisiae. Mol Genet Genomics 2002; 267(6):806-13.
51. Sykorova E, Lim KY, Kunicka Z et al. Telomere variability in the monocotyledonous plant order Asparagales. Proc Biol Sci 2003; 270(1527):1893-904.
52. Zappulla DC, Cech TR. Yeast telomerase RNA: a flexible scaffold for protein subunits. Proc Natl Acad Sci USA 2004; 101(27):10024-9.
53. Bhattacharyya A, Blackburn EH. A functional telomerase RNA swap in vivo reveals the importance of nontemplate RNA domains. Proc Natl Acad Sci USA 1997; 94(7):2823-7.
54. Lin J, Ly H, Hussain A et al. A universal telomerase RNA core structure includes structured motifs required for binding the telomerase reverse transcriptase protein. Proc Natl Acad Sci USA 2004; 101(41):14713-8.
55. Dandjinou AT, Levesque N, Larose S et al. A phylogenetically based secondary structure for the yeast telomerase RNA. Curr Biol 2004; 14(13):1148-58.
56. Wahlin J, Cohn M. Saccharomyces cerevisiae RAP1 binds to telomeric sequences with spatial flexibility. Nucleic Acids Res 2000; 28(12):2292-2301.
57. Nosek J, Kosa P, Tomaska L. On the origin of telomeres: a glimpse at the pretelomerase world. Bioessays 2006; 28(2):182-90.
58. Pimpinelli S. Drosophila telomeres. In: de Lange T, Lundblad V, Blackburn EH, eds. Telomeres. 2 ed. Cold Spring Harbor: Cold Spring Harbor Laboratory Press, 2006:433-463.
59. Casacuberta E, Pardue ML. HeT-A and TART, two Drosophila retrotransposons with a bona fide role in chromosome structure for more than 60 million years. Cytogenet Genome Res 2005; 110(1-4):152-9.
60. Fulton TB, Blackburn EH. Identification of Kluyveromyces lactis telomerase: discontinuous synthesis along the 30-nucleotide-long templating domain. Mol Cell Biol 1998; 18(9):4961-70.
61. Kosa P, Valach M, Tomaska L et al. Complete DNA sequences of the mitochondrial genomes of the pathogenic yeasts Candida orthopsilosis and Candida metapsilosis: insight into the evolution of linear DNA genomes from mitochondrial telomere mutants. Nucleic Acids Res 2006; 34(8):2472-81.
62. Cohn M, Liti G, Barton DBH. Telomeres in fungi. In: Sunnerhagen P, Piskur J, eds. Comparative genomics using fungi as models. Heidelberg: Springer Verlag, 2006; 100-130.
63. Moyzis RK, Buckingham JM, Cram LS et al. A highly conserved repetitive DNA sequence, (TTAGGG)n, present at the telomeres of human chromosomes. Proc Natl Acad Sci USA 1988; 85(18):6622-6.
64. Forney J, Henderson ER, Blackburn EH. Identification of the telomeric sequence of the acellular slime molds Didymium iridis and Physarum polycephalum. Nucleic Acids Res 1987; 15(22):9143-52.
65. Emery HS, Weiner AM. An irregular satellite sequence is found at the termini of the linear extrachromosomal rDNA in Dictyostelium discoideum. Cell 1981; 26(3 Pt 1):411-9.
66. Blackburn EH, Challoner PB. Identification of a telomeric DNA sequence in Trypanosoma brucei. Cell 1984; 36(2):447-57.
67. Van der Ploeg LH, Liu AY, Borst P. Structure of the growing telomeres of Trypanosomes. Cell 1984; 36(2):459-68.
68. Freitas-Junior LH, Porto RM, Pirrit LA et al. Identification of the telomere in Trypanosoma cruzi reveals highly heterogeneous telomere lengths in different parasite strains. Nucleic Acids Res 1999; 27(12):2451-6.
69. Chiurillo MA, Beck AE, Devos T et al. Cloning and characterization of Leishmania donovani telomeres. Exp Parasitol 2000; 94(4):248-58.
70. Le Blancq SM, Kase RS, Van der Ploeg LH. Analysis of a Giardia lamblia rRNA encoding telomere with [TAGGG]n as the telomere repeat. Nucleic Acids Res 1991; 19(20): 5790.
71. Baroin A, Prat A, Caron F. Telomeric site position heterogeneity in macronuclear DNA of Paramecium primaurelia. Nucleic Acids Res 1987; 15(4):1717-28.
72. Shippen-Lentz D, Blackburn EH. Telomere terminal transferase activity from Euplotes crassus adds large numbers of TTTTGGGG repeats onto telomeric primers. Mol Cell Biol 1989; 9(6):2761-2764.
73. Klobutcher LA, Swanton MT, Donini P et al. All gene-sized DNA molecules in four species of hypotrichs have the same terminal sequence and an unusual 3' terminus. Proc Natl Acad Sci USA 1981; 78(5):3015-9.
74. Ponzi M, Pace T, Dore E et al. Identification of a telomeric DNA sequence in Plasmodium berghei EMBO J 1985; 4(11):2991-5.
75. Weiss-Schneeweiss H, Riha K, Jang CG et al. Chromosome termini of the monocot plant Othocallis siberica are maintained by telomerase, which specifically synthesises vertebrate-type telomere sequences. Plant J 2004; 37(4):484-93.

76. Ganal MW, Lapitan NL, Tanksley SD. Macrostructure of the tomato telomeres. Plant Cell 1991; 3(1):87-94.
77. Richards EJ, Ausubel FM. Isolation of a higher eukaryotic telomere from Arabidopsis thaliana. Cell 1988; 53(1):127-36.
78. Higashiyama T, Maki S, Yamada T. Molecular organization of Chlorella vulgaris chromosome I: presence of telomeric repeats that are conserved in higher plants. Mol Gen Genet 1995; 246(1):29-36.
79. Petracek ME, Lefebvre PA, Silflow CD et al. Chlamydomonas telomere sequences are A + T-rich but contain three consecutive G-C base pairs. Proc Natl Acad Sci USA 1990; 87(21):8222-6.
80. Muller F, Wicky C, Spicher A et al. New telomere formation after developmentally regulated chromosomal breakage during the process of chromatin diminution in Ascaris lumbricoides. Cell 1991; 67(4):815-22.
81. Wicky C, Villeneuve AM, Lauper N et al. Telomeric repeats (TTAGGC)n are sufficient for chromosome capping function in Caenorhabditis elegans. Proc Natl Acad Sci USA 1996; 93(17):8983-8.
82. Teschke C, Solleder G, Moritz KB. The highly variable pentameric repeats of the AT-rich germline limited DNA in Parascaris univalens are the telomeric repeats of somatic chromosomes. Nucleic Acids Res 1991; 19(10):2677-84.
83. Okazaki S, Tsuchida K, Maekawa H et al. Identification of a pentanucleotide telomeric sequence, (TTAGG)n, in the silkworm Bombyx mori and in other insects. Mol Cell Biol 1993; 13(3):1424-1432.
84. Robertson HM, Gordon KH. Canonical TTAGG-repeat telomeres and telomerase in the honey bee, Apis mellifera. Genome Res 2006; 16(11):1345-51.
85. Schechtman MG. Characterization of telomere DNA from Neurospora crassa. Gene 1990; 88(2):159-65.
86. Powell WA, Kistler HC. In vivo rearrangement of foreign DNA by Fusarium oxysporum produces linear self-replicating plasmids. J Bacteriol 1990; 172(6):3163-71.
87. Woods JP, Goldman WE. In vivo generation of linear plasmids with addition of telomeric sequences by Histoplasma capsulatum. Mol Microbiol 1992; 6(23):3603-10.
88. Coleman MJ, McHale MT, Arnau J et al. Cloning and characterisation of telomeric DNA from Cladosporium fulvum. Gene 1993; 132(1):67-73.
89. Javerzat JP, Bhattacherjee V, Barreau C. Isolation of telomeric DNA from the filamentous fungus Podospora anserina and construction of a self-replicating linear plasmid showing high transformation frequency. Nucleic Acids Res 1993; 21(3):497-504.
90. Guzman PA, Sanchez JG. Characterization of telomeric regions from Ustilago maydis. Microbiology 1994; 140(Pt 3):551-7.
91. Farman ML, Leong SA. Genetic and physical mapping of telomeres in the rice blast fungus, Magnaporthe grisea Genetics 1995; 140(2):479-92.
92. Underwood AP, Louis EJ, Borts RH et al. Pneumocystis carinii telomere repeats are composed of TTAGGG and the subtelomeric sequence contains a gene encoding the major surface glycoprotein. Mol Microbiol 1996; 19(2):273-81.
93. Bhattacharyya A, Blackburn EH. Aspergillus nidulans maintains short telomeres throughout development. Nucleic Acids Res 1997; 25(7):1426-31.
94. Kusumoto KI, Suzuki S, Kashiwagi Y. Telomeric repeat sequence of Aspergillus oryzae consists of dodeca-nucleotides. Appl Microbiol Biotechnol 2003; 61(3):247-51.
95. Edman JC. Isolation of telomerelike sequences from Cryptococcus neoformans and their use in high-efficiency transformation. Mol Cell Biol 1992; 12(6):2777-83.
96. Matsumoto T, Fukui K, Niwa O et al. Identification of healed terminal DNA fragments in linear minichromosomes of Schizosaccharomyces pombe. Mol Cell Biol 1987; 7(12):4424-30.
97. Shampay J, Szostak JW, Blackburn EH. DNA sequences of telomeres maintained in yeast. Nature 1984; 310(5973):145-157.
98. McEachern MJ, Hicks JB. Unusually large telomeric repeats in the yeast Candida albicans. Mol Cell Biol 1993; 13(1):551-560.

CHAPTER 6

Evolution of Telomere Binding Proteins

Martin P. Horvath*

Abstract

Telomere binding proteins provide essential functions for chromosome maintenance in most eukaryotes and consequently are well suited for analysis in the context of evolution. This review focuses on patterns gleaned from structural and functional characterization of telomere proteins that reveal contrasting evolutionary histories for double-stranded and single-stranded DNA-binding protein families. The myb-like/homeodomain DNA-binding motif is ubiquitous among members of the double-stranded telomere DNA-binding protein family which includes Rap1 and Taz1 in yeast, TRF1 and TRF2 in vertebrates, as well as putative plant-specific telomere proteins, TBP1, TRP1 and Smh1. In this myb-motif family, strong purifying selection has preserved amino-acid sequence among distantly related lineages. Accessory domains linked with the myb-like DNA-binding domain define distinct lineages, indicating that the myb motif was probably recruited multiple times from a reserve of more ancestral forms functioning as transcription factors. The oligonucleotide/oligosaccharide/oligopeptide-binding (OB)-fold is universally found in members of the second class of telomere binding proteins that recognize and bind T/G-rich telomere sequences in the form of single-stranded DNA. Examples of proteins in this class include TEBP-α and TEBP-β from ciliated protozoa, Cdc13 from budding yeast, and Pot1 which is widely distributed in fission yeast, vertebrates and plants. For these OB-fold proteins, rapid divergence at the amino-acid sequence level has all but erased traces of common ancestry. Homology is apparent, however, when comparing three-dimensional structures and functional characters. Sequence alignments consistent with these structural comparisons provide a tentative glimpse of deeply rooted lineages that likely emerged from the general single-stranded DNA-binding proteins dedicated to DNA replication and repair. Interaction networks among telomere DNA-binding proteins and their associated interacting partners hint at further common patterns and innovations encountered during the evolution of telomere capping complexes.

Introduction

Telomere binding proteins recognize the short tandem repeats characterizing telomere DNA,[1] and are essential for stable chromosome maintenance as reflected in the precipitous genome-destabilizing outcomes following gene deletion and over-expression of mutant alleles (reviewed in ref. 2). The nucleoprotein complexes formed upon association of telomere binding proteins with telomere DNA distinguish natural chromosome ends from double-stranded breaks and thereby protect chromosome termini from inappropriate end-to-end ligation. Additionally, telomere binding proteins recruit and regulate telomerase to ensure an appropriate length of structural DNA is maintained as a buffer against loss of genetic information stored in genes close to the ends of linear chromosomes. Recent reviews of telomere binding proteins from yeast and vertebrates have focused on telomere homeostasis and the relationship between these proteins

*Martin P. Horvath—Biology Department, University of Utah, 257 S 1400 E, Salt Lake City UT 84112-0840, USA. Email: horvath@biology.utah.edu

Origin and Evolution of Telomeres, edited by Jozef Nosek and Ľubomír Tomáška.

and telomerase and DNA repair pathways.[2,3] The emphasis of this chapter will be on the evolution of these important proteins. Questions addressed include "What are the origins of telomere end binding proteins?" and "What are the functions and interactions among telomere proteins predicted by analogy with divergent systems?"

Telomere binding proteins can be broadly divided into two classes on the basis of double-stranded versus single-stranded DNA-binding specificity. DNA-binding specificity correlates with folding motifs particular to each class, with the myb-like/homeodomain motif dictating double-stranded DNA recognition and the OB-fold proteins directed towards single-stranded telomere DNA. This review begins with an introduction to the structures of the myb-like motif and the OB-fold since recent insights regarding evolution of telomere binding proteins rely on relationships inferred from three-dimensional structures of these DNA-binding elements. The origins of double-stranded and single-stranded telomere DNA-binding proteins are then pursued followed by a discussion of cooperative interaction networks characterizing telomere-capping systems.

Protein Folding Motifs for Binding Telomere DNA

Structural biology has been particularly important for inferring evolutionary relationships among telomere binding proteins.[4] Crystal structures of the DNA-binding domain of Rap1 from yeast and the telomere end binding protein (TEBP) from *Sterkiella nova* (formerly *Oxytricha nova*) in complex with cognate double-stranded and single-stranded telomere DNA, respectively, provided the first high-resolution views of telomere proteins in action.[5,6] Since then the structures of several DNA-binding domains or protein-protein interaction domains derived from telomere binding proteins have been determined (see Table 1). The structural architecture of telomere proteins with an emphasis on DNA-protein interactions has been previously reviewed.[7] In this section, the basic structural elements of telomere protein DNA-binding domains are summarized so as to provide a foundation for understanding evolutionary analysis of telomere binding proteins.

The Homeodomain/Myb-Like Motif

The homeodomain/myb-like motif is found in all telomere binding proteins that recognize double-stranded telomere DNA. The myb motif is named after the transcription factor c-Myb, a proto-oncogene which regulates differentiation and proliferation during hematopoiesis.[8] The myb-fold consists of three α-helices arranged in an orthogonal bundle around an independent hydrophobic core (Fig. 1). The third helix presents residues that make sequence-specific contacts with

Table 1. Telomere binding protein structures

Protein Family	Protein Complex	PDB Id	References
Myb/ homeodomain	*S. cerevisiae* Rap1–dsDNA, apo	1ign,	5
	human TRF1–dsDNA, apo	1w0t, 1ba5	10,107
	human TRF2–dsDNA, apo	1w0u, 1vf9	10,108
	human Rap1 (does not bind DNA)	1fex	67
	c-Myb–dsDNA	1mse, 1h89	9,109
OB-fold	*S. nova* TEBP-α/β–ssDNA, α–ssDNA, αN–ssDNA, αN (apo)	1otc, 1jb7, 2i0q, 1kix, 1k8g	6,73,105,106,110
	S. cerevisiae cdc13–ssDNA	1kxl, 1s40	32,41
	S. pombe Pot1–ssDNA	1qzg, 1qzh	34
	human Pot1–ssDNA	1xjv	33
	human Tpp1	2i46	35
	human RPA-70–ssDNA, apo; RPA trimer core; RPA-70 OB-N	1jmc, 1fgu, 1l1o, 1ewi, 2b29, 2b3g	30,40,111-113

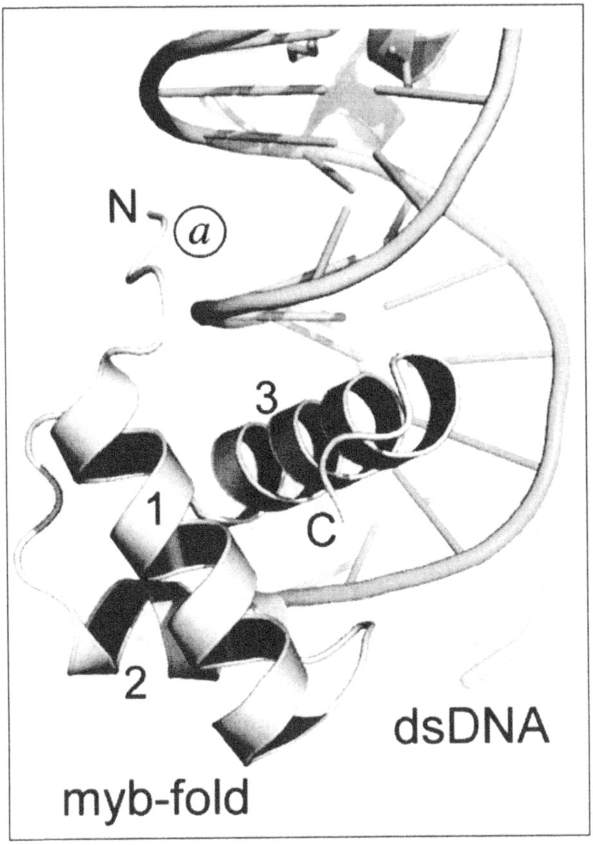

Figure 1. Myb/homeodomain-DNA complex. One of the two DNA-binding domains of the human TRF2 dimer (pdb id 1w0u) is represented with its three α-helices as ribbons. The C-terminal helix 3 fits within the major groove of double-stranded telomere repeat DNA where particular valine, lysine, aspartate and arginine residues make sequence-specific contacts. The N-terminal tail (labeled with *a*) crosses over to make interactions with the adjacent minor groove.

bases in the major groove of B-form DNA.[5,9,10] For telomere proteins, these DNA-recognition residues are especially well conserved and define the so-called telobox sequence feature.[11,12] In TRF1 and TRF2 the DNA-recognition helix is augmented by an N-terminal tail (*a* in Fig. 1), particular to the homeodomain superfamily, that makes DNA contacts in the adjacent minor groove.[10]

The different architectures of DNA-binding domains constructed from myb-like motifs (Fig. 2) define distinct myb-like protein families. In the DNA-binding domain of c-Myb, three imperfect tandem repeats of the three-helical orthogonal bundle contribute to DNA recognition, with repeats 2 and 3 making sequence specific DNA contacts and repeat 1 in close proximity with but not directly touching the DNA.[9] In Rap1, the principle double-stranded DNA-binding protein found at telomeres of budding yeast,[13] two imperfect tandem myb-like repeats make up the DNA-binding domain. Myb repeats in Rap1 are connected by a linker that is somewhat longer than the linker connecting repeats in c-Myb, a feature that may have been adaptive for recognition of telomere DNA repeats that are positioned with variable phasing in the degenerate sequences particular to telomeres of *Saccharomyces cerevisiae*.[5]

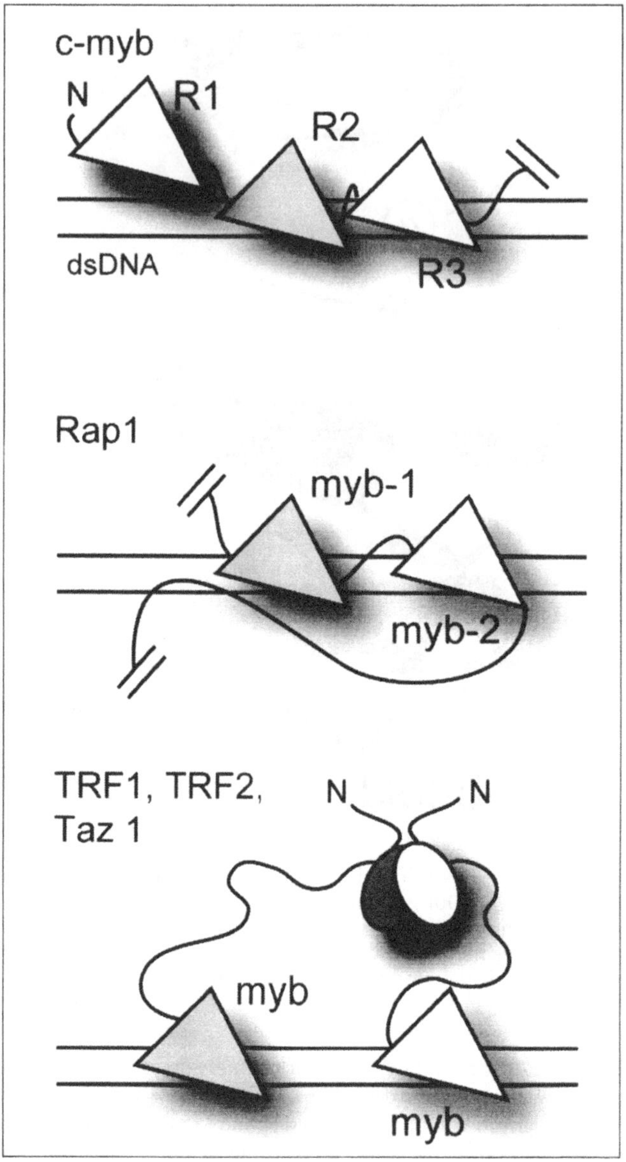

Figure 2. Myb-like domain and subunit organization. (top panel) c-Myb recognizes its cognate sequence using three imperfect tandem repeats (R1, R2, and R3; pdb id 1mse). Repeats R2 and R3 make direct contact with six basepairs of DNA whilst R1 stabilizes the complex through a favorable electrostatic contribution but does not touch the DNA.[9] (middle panel) The two myb-like imperfect tandem repeats found in the DNA-binding domain of Rap1 (pdb id 1ign) are arranged with increased spacing.[5] (bottom panel) Single myb-domain proteins TRF1, TRF2 and Taz1 bind DNA as dimers with two equal myb/homeodomains brought in proximity *via* flexible linker regions. Homodimerization may be an adaptive innovation that facilitates looping of telomere DNA[22,90] and telomere clustering during meiosis.[21]

TRF1, the first described human telomere protein,[14] binds telomere double-stranded DNA as a dimer,[15] with flexible regions connecting the conserved TRF-homology (TRFH) dimerization domain with a single C-terminal DNA-binding myb-like domain.[16,17] Taz1 in fission yeast[18] and TRF2, a paralog to TRF1 in vertebrates,[19,20] share the same TRFH/C-terminal single-myb domain structure seen in TRF1. Consideration of domain architecture in Rap1, Taz1, TRF1 and TRF2 supports the view that telomere double-stranded DNA-binding proteins are constrained to use two myb-like repeats for recognition of telomere DNA. The two myb-like units can be arranged either as tandem repeats as seen in Rap1 or as individual units that are brought together by protein dimerization. Dimerization may have been an adaptive innovation to facilitate telomere-pairing during meiosis in fission yeast[21] and looping of telomere DNA as observed for end-protection in vertebrates.[22]

Single myb-like repeats are found in numerous plant-specific proteins that bind with telomere repeat sequences in vitro.[23-26] Members of the plant-specific family defined by TBP1, TRP1 and several TRF-like proteins have a myb-like motif close to the C-terminus and interact with DNA as dimers, features that are similar with those characterizing the Taz1/TRF1/TRF2 family of fission yeast and vertebrates.[24-26] These putative telomere proteins have distinguishing plant-specific features, however, in that they lack a recognizable TRFH domain and in each case the myb-like motif is extended to include a fourth helix.[26,27] Protein engineering experiments showed that the fourth helix confers in vitro DNA-binding ability to canonical three-helix myb-like domains that otherwise do not bind telomere DNA.[26] The fourth helix contributes to a larger hydrophobic core and repositions the DNA-recognition helix slightly but does not appear to enlarge the DNA-interface as inferred from an NMR-determined structure and chemical shift perturbation studies.[27] The single myb-like histone-1 (Smh1) protein from maize introduced a second class of myb-like proteins in plants that are implicated in telomere structure on the basis of in vitro DNA-binding.[23] Proteins like Smh1 have yet another domain architecture that comprises an N-terminal myb-like domain, a middle domain derived from the conserved globular domain of linker histones and a C-terminal coiled-coil domain.[23] High redundancy expected for these plant proteins complicate interpretation of gene knockout and mutant allele expression studies meaning that positive identification of the telomere-specific factors in plants is still in progress. It seems likely, however, that double-stranded DNA-binding factors in plants will add distinct plant-specific domain architectures for this class of telomere proteins.

The OB-Fold

Murzin first described the OB-fold as an example of a homologous protein family for which evidence of common ancestry is preserved in the three-dimensional structures of extant family members but not readily apparent on the basis of amino-acid sequence similarity.[28] The X-ray cocrystal structure of a telomere end binding protein (TEBP) from *S. nova* complexed with telomere single-stranded DNA revealed three OB-folds devoted to recognition of DNA and a fourth OB-fold involved in protein-protein interactions between TEBP-α and TEBP-β proteins.[6] These OB-folds placed the otherwise orphaned TEBP-α and TEBP-β proteins in a large family with diverse members that include single-stranded DNA-binding proteins from bacteria,[29] replication protein A (RPA),[30] and the anti-codon binding domain of many amino-acyl-tRNA synthetases.[28,31] Subsequent structure determinations of the single-stranded DNA-binding domains derived from several other telomere binding proteins (Table 1) has led to the conclusion that use of OB-folds is universally conserved among this class of proteins.[32-36]

OB-folds have been reviewed previously with an emphasis on evolution in host/pathogen contexts,[37] recognition of DNA and RNA,[31] and superfamily inter-relations inferred from sequence profile-based phylogenetics.[38] The OB-fold (Fig. 3) comprises two orthogonally packed anti-parallel beta sheets with s1:s4:s5 strand topology in sheet *a* and s1':s2:s3 topology in sheet *b*, with the N-terminal stand s1/s1' continuing as the outer edge of both sheets. Strands s4 and s5 often continue past sheet *a* and fold over to extend sheet *b* and thus complete a closed beta-barrel-like structure. The segment joining strands s3 and s4 often includes a short helix, but this region adopts an extended coil-like structure in many counter-examples including an SSB protein from archaea[39]

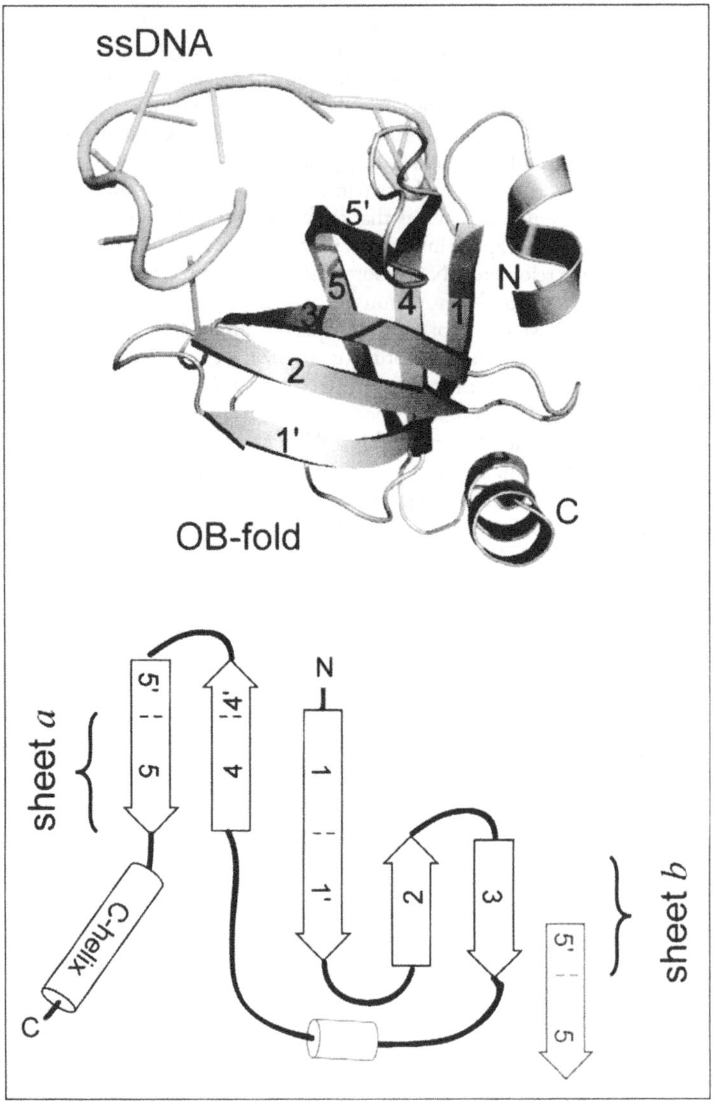

Figure 3. OB-fold. (Top panel) The second OB-fold from the DNA-binding domain of *S. nova* TEBP-α in complex with single-stranded telomere DNA (pdb id 2i0q) is shown as a ribbon representation. (Bottom panel) Strand topology defining the OB-fold is shown with strands s1:s4:s5 making up sheet *a* and strands s1′:s2:s3:s5′ making up sheet *b*. Loops connecting beta strands, particularly loops L_{12} and L_{45}, form the DNA-interaction surface. A C-terminal helix included with many, but not all, OB-folds mediates protein-protein or domain-domain interactions.

and certain OB-folds from RPA.[30,40] Most telomere-associated OB-folds are further characterized by a C-terminal alpha-helix but this structural element is neither absolutely conserved among telomere-specific OB-folds nor excluded from nontelomere protein examples. The loops connecting beta strands of the OB-fold are variable in length and these insertions/deletions account for the unreliability of current bioinformatics tools for positively identifying OB-folds.

The domain and subunit organization of OB-fold telomere components is relevant for understanding inferences regarding molecular evolution. A single OB-fold makes up the DNA-binding domain of Cdc13 from budding yeast.[32,41] As far as we currently know from structural biology, the DNA-binding domain of Pot1 from fission yeast contains one OB-fold;[34] however, the binding behaviors recently measured for full-length and truncated yeast Pot 1 proteins strongly suggest that there could be two OB-folds involved with DNA binding.[42,43] As seen in crystal structures, two OB-folds make up the DNA-binding domain in human Pot1[33] and the DNA-binding domain of TEBP-α from *S. nova*.[6] In the case of the *S. nova* telomere complex, an OB-fold contributed by TEBP-β extends the DNA-protein interface so that the complete DNA-binding portion of the α/β hetero-dimer consists of three OB-folds (Fig. 4). The interactions responsible for α/β association are established by means of an extended peptide loop contributed by TEBP-β that wraps around the C-terminal domain of α which comprises an OB-fold as well.[6]

Hetero-oligomerization of single-stranded telomere DNA-binding proteins, as first characterized in the *S. nova* telomere-capping complex,[6,44,45] has recently emerged as a theme that is conserved in vertebrates and yeast. Pot1 proteins were identified on the basis of weak but nonetheless significant sequence similarity with the N-terminal OB-fold (αOB-1) of TEBP-α,[46] but corresponding homologs for TEBP-β remained elusive until recently. Bioinformatics approaches and structure determination now confirm that the human Pot1-interacting protein Tpp1 (formerly called PTOP, PIP1, or TINT1)[47-49] is an OB-fold protein and a homolog of TEBP-β,[35,50] suggesting that subunit organization as well as OB-folds are conserved among deeply rooted eukaryotic lineages.

A similar case of hetero-oligomerization involving OB-folds with profound implications for our understanding of telomere protein evolution was recently described for telomere components from budding yeast.[51] Cdc13 from *S. cerevisiae* associates with single-stranded telomere DNA[32,52,53] and carries out essential and separable functions for telomere length homeostasis and end-protection by recruiting additional components to telomere ends.[54] For end protection, Cdc13 suppresses inappropriate DNA-damage response signals by recruiting Stn1[55] and Ten1.[56] Sequence comparisons, in vitro DNA-binding experiments, yeast two-hybrid tests and complementation by domain swapping in yeast indicate that Stn1 and Ten1 are paralogs of RPA subunits with predicted OB-folds of their own.[51] RPA is the eukaryotic single-stranded DNA-binding (SSB) protein with essential functions in semi-conservative DNA replication, recombination and repair.[57,58] RPA is a multi-OB-fold protein made up of three different subunits (see Fig. 4).[57,58] Sequence similarity, analogous biochemical properties among telomere associated factors and RPA subunits and the ability of an OB-fold derived from Stn1 to functionally replace the OB-fold of an RPA subunit in vivo strongly suggest that a complex consisting of Cdc13, Stn1 and Ten1 constitutes a multi-OB-fold, RPA-like complex dedicated to telomere end protection in budding yeast.[51]

Origins of Telomere Binding Proteins

Myb-Domain Telomere Proteins Are Derived from Transcription Factors

The ancestral form of the myb-like telomere DNA-binding motif is likely the DNA-binding region of a trans acting factor devoted to regulation of gene expression. Repressors and activators bearing similarity to the myb-motif are widespread among eubacteria, archaea and eukarya indicating that this protein folding unit was well established in the last common ancestor. Transcription factors with multiple myb repeats have experienced explosive expansion especially in animals and plants.[59] Rap1, with its two myb-repeat DNA-binding domain, probably represents the descendent of one such protein that aquired a telomeric role even while retaining its gene regulatory functions.[60]

Ubiquitous use of myb-like domains in the DNA-binding regions of telomere binding proteins suggests a common origin. Domains outside of the DNA-binding region are distinct and different, however and indicate that fixation during the course of evolution involved functions in addition to simply binding telomere sequence repeats. Two added functions apparent in extant proteins are (1) recruitment and coordinated interaction with other telomere factors, some of which have

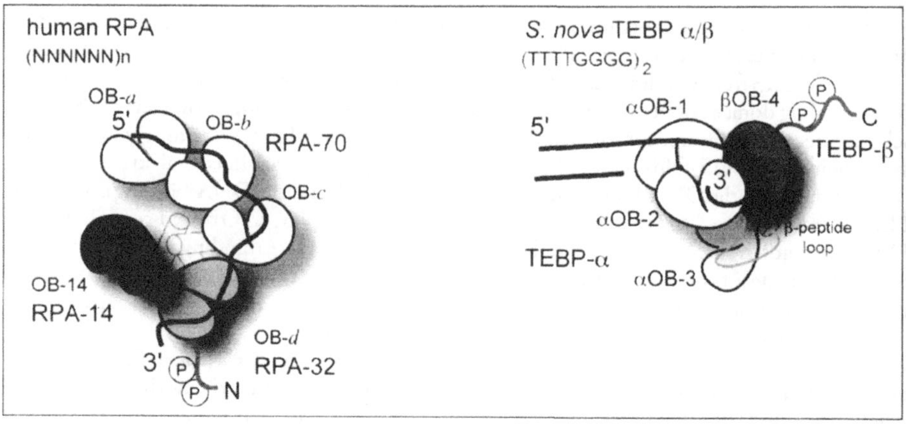

Figure 4. OB-fold domain and subunit organization. (left-hand panel) OB-folds contained within the human RPA hetero-trimer form a nonsequence specific complex with single-stranded DNA (adapted from multiple structures; pdb ids 1jmc and 1l1o). Subunit RPA-70 contributes the major single-stranded DNA-binding domain comprising OB-folds OB-*a* and OB-*b*.[30] OB-*c*, also from the RPA-70 subunit, and OB-*d* from the middle domain of RPA-32 extend the DNA interaction surface.[114] Three helices contributed by each of the three subunits form the trimerization core.[40] (right-hand panel) OB-folds contained within the *S. nova* TEBP α/β hetero-dimer from a sequence and structure-specific complex with single-stranded telomere DNA (adapted from pdb 2i0q). The N-terminal domain of TEBP-α comprising αOB-1 and αOB-2 constitutes the major telomere single-stranded DNA-binding domain.[100] TEBP-β contributes a third OB-fold,[6] βOB-4, to substantially reorganize the DNA-protein interface relative to that seen for DNA-complexes with TEBP-α[105] or the N-terminal domain of TEBP-α.[106] Subunit interactions within TEBP are mediated by a helix and peptide loop from TEBP-β that make extensive contacts with αOB-3 contributed by the C-terminal domain of TEBP-α.[6] Both TEBP-β and RPA-32 are phosphorylated in peripheral regions, the N-terminal region of RPA-32[87] and the C-terminal tail of TEBP-β.[85,86]

DNA-binding domains of their own and (2) oligomerization of the telomere binding protein through homotypic interactions. Figure 5 shows a phylogenetic tree constructed from amino-acid sequence alignment of homeodomain/myb-like domains from TRF1, TRF2, Taz1, Rap1, as well as TRF-like single-myb proteins from amoeba and plants. In this analysis the TRF1/TRF2 family appears well separated from the single-myb motifs found in other contexts.

Figure 5, viewed on next page. Amino-acid sequence alignment and phylogenetic tree constructed for myb-like domains from telomere binding proteins. Residue positions and helix boundaries are those derived from the crystal structure of human TRF2 (pdb id 1w0u). Residues within helix 3 that make sequence specific DNA contacts are indicated with a triangle (▼). TRF1/TRF2 paralogy extends to the root of vertebrates. The TRF-like sequences predicted from the Zebra fish (*Danio rerio*) genome are the least well resolved. Sequences belonging to the TRF1 and TRF2 clades, except Taz1, were identified and aligned with a *BLASTp* search at the NCBI site (http://www.ncbi.nlm.nih.gov/BLAST/). The same search yielded TRF-like sequences Entamoeba-1 (gi|67467563) and Entamoeba-2 (gi|67484616), and MtSmh1, the single myb histone H1 protein from *Medicago truncatula* (gi|92872731). The later served as outgroup for tree reconstruction. Sequences for Taz1, ScRap1 myb repeats R1 and R2 and the single-myb domains from *Arabidopsis* proteins AtTRP1, AtTBP1, and AtTRFL4 were added on the basis of previously published sequence alignments.[23-26] Tree topology was obtained by parsimony analysis and tested by the bootstrap method using the program *Phylip*.[115] Gapped positions were excluded from consideration. Nodes observed in greater than 50% of 1000 simulations are annotated with bootstrap support (%).

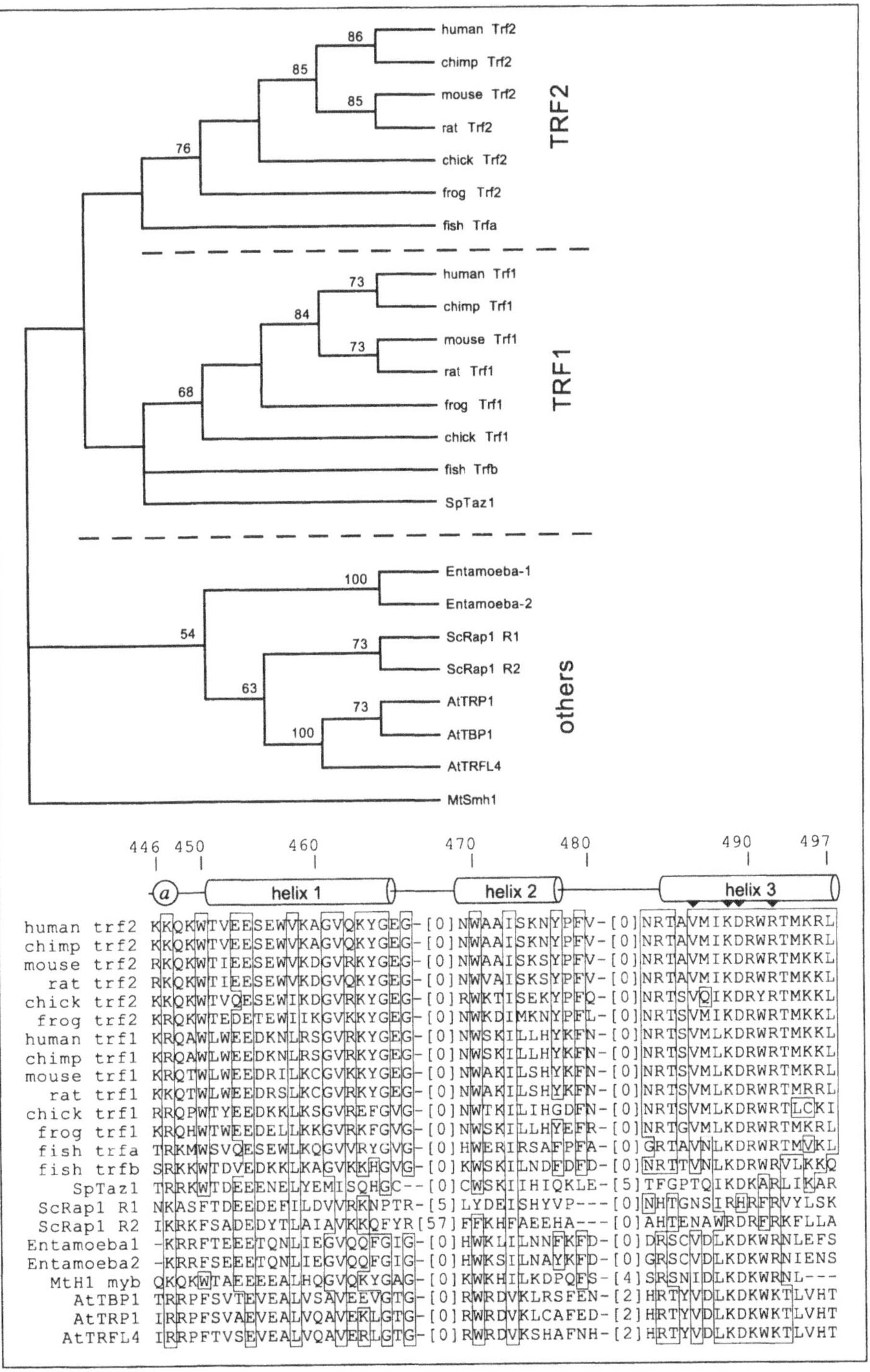

Figure 5. Figure legend on previous page.

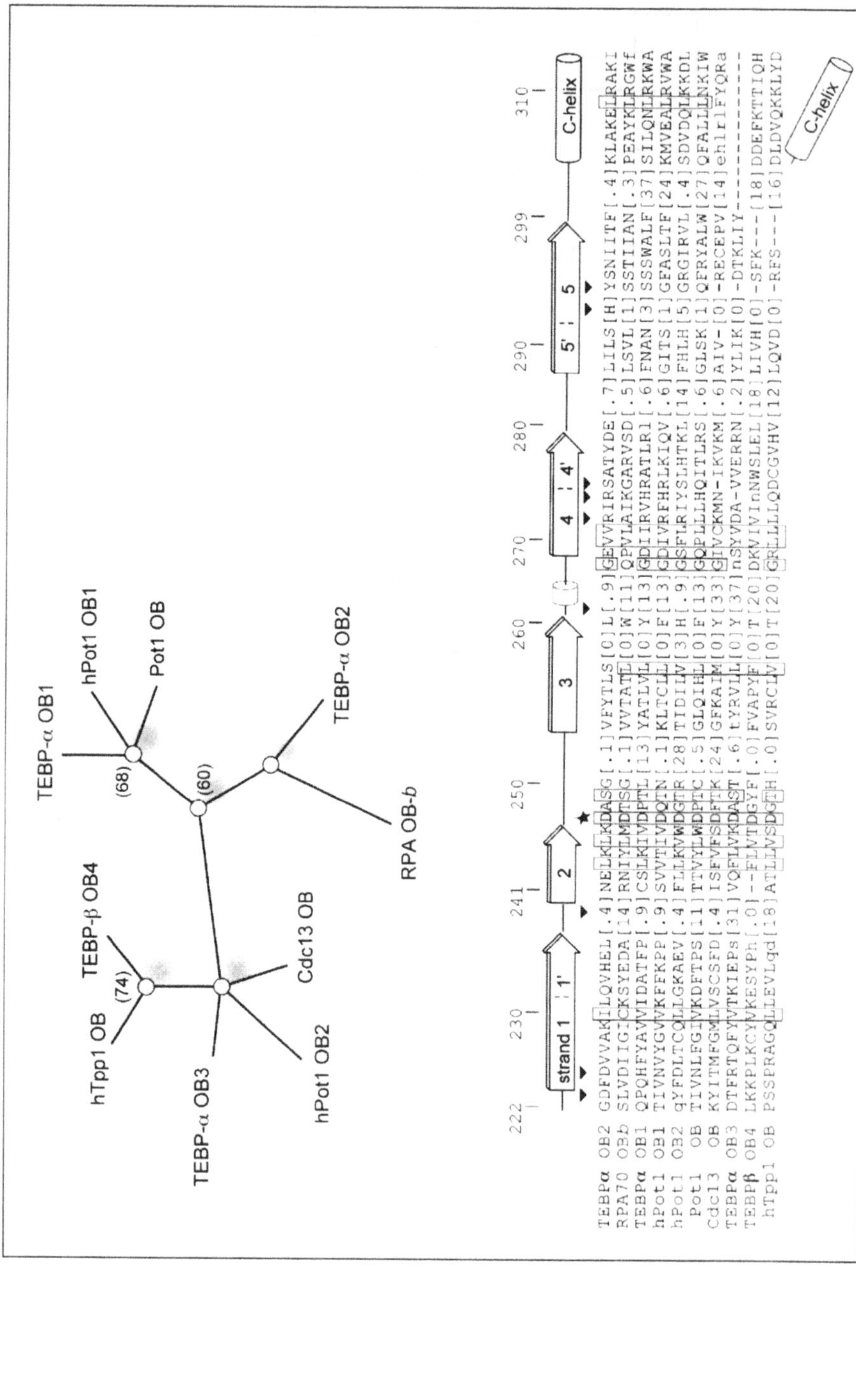

Figure 6. Figure legend on next page.

The single-myb telomere proteins may have arisen via separate gene duplication events coupled with acquisition of different dimerization domains. The high degree of amino-acid sequence conservation among myb-like motifs argues for a common ancestral protein. It seems also possible, however, that two or more myb motifs derived from different proteins functioning in the binding of T/G-rich gene regulatory sequences would experience similar selective pressures as a role in telomere biology became established. Myb-domains resident in transcription factors may have been predisposed for recruitment to a telomere context since myb-type DNA-binding domains constitute tandem protein repeats suitable for the recognition of tandemly repeated telomere sequences. Convergent evolution following separate gene duplication events and acquisition of accessory domains would account for both the common sequence signature (telobox) of current myb-motifs and the diverse nature of accessory domains involved in dimerization and protein recruitment.

Competition among incipient telomere binding proteins seems to have occurred with generally only one or two closely related proteins retaining the role of direct DNA-binding as is the case for Taz1 in fission yeast,[18] TRF1/TRF2 in vertebrates[20,61] and Rap1 in budding yeast.[62] Rap1 homologs were retained alongside Taz1/TRF lineages of fission yeast and humans,[63-66] but these Rap1 homologs lack the positive electrostatic character necessary for DNA-binding[67] and currently associate with telomeres via protein-protein interactions.[63-66,68] Intriguingly, there is no evidence for converse retention of a TRF-like homolog in budding yeast. Several reports and reviews have noted this puzzling situation and speculate that a telomerase RNA template crisis occurred in the hemiascomycetes leading to TRF loss in this group.[63,69]

OB-Fold Telomere Proteins Are Derived from SSB

Structural homology first pointed to an evolutionary relationship among OB-fold telomere proteins and RPA,[6,30] the eukaryotic SSB devoted to binding single-stranded DNA during DNA replication and repair.[57,58] Evidence for such a relationship also comes from studies showing that telomere binding proteins such as TEBP-α and Pot1 have biochemical activities consistent with a role in preventing or resolving folded forms of single-stranded DNA,[70-73] a functional role also played by RPA. Additionally, human telomere protein Pot1 activates in vitro unwinding of DNA structures by WRN and BLM helicases[74] and together with Tpp1 potentiates telomerase primer-extension reactions,[35] properties that are strikingly similar with how RPA potentiates in vitro DNA replication reactions.[75,76]

SSB proteins constructed with OB-folds and carrying out analogous functions in DNA-replication are ubiquitous among eubacteria and archaea. Subunit organization for bacterial and archaeal SSB proteins is relatively simple with tetramers of a single protein subunit generally found in the SSB proteins of bacteria[29] and mitochondria[77,78] and a monomeric RPA-like SSB found in *Sulfolobus solfataricus*.[39] A hetero-trimer comprising 70 kDa, 32 kDa and 14 kDa

Figure 6, viewed on previous page. Alignment and phylogentic tree contructed for OB-folds in single-stranded telomere DNA-binding proteins. Secondary structure and residue numbers annotating the alignment were derived from the second OB-fold of TEBP-α, α-OB2. DNA contacts for this protein are indicated with a triangle (▼). Sequences were aligned on the basis of three-dimensional superpositions as facilitated by the *VAST* server (http://www.ncbi.nlm.nih.gov/Structure/VAST/vast.shtml).[116] Only nongap positions beginning with strand s1 and ending with strand s5 were included in most parsimonious tree construction. OB-*b* from RPA-70 served as a nontelomere outgroup. Bootstrap support is indicated for nodes observed in more than 50% of 1000 simulations. The same tree topology was obtained through Bayesian analysis of the alignment by *MrBayes*[117,118] (D. L. Theobald, personal communication). Although sequence similarity among OB-fold members is tenuous, the analysis appears reasonable in that close relatives identified on the basis of amino-acid sequence comparisons match expectations based on structural similarity, e.g., the Pot1/TEBP-α OB1 clade. Additionally, a previously identified sequence motif, [D-X-(T/S/Y)],[39] is apparent in all members (starred). Had structural superposition resulted in miss-aligned sequences it is unlikely that such a motif would be so consistently represented.

protein subunits is conserved for RPA in yeast and humans.[75,79-81] As reviewed above, the three proteins Cdc13, Stn1 and Ten1, each with predicted OB-folds, operate together in budding yeast to suppress inappropriate DNA-damage responses directed at telomeres.[51,55,56] The idea emerging is that telomeres are protected by RPA-like complexes, which co-evolved with RPA from an ancestral OB-fold SSB-like protein. Recruitment of an SSB-like protein for chromosome end protection also likely occurred separately for protection of linear chromosomes particular to certain mitochondria.[82,83]

Reconstruction of the evolutionary history of OB-folds is complicated by rapid divergence among family members. While the myb-like motif is recognizable on the basis of amino-acid sequence similarity, OB-folds are reliably identified only by structure determination. In certain cases, closely related OB-folds retain a degree of sequence similarity; however, even under these circumstances the first OB-fold of the group has been identified from its X-ray or NMR structure. The contrasting pattern of sequence conservation observed for myb-motifs and sequence divergence in the case of OB-fold proteins suggest that strong purifying selection and perhaps convergent evolution have been acting on myb-like domains while comparatively relaxed constraints and divergent evolution operate for the OB-fold proteins.

An amino-acid sequence alignment and phylogenetic tree constructed for OB-folds derived from telomere proteins is shown in Figure 6. In this analysis, equivalent positions were inferred from superposition of three-dimensional structures and amino-acid characters were then treated as if derived from a normal multiple-sequence alignment. Structure-based phylogenetic reconstructions like the one presented here are increasingly being applied in order to examine very deep phylogenies.[84] Although most pairwise sequence comparisons have less than 20% sequence identity, the alignment does show one strongly conserved sequence motif [D X (T/S/Y)] in the region connecting strands s2 and s3. Had the sequences been grossly mis-aligned by structural superposition, this motif would not be so consistently represented in the current analysis. Recognized previously as conserved among eukaryotic SSB proteins but absent in bacterial SSB, this motif establishes intra-domain contacts that are incompatible with homo-oligomerization.[39] Seeing this [D X (T/S/Y)] motif in telomere binding protein OB-folds, RPA and an archaeal SSB[39] provides further support for the idea that telomere proteins co-evolved with RPAs as the archaea and eukarya lineages emerged. The OB-fold phylogenetic tree indicates that partial gene duplications accounting for the multiple OB-folds and subunits characteristic of telomere proteins occurred very early, likely within the last common ancestor of all eukaryotes just as RPA (and RPA-like) complexes were assuming roles in DNA replication.

The major single-stranded DNA-binding activity associated with the TEBP of *S. nova* resides in the N-terminal domain of TEBP-α. The second OB-fold from this protein, α-OB2, appears to be a close relative to OB-*b*, the second OB-fold contained within the major single-stranded DNA-binding domain of RPA-70 (Fig. 6). Extending the TEBP-RPA comparison makes TEBP-α and TEBP-β analogous and perhaps homologous with RPA-70 and RPA-32 subunits. Phosphorylation of protein regions in TEBP-β[85,86] and RPA-32[87] provides further support for the suggestion that these proteins share a common function and origin. All current evidence indicates that the ciliate TEBP-α/β protein is a hetero-dimer, not a trimer, apparently at odds with descent from an RPA-like complex. The sequence alignment presented in Figure 6 is consistent with loss of a third subunit accomplished through truncation of TEBP-α. Subunit interactions in RPA are established by means of three helices located C-terminal to each of the three OB-folds constituting the RPA trimerization core.[40] While TEBP-β makes use of an analogous helix for hetero-dimerization, the C-terminal helix expected to follow the third OB-fold of TEBP-α is missing; the polypeptide chain terminates immediately at the end of strand s5 for this OB-fold (α-OB3 in Fig. 6). Loss of the C-terminal helix and concomitant loss of a third RPA-like subunit may have allowed for more efficient allocation of resources since there are so many telomeres (~100 million per macronucleus) in this organism,[88] each of which is capped by a TEBP-α/β protein complex.

Evolution of Cooperative Telomere Systems

Evolution of telomere proteins appears to have been driven by forces selecting for cooperative systems. Emerging cooperative systems were likely adaptive since these afforded more finely tuned segregation and regulation of two principle telomere functions, telomere homeostasis and telomere end protection. This idea is exemplified by gene duplication apparent for TRF1 and TRF2, which occurred early in the vertebrate lineage (Fig. 5). Gene duplication of the TRF ancestor likely relaxed functional constraints so as to enable specialization and novel biochemical innovations in TRF1 and TRF2. Currently, TRF1 appears most closely associated with telomere length homeostasis whilst TRF2 is essential for sequestering the 3'-terminal DNA and preventing an inappropriate DNA-damage response (reviewed in ref. 2). Fixation of TRF2 may have been driven by its unique ability to condense and supercoil DNA,[89] which could facilitate t-loop formation.[22,90] Communication between the emerging TRF1 and TRF2 systems was probably crucial for harmonious co-existence of the paralogs. Duplication of the TRF-encoding gene, therefore, likely imposed strong selective pressure to develop inter-subunit communication as currently mediated by Tpp1 and Ten2 within the cooperative telomere complex shelterin.[47,91,92] Similar mechanisms may have driven fixation of duplicated Pot1-encoding genes in *Tetrahymena*,[93] plants[94] and rodents.[95-97] Although the picture is still coming into focus,[98,99] it seems that Pot1 paralogy allows for more specialized telomere functions as is the case for TRF1 and TRF2.

How does protein-protein communication work within complexes of interacting telomere components? Cooperativity was first characterized in the telomere complex from *S. nova*.[44,45,100,101] This telomere system was originally chosen because of natural protein abundance considerations;[102-104] however, it still affords unique opportunities to understand protein-protein communication since three-dimensional structures for the entire DNA–TEBP-α–TEBP-β complex[6,73] as well as DNA cocomplexes with individual subunits[105] and domains[106] have been solved. Additionally, protein subunits and domains derived from this system are readily expressed in high yields as required for isothermal titration calorimetry, a method that is revealing the thermodynamic underpinnings for cooperativity and allostery.[72,73] These studies show that, despite its apparent simplicity (two proteins plus 16-mer single-stranded DNA), the telomere end complex is remarkably responsive and suggest that allosteric trigger points likely coordinate the hand-off of DNA from a tenacious protective complex to an extension-competent complex with telomerase. Relationships becoming increasingly apparent in telomere systems from protozoa, yeast and humans predict that similar thermodynamic gears and levers have also evolved in the more complicated telomere systems.

Acknowledgements

This work was supported through a grant from the NIH (R01 GM067994).

References

1. Zakian VA. Telomeres: beginning to understand the end. Science 1995; 270:1601-1607.
2. Smogorzewska A, de Lange T. Regulation of telomerase by telomeric proteins. Annu Rev Biochem 2004; 73:177-208.
3. d'Adda di Fagagna F, Teo SH, Jackson SP. Functional links between telomeres and proteins of the DNA-damage response. Genes Dev 2004; 18:1781-1799.
4. Theobald DL, Cervantes RB, Lundblad V et al. Homology among telomeric end-protection proteins. Structure (Camb) 2003; 11:1049-1050.
5. Konig P, Giraldo R, Chapman L et al. The crystal structure of the DNA-binding domain of yeast RAP1 in complex with telomeric DNA. Cell 1996; 85:125-136.
6. Horvath MP, Schweiker VL, Bevilacqua JM et al. Crystal structure of the Oxytricha nova telomere end binding protein complexed with single strand DNA. Cell 1998; 95:963-974.
7. Rhodes D, Fairall L, Simonsson T et al. Telomere architecture. EMBO Rep 2002; 3:1139-1145.
8. Friedman AD. Runx1, c-Myb and C/EBPalpha couple differentiation to proliferation or growth arrest during hematopoiesis. J Cell Biochem 2002; 86:624-629.
9. Ogata K, Morikawa S, Nakamura H et al. Solution structure of a specific DNA complex of the Myb DNA-binding domain with cooperative recognition helices. Cell 1994; 79:639-648.
10. Court R, Chapman L, Fairall L et al. How the human telomeric proteins TRF1 and TRF2 recognize telomeric DNA: a view from high-resolution crystal structures. EMBO Rep 2005; 6:39-45.

11. Bilaud T, Koering CE, Binet-Brasselet E et al. The telobox, a Myb-related telomeric DNA binding motif found in proteins from yeast, plants and human. Nucleic Acids Res 1996; 24:1294-1303.
12. Vassetzky NS, Gaden F, Brun C et al. Taz1p and Teb1p, two telobox proteins in Schizosaccharomyces pombe, recognize different telomere-related DNA sequences. Nucleic Acids Res 1999; 27:4687-4694.
13. Marcand S, Gilson E, Shore D. A protein-counting mechanism for telomere length regulation in yeast. Science 1997; 275:986-990.
14. Chong L, van Steensel B, Broccoli D et al. A human telomeric protein. Science 1995; 270:1663-1667.
15. Bianchi A, Smith S, Chong L et al. TRF1 is a dimer and bends telomeric DNA. EMBO J 1997; 16:1785-1794.
16. Bianchi A, Stansel RM, Fairall L et al. TRF1 binds a bipartite telomeric site with extreme spatial flexibility. EMBO J 1999; 18:5735-5744.
17. Fairall L, Chapman L, Moss H et al. Structure of the TRFH dimerization domain of the human telomeric proteins TRF1 and TRF2. Mol Cell 2001; 8:351-361.
18. Cooper JP, Nimmo ER, Allshire RC et al. Regulation of telomere length and function by a Myb-domain protein in fission yeast. Nature 1997; 385:744-747.
19. Bilaud T, Brun C, Ancelin K et al. Telomeric localization of TRF2, a novel human telobox protein. Nat Genet 1997; 17:236-239.
20. Broccoli D, Smogorzewska A, Chong L et al. Human telomeres contain two distinct Myb-related proteins, TRF1 and TRF2. Nat Genet 1997; 17:231-235.
21. Cooper JP, Watanabe Y, Nurse P. Fission yeast Taz1 protein is required for meiotic telomere clustering and recombination. Nature 1998; 392:828-831.
22. Stansel RM, de Lange T, Griffith JD. T-loop assembly in vitro involves binding of TRF2 near the 3□ telomeric overhang. EMBO J 2001; 20:5532-5540.
23. Marian CO, Bordoli SJ, Goltz M et al. The maize Single myb histone 1 gene, Smh1, belongs to a novel gene family and encodes a protein that binds telomere DNA repeats in vitro. Plant Physiol 2003; 133:1336-1350.
24. Yu EY, Kim SE, Kim JH et al. Sequence-specific DNA recognition by the Myb-like domain of plant telomeric protein RTBP1. J Biol Chem 2000; 275:24208-24214.
25. Chen CM, Wang CT, Ho CH. A plant gene encoding a Myb-like protein that binds telomeric GGTTAG repeats in vitro. J Biol Chem 2001; 276:16511-16519.
26. Karamysheva ZN, Surovtseva YV, Vespa L et al. A C-terminal Myb extension domain defines a novel family of double-strand telomeric DNA-binding proteins in Arabidopsis. J Biol Chem 2004; 279:47799-47807.
27. Sue SC, Hsiao HH, Chung BC et al. Solution structure of the Arabidopsis thaliana telomeric repeat-binding protein DNA binding domain: a new fold with an additional C-terminal helix. J Mol Biol 2006; 356:72-85.
28. Murzin AG. OB(oligonucleotide/oligosaccharide binding)-fold: common structural and functional solution for nonhomologous sequences. EMBO J 1993; 12:861-867.
29. Raghunathan S, Kozlov AG, Lohman TM et al. Structure of the DNA binding domain of E. coli SSB bound to ssDNA. Nat Struct Biol 2000; 7:648-652.
30. Bochkarev A, Pfuetzner RA, Edwards AM et al. Structure of the single-stranded-DNA-binding domain of replication protein A bound to DNA. Nature 1997; 385:176-181.
31. Theobald DL, Mitton-Fry RM, Wuttke DS. Nucleic acid recognition by OB-fold proteins. Annu Rev Biophys Biomol Struct 2003; 32:115-133.
32. Mitton-Fry RM, Anderson EM, Hughes TR et al. Conserved structure for single-stranded telomeric DNA recognition. Science 2002; 296:145-147.
33. Lei M, Podell ER, Cech TR. Structure of human POT1 bound to telomeric single-stranded DNA provides a model for chromosome end-protection. Nat Struct Mol Biol 2004; 11:1223-1229.
34. Lei M, Podell ER, Baumann P et al. DNA self-recognition in the structure of Pot1 bound to telomeric single-stranded DNA. Nature 2003; 426:198-203.
35. Wang F, Podell ER, Zaug AJ et al. The POT1-TPP1 telomere complex is a telomerase processivity factor. Nature 2007; 445:506-510.
36. Croy JE, Wuttke DS. Themes in ssDNA recognition by telomere-end protection proteins. Trends Biochem Sci 2006; 31:516-525.
37. Arcus V. OB-fold domains: a snapshot of the evolution of sequence, structure and function. Curr Opin Struct Biol 2002; 12:794-801.
38. Theobald DL, Wuttke DS. Divergent evolution within protein superfolds inferred from profile-based phylogenetics. J Mol Biol 2005; 354:722-737.
39. Kerr ID, Wadsworth RI, Cubeddu L et al. Insights into ssDNA recognition by the OB fold from a structural and thermodynamic study of Sulfolobus SSB protein. EMBO J 2003; 22:2561-2570.

40. Bochkareva E, Korolev S, Lees-Miller SP et al. Structure of the RPA trimerization core and its role in the multistep DNA-binding mechanism of RPA. EMBO J 2002; 21:1855-1863.
41. Mitton-Fry RM, Anderson EM, Theobald DL et al. Structural basis for telomeric single-stranded DNA recognition by yeast Cdc13. J Mol Biol 2004; 338:241-255.
42. Trujillo KM, Bunch JT, Baumann P. Extended DNA binding site in Pot1 broadens sequence specificity to allow recognition of heterogeneous fission yeast telomeres. J Biol Chem 2005; 280:9119-9128.
43. Croy JE, Podell ER, Wuttke DS. A new model for Schizosaccharomyces pombe telomere recognition: the telomeric single-stranded DNA-binding activity of Pot11-389. J Mol Biol 2006; 361:80-93.
44. Gray JT, Celander DW, Price CM et al. Cloning and expression of genes for the Oxytricha telomere-binding protein: specific subunit interactions in the telomeric complex. Cell 1991; 67:807-814.
45. Fang G, Cech TR. Oxytricha telomere-binding protein: DNA-dependent dimerization of the alpha and beta subunits. Proc Natl Acad Sci USA 1993; 90:6056-6060.
46. Baumann P, Cech TR. Pot1, the putative telomere end-binding protein in fission yeast and humans. Science 2001; 292:1171-1175.
47. O'Connor MS, Safari A, Xin H et al. A critical role for TPP1 and TIN2 interaction in high-order telomeric complex assembly. Proc Natl Acad Sci USA 2006; 103:11874-11879.
48. Ye JZ, Hockemeyer D, Krutchinsky AN et al. POT1-interacting protein PIP1: a telomere length regulator that recruits POT1 to the TIN2/TRF1 complex. Genes Dev 2004; 18:1649-1654.
49. Liu D, Safari A, O'Connor MS et al. PTOP interacts with POT1 and regulates its localization to telomeres. Nat Cell Biol 2004; 6:673-680.
50. Xin H, Liu D, Wan M et al. TPP1 is a homologue of ciliate TEBP-beta and interacts with POT1 to recruit telomerase. Nature 2007; 445:559-562.
51. Gao H, Cervantes RB, Mandell EK et al. RPA-like proteins mediate yeast telomere function. Nat Struct Mol Biol 2007; 14:208-214.
52. Nugent CI, Hughes TR, Lue NF et al. Cdc13p: a single-strand telomeric DNA-binding protein with a dual role in yeast telomere maintenance. Science 1996; 274:249-252.
53. Hughes TR, Weilbaecher RG, Walterscheid M et al. Identification of the single-strand telomeric DNA binding domain of the Saccharomyces cerevisiae Cdc13 protein. Proc Natl Acad Sci USA 2000; 97:6457-6462.
54. Pennock E, Buckley K, Lundblad V. Cdc13 delivers separate complexes to the telomere for end protection and replication. Cell 2001; 104:387-396.
55. Grandin N, Reed SI, Charbonneau M. Stn1, a new Saccharomyces cerevisiae protein, is implicated in telomere size regulation in association with Cdc13. Genes Dev 1997; 11:512-527.
56. Grandin N, Damon C, Charbonneau M. Ten1 functions in telomere end protection and length regulation in association with Stn1 and Cdc13. EMBO J 2001; 20:1173-1183.
57. Wold MS. Replication protein A: a heterotrimeric, single-stranded DNA-binding protein required for eukaryotic DNA metabolism. Annu Rev Biochem 1997; 66:61-92.
58. Iftode C, Daniely Y, Borowiec JA. Replication protein A (RPA): the eukaryotic SSB. Crit Rev Biochem Mol Biol 1999; 34:141-180.
59. Lipsick JS. One billion years of Myb. Oncogene 1996; 13:223-235.
60. Shore D, Nasmyth K. Purification and cloning of a DNA binding protein from yeast that binds to both silencer and activator elements. Cell 1987; 51:721-732.
61. Zhong Z, Shiue L, Kaplan S et al. A mammalian factor that binds telomeric TTAGGG repeats in vitro. Mol Cell Biol 1992; 12:4834-4843.
62. Shore D. Telomere length regulation: getting the measure of chromosome ends. Biol Chem 1997; 378:591-597.
63. Li B, Oestreich S, de Lange T. Identification of human Rap1: implications for telomere evolution. Cell 2000; 101:471-483.
64. Kanoh J, Ishikawa F. spRap1 and spRif1, recruited to telomeres by Taz1, are essential for telomere function in fission yeast. Curr Biol 2001; 11:1624-1630.
65. Chikashige Y, Hiraoka Y. Telomere binding of the Rap1 protein is required for meiosis in fission yeast. Curr Biol 2001; 11:1618-1623.
66. Park MJ, Jang YK, Choi ES et al. Fission yeast Rap1 homolog is a telomere-specific silencing factor and interacts with Taz1p. Mol Cells 2002; 13:327-333.
67. Hanaoka S, Nagadoi A, Yoshimura S et al. NMR structure of the hRap1 Myb motif reveals a canonical three-helix bundle lacking the positive surface charge typical of Myb DNA-binding domains. J Mol Biol 2001; 312:167-175.
68. Miller KM, Ferreira MG, Cooper JP. Taz1, Rap1 and Rif1 act both interdependently and independently to maintain telomeres. EMBO J 2005; 24:3128-3135.
69. Teixeira MT, Gilson E. Telomere maintenance, function and evolution: the yeast paradigm. Chromosome Res 2005; 13:535-548.

70. Lei M, Zaug AJ, Podell ER et al. Switching human telomerase on and off with hPOT1 protein in vitro. J Biol Chem 2005; 280:20449-20456.
71. Zaug AJ, Podell ER et al. Human POT1 disrupts telomeric G-quadruplexes allowing telomerase extension in vitro. Proc Natl Acad Sci USA 2005; 102:10864-10869.
72. Buczek P, Horvath MP. Thermodynamic characterization of binding Oxytricha nova single strand telomere DNA with the alpha protein N-terminal domain. J Mol Biol 2006; 359:1217-1234.
73. Buczek P, Horvath MP. Structural reorganization and the cooperative binding of single-stranded telomere DNA in Sterkiella nova. J Biol Chem 2006; 281:40124-40134.
74. Opresko PL, Mason PA, Podell ER et al. POT1 stimulates RecQ helicases WRN and BLM to unwind telomeric DNA substrates. J Biol Chem 2005; 280:32069-32080.
75. Fairman MP, Stillman B. Cellular factors required for multiple stages of SV40 DNA replication in vitro. EMBO J 1988; 7:1211-1218.
76. Brill SJ, Stillman B. Yeast replication factor-A functions in the unwinding of the SV40 origin of DNA replication. Nature 1989; 342:92-95.
77. Yang C, Curth U, Urbanke C et al. Crystal structure of human mitochondrial single-stranded DNA binding protein at 2.4 A resolution. Nat Struct Biol 1997; 4:153-157.
78. Curth U, Urbanke C, Greipel J et al. Single-stranded-DNA-binding proteins from human mitochondria and Escherichia coli have analogous physicochemical properties. Eur J Biochem 1994; 221:435-443.
79. Wold MS, Kelly T. Purification and characterization of replication protein A, a cellular protein required for in vitro replication of simian virus 40 DNA. Proc Natl Acad Sci USA 1988; 85:2523-2527.
80. Brill SJ, Stillman B. Replication factor-A from Saccharomyces cerevisiae is encoded by three essential genes coordinately expressed at S phase. Genes Dev 1991; 5:1589-1600.
81. Henricksen LA, Umbricht CB, Wold MS. Recombinant replication protein A: expression, complex formation and functional characterization. J Biol Chem 1994; 269:11121-11132.
82. Nosek J, Tomaska L, Pagacova B et al. Mitochondrial telomere-binding protein from Candida parapsilosis suggests an evolutionary adaptation of a nonspecific single-stranded DNA-binding protein. J Biol Chem 1999; 274:8850-8857.
83. Tomaska L, Nosek J, Kucejova B. Mitochondrial single-stranded DNA-binding proteins: in search for new functions. Biol Chem 2001; 382:179-186.
84. Hall BG, Barlow M. Structure-based phylogenies of the serine beta-lactamases. J Mol Evol 2003; 57:255-260.
85. Hicke B, Rempel R, Maller J et al. Phosphorylation of the Oxytricha telomere protein: possible cell cycle regulation. Nucleic Acids Res 1995; 23:1887-1893.
86. Paeschke K, Simonsson T, Postberg J et al. Telomere end-binding proteins control the formation of G-quadruplex DNA structures in vivo. Nat Struct Mol Biol 2005; 12:847-854.
87. Binz SK, Sheehan AM, Wold MS. Replication protein A phosphorylation and the cellular response to DNA damage. DNA Repair (Amst) 2004; 3:1015-1024.
88. Prescott DM, Murti KG. Chromosome structure in ciliated protozoans. Cold Spring Harb Symp Quant Biol 1973; 38:609-618.
89. Amiard S, Doudeau M, Pinte S et al. A topological mechanism for TRF2-enhanced strand invasion. Nat Struct Mol Biol 2007; 14:147-154.
90. Griffith JD, Comeau L, Rosenfield S et al. Mammalian telomeres end in a large duplex loop. Cell 1999; 97:503-514.
91. Liu D, O'Connor MS, Qin J et al. Telosome, a mammalian telomere-associated complex formed by multiple telomeric proteins. J Biol Chem 2004; 279:51338-51342.
92. de Lange T. Shelterin: the protein complex that shapes and safeguards human telomeres. Genes Dev 2005; 19:2100-2110.
93. Jacob NK, Lescasse R, Linger BR et al. Tetrahymena POT1a regulates telomere length and prevents activation of a cell cycle checkpoint. Mol Cell Biol 2007; 27:1592-1601.
94. Shakirov EV, Surovtseva YV, Osbun N et al. The Arabidopsis Pot1 and Pot2 proteins function in telomere length homeostasis and chromosome end protection. Mol Cell Biol 2005; 25:7725-7733.
95. Hockemeyer D, Daniels JP, Takai H et al. Recent expansion of the telomeric complex in rodents: Two distinct POT1 proteins protect mouse telomeres. Cell 2006; 126:63-77.
96. He H, Multani AS, Cosme-Blanco W et al. POT1b protects telomeres from end-to-end chromosomal fusions and aberrant homologous recombination. EMBO J 2006; 25:5180-5190.
97. Wu L, Multani AS, He H et al. Pot1 deficiency initiates DNA damage checkpoint activation and aberrant homologous recombination at telomeres. Cell 2006; 126:49-62.
98. Price CM. Stirring the POT1: surprises in telomere protection. Nat Struct Mol Biol 2006; 13:673-674.
99. Baumann P. Are mouse telomeres going to pot? Cell 2006; 126:33-36.

100. Fang G, Gray JT, Cech TR. Oxytricha telomere-binding protein: separable DNA-binding and dimerization domains of the alpha-subunit. Genes Dev 1993; 7:870-882.
101. Buczek P, Orr RS, Pyper SR et al. Binding linkage in a telomere DNA-protein complex at the ends of Oxytricha nova chromosomes. J Mol Biol 2005; 350:938-952.
102. Gottschling DE, Zakian VA. Telomere proteins: specific recognition and protection of the natural termini of Oxytricha macronuclear DNA. Cell 1986; 47:195-205.
103. Price CM, Cech TR. Telomeric DNA-protein interactions of Oxytricha macronuclear DNA. Genes Dev 1987; 1:783-793.
104. Price CM, Cech TR. Properties of the telomeric DNA-binding protein from Oxytricha nova. Biochemistry 1989; 28:769-774.
105. Peersen OB, Ruggles JA, Schultz SC. Dimeric structure of the Oxytricha nova telomere end-binding protein alpha-subunit bound to ssDNA. Nat Struct Biol 2002; 9:182-187.
106. Classen S, Ruggles JA, Schultz SC. Crystal structure of the N-terminal domain of Oxytricha nova telomere end-binding protein alpha subunit both uncomplexed and complexed with telomeric ssDNA. J Mol Biol 2001; 314:1113-1125.
107. Nishikawa T, Nagadoi A, Yoshimura S et al. Solution structure of the DNA-binding domain of human telomeric protein, hTRF1. Structure 1998; 6:1057-1065.
108. Hanaoka S, Nagadoi A, Nishimura Y. Comparison between TRF2 and TRF1 of their telomeric DNA-bound structures and DNA-binding activities. Protein Sci 2005; 14:119-130.
109. Tahirov TH, Sato K, Ichikawa-Iwata E et al. Mechanism of c-Myb-C/EBP beta cooperation from separated sites on a promoter. Cell 2002; 108:57-70.
110. Horvath MP, Schultz SC. DNA G-quartets in a 1.86 A resolution structure of an Oxytricha nova telomeric protein-DNA complex. J Mol Biol 2001; 310:367-377.
111. Bochkareva E, Belegu V, Korolev S et al. Structure of the major single-stranded DNA-binding domain of replication protein A suggests a dynamic mechanism for DNA binding. EMBO J 2001; 20:612-618.
112. Jacobs DM, Lipton AS, Isern NG et al. Human replication protein A: global fold of the N-terminal RPA-70 domain reveals a basic cleft and flexible C-terminal linker. J Biomol NMR 1999; 14:321-331.
113. Bochkareva E, Kaustov L, Ayed A et al. Single-stranded DNA mimicry in the p53 transactivation domain interaction with replication protein A. Proc Natl Acad Sci USA 2005; 102:15412-15417.
114. Philipova D, Mullen JR, Maniar HS et al. A hierarchy of SSB protomers in replication protein A. Genes Dev 1996; 10:2222-2233.
115. PHYLIP-Phylogeny inference package (version 3.6) [computer program]. 2004.
116. Gibrat JF, Madej T, Bryant SH. Surprising similarities in structure comparison. Curr Opin Struct Biol 1996; 6:377-385.
117. Huelsenbeck JP, Ronquist F. MRBAYES: Bayesian inference of phylogenetic trees. Bioinformatics 2001; 17:754-755.
118. Ronquist F, Huelsenbeck JP. MrBayes 3: Bayesian phylogenetic inference under mixed models. Bioinformatics 2003; 19:1572-1574.

Chapter 7

Telomeres: Guardians of Genomic Integrity or Double Agents of Evolution?

Michael J. McEachern*

Abstract

Telomeres are crucial components required for genomic stability. Telomere dysfunction can result in enormously elevated rates of chromosomal alterations, particularly in subtelomeric regions. Interestingly, the chromosomal regions in the vicinity of telomeres are often among the most rapidly evolving in the genome. These facts could suggest that the protective capping function of telomeres has not evolved to be fail-safe but instead to permit a certain rate of failure that can foster evolution through subtelomeric rearrangements.

Introduction

Chromosome Ends: The Wild West of the Genome?

Just as the geography of the earth has influenced the nature and evolution of human communities, so too has the geography of chromosomes influenced the nature and evolution of genes. A prominent example of this is subtelomeric DNA. Subtelomeric regions are the frontier outposts of the genomes and existence for genes there can be precarious and not always governed by the laws that regulate genes in other parts of the chromosome. Like the inhabitants of frontier towns, the genes in subtelomeric regions are often not representative of those found elsewhere. Instead, subtelomeric regions are often enriched for genes that allow rapid adaptation to new environments. In this chapter, I examine the special consequences that can result from a location near a chromosome end and how this may relate to the structure, function and evolution of subtelomeric sequences.

Telomeric and Broken DNA Ends and the Processes That Act on Them

The long linear DNA molecules that constitute eukaryotic chromosomes can have two types of DNA ends, telomeric ends and broken ends.[1] The fundamental difference between a telomere and a broken DNA end is that the former is the natural stable end of the chromosome and the latter is something that typically occurs from damage and that the cell will usually rapidly repair. Telomeres and broken ends are therefore usually treated by the cell in opposite ways.

Telomeric ends are the normal targets for sequence addition by telomerase, which counteracts the gradual sequence loss from ends that occurs as a consequence of the 'end replication problem'.[2] Telomeric ends also function to prevent chromosome ends from acting like broken DNA ends. This protective role is commonly known as capping. In the great majority of eukaryotes, the distinctive features that distinguish telomeres from broken ends result from the specific DNA sequences that make up telomeres as well as the specific proteins that bind them. In the large

*Michael J. McEachern—Department of Genetics, Fred Davison Life Sciences Complex, C-324, University of Georgia, Athens, GA 30605, USA. Email: mjm@arches.uga.edu

Origin and Evolution of Telomeres, edited by Jozef Nosek and Ľubomír Tomáška.

majority of eukaryotes, telomeric DNA is composed of tandem arrays of a short repeat (5-26 bp) the sequence of which is specified by the template region of the telomerase RNA.[3,4] Telomere capping generally involves telomere-specific DNA binding proteins. These include proteins that bind double stranded telomeric repeats as well as those that bind the single stranded 3' overhangs at telomeric termini.[5] Paradoxically, some proteins that bind to broken DNA ends also make critical contributions to telomere capping. Additionally, some organisms appear to use t-loops, structures where the 3' overhang is thought to be strand invaded into more internal telomeric repeats, as a means to help cap telomeres.[6]

Telomeric repeat sequences are not always limited to being at chromosome ends. Many species have telomere-like repeat arrays present at interstitial locations in chromosomes. These likely arise from a number of mechanisms including telomere-telomere fusions and aberrant repair of DNA double strand breaks (DSBs).[7] In at least some cases, these interstitial telomeric repeats can influence chromosome structure and stability by being hotspots for chromosome breakage or seed sequences for formation of a new telomere. The incidence and effects of interstitial telomeric repeats are reviewed elsewhere.[8]

In contrast to telomeric ends, broken DNA ends (such as produced by DSBs) are severe forms of DNA damage and are precursors of many if not most chromosomal rearrangements. DSBs are typically repaired by either of two repair pathways, nonhomologous end joining (NHEJ) or homologous recombination (HR) and are not normally substrates for addition by telomerase.[9] NHEJ is a ligation reaction requiring the specialized ligase IV enzyme plus certain additional proteins.[10,11] Mitotic DSB repair by HR is thought to commonly proceed through a synthesis-dependent strand annealing mechanism that results in localized gene conversion but not in cross-overs. It requires an intact homologous sequence as a template and this is thought to be most commonly supplied by a sister chromatid. DSB repair through either NHEJ or HR will generally result in the two broken arms of a chromosome being rejoined. HR will normally bring about precise repair (albeit sometimes with gene conversions) while NHEJ is imprecise and will often incorporate small insertions or deletions at the junction point.

Telomeres therefore serve as key guardians of chromosomal integrity. By blocking chromosome ends from being subjected to NHEJ or HR events, telomeres act to preserve the integrity of the genome, particularly in and around subtelomeric regions. As will be discussed below, DNA repair events triggered by telomere failure or by DSBs near telomeres can be crucial factors in shaping the sequences that are present in subtelomeric regions.

Immediate Subtelomeric Regions and Their Possible Functions

Subtelomeric regions are the DNA sequences in the vicinity of chromosome ends. An exact definition is not possible but an approximate definition would be those sequences adjacent to the telomeres that have features that differentiate them from other regions of chromosomes. These regions are known from many organisms for unusual characteristics that include a complex repetitive structure, rapid evolution and a frequently heterochromatic nature.[12-14]

Subtelomeric sequences can loosely be grouped into two categories; regions immediately adjacent to the telomeric repeats that often lack genes and are present at a large percentage of the chromosome ends and more internal gene-containing regions that are present at smaller subsets of chromosome ends. When subtelomeric sequences are present at more than one chromosome end, they strongly tend to have the same orientation with respect to the telomeres where they are present. The functions of the generally gene-free, immediate subtelomeric sequences are not fully clear and may vary between species. Although they do not appear to be vital to the basic protective capping function of telomeres, there is evidence that they can contribute to telomere length regulation.[15-19] Subtelomeric sequences can also contribute to telomere position effect (TPE), the transcriptional repression that occurs next to telomeres.[20-22]

Another function of immediate subtelomeric elements can sometimes be to serve as templates for telomere repair through homologous recombination. In *S. cerevisiae*, mutants lacking telomerase frequently amplify subtelomeric Y' sequences and spread them to chromosome ends where they were not originally present.[23] Y' elements are DNA sequences of several kilobases in size that are present in one or two copies next to many but not all telomeres in wild type cells. In the related yeast, *Kluyveromyces lactis*, subtelomeric sequences can also be spread to other telomeres when telomeres become short and prone to recombination.[24] These recombination events are thought to represent break-induced replication (BIR), a nonreciprocal HR event copying a sequence from one chromosome arm to another that acts to restore telomeric sequences to chromosome ends that have lost most or all of their telomeric repeats.[25]

It is interesting to speculate that immediate subtelomeric repeats could sometimes have characteristics of selfish DNA elements. Replacement of part or all of such an element at one telomere by a BIR event would clearly involve competition between other such subtelomeric elements in the cell. If subtelomeric BIR events are not strictly limited to mutants with compromised telomere function, this would have the potential to select for elements particularly able to promote their own spread by BIR events. The common presence of families of small direct repeats in immediate subtelomeric sequences conceivably might act to facilitate the homology search of a Rad51-coated DNA filament that would be expected to initiate most BIR events.

Telomeres have been shown to be involved in the early stages of homologous chromosome pairing during meiosis.[26,27] However, yeast subtelomeric sequences appear to be relatively resistant to being sites of meiotic cross over events. Work in *S. cerevisiae* has shown that subtelomeric sequences, even if moved away from a telomere, are relatively resistant to crossover formation.[28] Those crossovers that do occur in subtelomeric DNA appear poorly able to bring about proper chromosome segregation at meiosis I.[29] Consistent with these data, cleavage by Spo11, the nuclease that makes the DNA double strand breaks that initiate meiotic recombination, is infrequent in subtelomeric DNA.[30,31] These data suggest that one function of subtelomeric DNA in yeast is to prevent Spo11-induced meiotic crossovers from occurring in regions of the chromosome where they would not be able to properly function. Meiotic recombination maps in humans, on the other hand, show an increase in recombination at the most distal markers.[32,33]

Subtelomeric Regions Are often Enriched in Contingency Genes

The genes in subtelomeric regions often include rapidly evolving gene families and "contingency genes"[34] involved in adaptation to changing environments. The most notable examples of this come from a number of eukaryotic pathogens where variant surface antigens or other gene families associated with rapid adaptation to the host's immune system are typically found at subtelomeric locations.[35] Organisms where this occurs include *Typanosoma brucei*,[36,37] *Plasmodium falciparum*,[38] *Pneumocystis carinii*,[39] and *Candida glabrata*.[40] In *T. brucei*, for example, there are hundreds of silent variant surface glycoprotein (VSG) genes but only one of these is transcriptionally expressed in a given cell at a time. Switching between different expressed copies of the gene occurs at high frequency in the mammalian host through a mechanism that generally involves being duplicated into one of 20-30 bloodstream expression sites residing next to a telomere.[36,37]

Multiple lines of evidence indicate that human subtelomeric regions evolve rapidly. An estimated half of known human subtelomeric DNA, which exists as patchworks of interchromosomal segmental duplications, is estimated to have arisen since the human-chimp split.[14,41,42] Among individuals, particular subtelomeric regions are often highly polymorphic. Subtelomeric mutations are thought to underlie ~5% of idiopathic mental retardation.[43] Human subtelomeric regions are also observed to undergo high rates of sister chromatid exchange.[44,45]

Rapid subtelomeric evolution is also apparent in Saccharomyces. A number of gene families relevant to adaptation to novel environments are present at subtelomeric locations in *S. cerevisiae.* The subtelomeric *SUC* and *RTM* gene families (encoding invertase and a resistance to molasses determinant, respectively) vary widely in copy number between yeast strains.[46-49] A comparative study of the genomes of four closely related Saccharomyces species found that the majority of genes without clear corresponding genes in other species were subtelomeric.[50] Yeast subtelomeric regions are also enriched for mutationally-inactivated genes.[50-52]

Although *S. cerevisiae* is normally free living, its subtelomeric gene families might be subject to changing selective pressures similar to those experienced by the cell surface protein gene families of parasites such as *T. brucei*. A number of the subtelomeric gene families in *S. cerevisiae* encode proteins that are secreted. These provide the potential for "cheaters", that produce little or none of the communal resource (the secreted enzyme), to be favored by natural selection at high cell concentrations. Recent work has demonstrated the plausibility of this idea for the subtelomeric *SUC* family in yeast.[53] The *SUC2* genes encode invertase, a secreted enzyme required for utilization of sucrose. Isogenic strains either lacking *SUC2* or containing two *SUC2* genes near different telomeres were passaged together at different cell densities on medium containing sucrose as a carbon source. While the cells lacking *SUC2* were less fit at low cell densities, they were more fit at high cell densities than the strain with *SUC2* genes. These fitness differences were specific to growth on sucrose. Thus, *S. cerevisiae* cells are likely to sometimes experience fluctuating environmental conditions that select for changing expression levels or copy number of *SUC2* genes.

Subtelomeric DNA Is Intrinsically Tolerant of Rearrangement

What forces help govern the often repetitive structure and rapid evolution of subtelomeric sequences? A key factor is simply their location near chromosome ends. To understand why this is true, it is useful to first discuss the kinds of aberrant repair events that can occur at a DSB. I will refer to any DSB repair that rejoins the original two chromosome arms, regardless of whether mutations arise at the junction point, as 'conservative' repair and any other outcomes as 'nonconservative' repair. To restore the integrity of a chromosome suffering a DSB it is necessary for broken ends to be eliminated and for the chromosome to terminate with telomeres. There are only two general routes to eliminate a broken end. Either it can be joined to a second end, thereby eliminating both, or it can be directly or indirectly extended to acquire a telomere. End joinings that create dicentric chromosomes (such as fusions between sister chromatids lacking a telomere) will be unstable and will likely lead to new DSBs through breakage-fusion-bridge cycles.[54-56] End joinings circularizing a chromosome fragment (which would require loss or failure of the other telomere) can function in mitotic cells but will lead to dicentric chromosomes from crossover formation in meiosis. Hence they are not likely to be maintained in populations. The type of nonconservative NHEJ event most likely to be tolerated would probably be mis-joining two simultaneous DSBs to produce a reciprocal translocation.

The second route to eliminate a broken end, acquiring a telomere, could occur in at least two ways. De novo telomere formation could occur by direct addition of telomeric repeats to a broken end by telomerase. There is evidence that this occasionally occur at broken ends, particularly at ends with coincidental similarity to telomeric sequence.[57-60] The most likely result of de novo telomere addition would be a terminal truncation of the affected chromosome and loss of the acentric fragment. A second way for a broken end to acquire a telomere would be for a BIR event to extend the sequence at a broken end by copying another chromosome arm all the way to the telomere. This would not only add telomeric repeats but also some amount, perhaps very considerable, of additional DNA and result in a duplication (nonreciprocal translocation) of the copied chromosome arm.

In addition to restoring the presence of a telomere, the nonconservative DSB repair events discussed above tend to have other features in common. For one, in moving or duplicating DNA regions, they typically maintain sequence orientation relative to the telomere. This is significant because, as mentioned above, sequences present at multiple subtelomeric regions within an organism commonly have the same orientation relative to the telomere. Second, there is less selective pressure for conservative repair of DSBs close to telomeres than there is for DSBs far from telomeres. At most positions within a chromosome, there will be strong selection for conservative DSB repair. This is because the aneuploidies that would result from the nonconservative repair would be smaller the closer the DSB had been to a telomere. Simply put, small terminal deletions or duplications would be less detrimental than large ones (Fig. 1). The important implication of this is that nonconservative DSB repair is more likely to be survivable the closer it occurs to

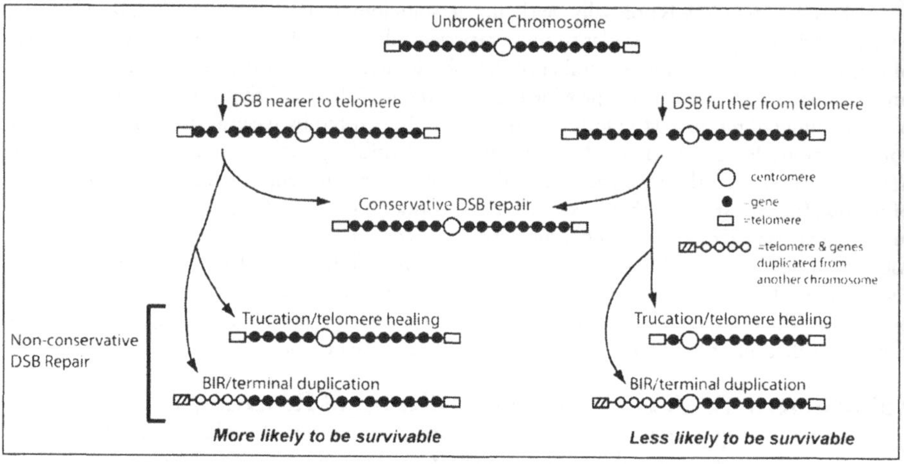

Figure 1. Subtelomeric regions are more tolerant of nonconservative DSB repair. See text for details.

telomeres. The corollary of this is that subtelomeric DNA will intrinsically be more tolerant of accumulating DNA rearrangements formed by nonconservative DSB repair than will other parts of the chromosome.

Direct evidence for the tolerance of subtelomeric regions to nonconservative DSB repair has come from work in *S. cerevisiae*. Yeast strains engineered to have a DSB at any of multiple places along a single chromosome were examined for their rate of survival and mode of DNA repair.[61] As the strains were haploid, no HR-mediated gene conversion using a homologue as template was possible. Also, because the chromosome break was actually a pair of closely spaced DSBs with noncompatible overhangs, repair by simple NHEJ was inefficient. Results from this study found that DSBs in subtelomeric regions were more frequently able to produce surviving cells. The reason for this was an increase in the number of recovered cells that had undergone nonconservative DSB repair. Most prevalent were BIR events that copied terminal regions of other chromosomal arms. However, instances of de novo telomere formation were also found.[61] Interestingly, repair of subtelomeric DSBs (but not DSBs at other locations), by both conservative NHEJ and other mechanisms, was found to mostly depend on telomeric tethering to the nuclear periphery.[62] This suggests the possibility that repair of subtelomeric DNA could be subject to regulation in a manner non-identical to that of DSBs elsewhere in the genome.

There is expected to be a synergistic evolutionary relationship between the tolerance of subtelomeric terminal deletions/duplications and the occurrence of nonconservative DSB repair events. Once a chromosomal terminus has been duplicated onto another chromosome end, this could greatly increase the likelihood that future repair of DSBs in or near the duplicated region will involve homologous recombination between those different chromosomes.

The Differences between Uncapped Telomeres and Broken DNA Ends

As described above, broken DNA ends formed occasionally near telomeres have clear potential to influence subtelomeric evolution. Another factor of potentially even greater importance in subtelomere structure and evolution is the influence of telomeres. Although the prime function of telomeres is to protect chromosome ends from acting like broken ends, this function can fail for a variety of reasons and such telomere uncapping can be a powerful engine for generating chromosomal rearrangements, particularly those affecting subtelomeric regions. Before discussing how this could happen it is useful to review certain features of uncapped telomeres and how these differ from broken ends.

Disruption of telomere capping can subject telomeres to NHEJ and HR, the same repair pathways that act on broken ends. However, uncapped telomeres differ from broken ends in a number of important ways. An obvious difference is that uncapped telomeres are limited to occurring at chromosome ends while broken ends can occur anywhere in the genome. This can strongly bias DNA rearrangements derived from aberrant repair of uncapped telomeres to occur in subtelomeric regions.

The causes of broken ends and uncapped telomeres are also typically different. Uncapped telomeres are not after all, DNA breaks, but rather malfunctioning natural DNA ends. Telomere uncapping can occur from either of two basic causes. Either telomeric DNA repeat arrays become too short to properly function or the function of one or more telomeric capping proteins are disrupted in some way. The former is observed when cells divide repeatedly in the absence of telomerase such as is the case in human somatic cells naturally lacking telomerase and in yeast mutants deleted for telomerase.[25,63] Although multiple telomeres in such cells can acquire capping defects more or less simultaneously, this type of capping defect appears to act at individual telomeres independently. Thus, the shortest telomeres in a cell may become uncapped at a time when other, longer telomeres do not. In contrast, capping defects occurring through mutation or imbalance of a telomeric protein will commonly affect all telomeres in the cell simultaneously. An additional important point is that, because of the varied roles of telomeric proteins, capping defects resulting from protein defects tend to be more varied in their phenotype than capping defects caused by telomere shortening.

Unlike broken DNA ends, which are an essentially all or nothing phenomenon, telomere uncapping is more variable. In some cases it might be total, where a telomere behaves indistinguishably from a broken end. However, in most cases it is more likely to be partial, where a telomere might have properties intermediate between a broken end and a normally functioning telomere. Different kinds of partial capping defects can easily be imagined. For example, telomere failure might occur stochastically, rendering a chromosome end subject to DNA repair pathways is some, but not all, cell divisions. Alternatively, a capping defect might cause a telomere to be prone to one type of DSB repair but not to the other. NHEJ favors ends that are blunt or with minimal overhangs while HR is favored by the long 3' single stranded tails that can be produced by exonucleolytic processing. Protection against these two processes is thus likely to require at least partially different mechanisms that might be differentially affected by a particular capping defect. To summarize, the effects of telomere capping defects are potentially more subtle and variable than those of broken ends. This gives them appreciable flexibility for influencing subtelomeric evolution.

Other important features of broken ends and uncapped telomeres concern both their structure and their manner of repair. Broken ends occur in two basic types. Those that occur in pairs from a DSB and those that occur singly from collapsed or stalled replication forks. The former are normally repaired conservatively by HR or NHEJ resulting in rejoining of the two ends. The latter are repaired through a HR-dependent process that leads to restarting the replication fork.[64] Uncapped telomeres, in contrast, can occur singly or in multiple and cannot be conservatively repaired by being joined to another end. Uncapped telomeres also typically contain some residual number of telomeric repeats. These have the potential to greatly bias DNA repair activity. They would, for example, certainly be expected to enhance the likelihood that telomerase (if present in the cell) would act on an uncapped telomere but they might also affect the ability of HR or NHEJ to act as well.

How uncapped telomeres are repaired varies depending upon circumstances and organism. The most simple repair would be to restore a functional telomere. If uncapping is due to telomere shortening, this can be straightforward. If telomerase is present, it should be able to directly elongate the telomere to a size that will permit capping. If telomerase is not present, an uncapped telomere may undergo HR reactions that can lengthen or replace it.[23,65] The simplest mechanism for this would be for the short telomere to strand invade a longer telomere and copy its sequence in a BIR event. Such events have been shown to be very efficient in yeast when a long telomere is available as a substrate.[66] However, this mechanism may not be feasible when all telomeres are becoming critically short together, such as in a typical senescing yeast telomerase deletion mutant. In such cases though, HR processes can still occasionally generate elongated telomeres, apparently by first producing a small telomeric circle and then using it as a template for a BIR event to make an elongated telomere.[67,68] Other cells, such as most senescing human somatic cells, do not appear to be able to lengthen telomeres using HR. As the replicative senescence caused by telomere shortening in human cells is thought to help prevent carcinogenesis, this failure of HR to act at short telomeres might be an adaptive trait. In the event that telomere uncapping is due to a defect in a telomeric protein, lengthening the telomere by telomerase or HR may not be able to restore a capped state. In such cases, telomeres might persist in a chronically uncapped state.

NHEJ has been shown to act on uncapped telomeres in a number of circumstances.[69-75] The most common outcome of this is for fusion to occur between two uncapped telomeres. This can occur between two telomeres of different chromosome arms or between identical replicated sister chromosome arms, the latter favored when only one uncapped telomere is present. In some cases fusions may occur between chromosome ends that retain telomeric repeats and in others it may occur only after all telomeric repeats have been lost. In extreme cases, all chromosome ends in the cell may undergo fusions.[76-78] Through eliminating chromosome ends altogether, telomere fusions can at least transiently eliminate any type of telomere capping defect. However, the dicentric chromosomes created by such fusions promote recurrent breakage-fusion-bridge cycles that tear apart and reform chromosomes.

Disruption of Telomere Capping Can Trigger High Rates of Subtelomeric Change

The consequences of telomere dysfunction have been documented in numerous studies done in multiple organisms. In some of these, dramatic effects on subtelomeric sequences have been observed. In an early example, a temperature-sensitive defect in *S. cerevisiae* Cdc13, (a protein that protects 3' overhangs and helps recruit telomerase) was found to lead to a large increase in mitotic recombination in subtelomeric regions after a period at the restrictive temperature.[79] The increase was greatest near the telomere but extended ~50 kb distance from it. Amplification of subtelomeric Y' elements were found to characterize the major type of post-senescence survivors that occur through HR in *S. cerevisiae* mutants lacking telomerase.[23,80] Y' element amplification in 'Type I' survivors generates large tandem arrays of the elements and spreads them to most or all chromosome ends. The amplified DNA can be up to several percent of total genomic DNA in these survivors. This subtelomeric amplification likely depends on the presence of short blocks of telomeric repeats that are typically present just internal to the Y' elements in both wild type cells and the Type I survivors. The shortening terminal telomeric repeat arrays likely become prone to HR and strand invade into the subtelomeric blocks of telomere repeats and thereby initiate the events leading to the large scale amplification. In *K. lactis*, where wild type cells have telomeric repeats only at chromosome ends, telomerase deletion postsenescence survivors do not exhibit subtelomeric amplification.[65] However, if even a single block of telomeric repeats is engineered to be internal to a subtelomeric marker gene, postsenescence survivors can readily form long tandem arrays of the alternating telomeric and marker gene sequences.[81]

Other types of subtelomeric rearrangements have also been observed at highly elevated frequencies in yeast telomerase deletion mutants during or after the period when telomeres were very short and malfunctioning.[82-84] These included terminal deletions, BIR events that replaced one subtelomeric region with duplicated subtelomeric sequence from another chromosome arm and fusions between chromosome ends that had lost telomeric and some subtelomeric sequence. Very recently, base substitution and frame shift mutations in a subtelomeric gene (but not an internal gene) were also found to be significantly elevated in senescing *S. cerevisiae* cells without telomerase.[85] Both the base mutations and the gross chromosomal rearrangements were found to be largely dependent on the Rev1 and Pol ζ DNA polymerases that are involved in the error-prone bypass of DNA damage. Other examples of large increases in the rate of subtelomeric BIR events have been observed in *K. lactis*. Telomerase deletion mutants with short dysfunctional telomeres can readily spread a subtelomeric marker gene to most or all other telomeres in the cell via shared subtelomeric homology.[24] In this manner, a "gene family" with copies of the marker gene spread to most or all subtelomeres can be generated very rapidly in the laboratory. Elevated subtelomeric BIR has also been seen in other situations when telomere function has been compromised in *K. lactis*. Most notably, cells with telomeres stable at appreciably shorter than normal lengths and that do not undergo growth senescence can experience subtelomeric BIR rates that can be 40-200 fold elevated relative to wild type cells.[24] This elevated BIR rate likely stems from the telomeres in these cells frequently dropping below a critical length that is able to prevent the telomere from engaging in HR. Work using tagged telomeric sequence in telomerase deletion mutants has shown that *K. lactis* telomeres can undergo recombination once they drop below about 100 bp in size.[66] How a chromosome end retaining telomeric repeats could lead to subtelomeric gene conversions is not certain. One possibility is that the uncapped chromosome end is processed to have a 3' single strand tail that becomes bound by the Rad51 protein and the resulting strand invasion into another telomere could be followed by branch migration into subtelomeric regions. Certain resolutions of such structures might then be easily imagined to cause subtelomeric gene conversions.

Data showing that telomere dysfunction can promote subtelomeric recombination is not limited to yeast. In *Trypanosoma brucei*, telomere shortening caused by loss of telomerase has been shown to promote a duplicative gene conversion (thought to be BIR) that replaced the expressed variant surface gene at the expression site with another VSG gene that was previously unexpressed.[86] This had led to the suggestion that short telomere lengths in trypanosomes may be associated with a higher rate of antigenic switching.[87]

Adaptive Telomere Failure: A Fast Track for Subtelomeric Evolution?

Telomeres are clearly of huge importance in guarding the stability of subtelomeric regions. However, this very property makes them potentially ideal agents for regulating and promoting subtelomeric evolution. I suggest that regulated failure of the telomere capping function could be an adaptive trait that can facilitate evolution of subtelomeric sequences. In essence, this hypothesis states that telomere capping in some, perhaps many, organisms has evolved to be less than 100% effective at protecting telomeric ends from engaging in DNA repair processes. It is reasonable to further predict that the repair process that would be primarily favored by adaptive telomere failure would be homologous recombination. HR would have the ability to recombine or amplify subtelomeric gene families with less risk of detrimental formation of dicentric chromosomes.

Several variants of this basic adaptive telomere failure model are conceivable. A simple possibility is that telomere capping function in some species might have evolved to have a finite and perhaps appreciable failure rate under all circumstances. This would make the testable prediction that telomere capping in such a species could be improved by mutation or overexpression of capping components in ways that would reduce the background rates of subtelomeric rearrangements.

An especially intriguing possibility is that an increase in the rate of telomere failure might be inducible. In this scenario, telomeres might function with a very low failure rate in most situations but with an appreciably higher failure rate under certain other circumstances. Unicellular eukaryotes subjected to starvation or other stress conditions could represent an example of a situation where increased rates of subtelomeric rearrangements might be evolutionarily favorable. Inducing an increased telomere failure rate would be trivial mechanistically. The decreased expression, or modification of a single telomere capping component might easily be sufficient and could affect all telomeres equally. It is also possible that an inducible telomere capping defect could evolve that was specific to a single telomere, perhaps through affecting transcription or a DNA binding protein located next to a particular telomere.

Increased telomere failure rates could obviously also arise through mutation. Sufficiently large populations, such as occur in many microbes, that are subject to strong selective pressures can favor the emergence of mutator phenotypes that produce elevated mutations rates.[88] A mutation arising in a telomere capping component that elevated the rate of formation subtelomeric rearrangements would serve as a very effective region-specific mutator. Once the original selective pressure was relaxed, reversion or suppression of the telomeric capping mutation would likely occur. In higher eukaryotes, it is interesting to speculate that alleles that cause increased rates of telomere failure might be present in a subset of individuals in populations. These alleles might boost the rates of subtelomeric rearrangements in the individuals where they occurred. Such alleles might affect all cells within individuals carrying them or they could be imagined to act in a germline-specific manner.

An obvious potential example of adaptive telomere failure identified to date is the telomere shortening that occurs in most human somatic cells because of the transcriptional repression of telomerase in embryonic development. Human cells containing some small number of telomeres shortened to a critical degree are subject to a permanent growth arrest called replicative senescence.[89] This phenomenon is widely believed to be an adaptation that blocks the unlimited replication of cancerous or precancerous cells. In this case, the adaptive role of the telomere dysfunction is limited to somatic cells and is thought to involve fostering genetic stability rather than genetic change. Interestingly, human cells lacking the pathway that arrests growth in response to short telomeres eventually experience massive cell death, apparently from the deleterious consequences of high rates of telomere fusions. The rare cells that survive beyond this crisis emerge with an active telomere maintenance pathway (either telomerase or HR-mediated ALT) and often exhibit considerable karyotypic abnormality.[56,89] In such cases, the telomere dysfunction is thought to potentially serve as a mutator mechanism that can help promote cancer cell formation.

Telomere Position Effect Furthers the Adaptive Plasticity of Subtelomeric DNA

In addition to enhanced genetic plasticity, subtelomeric regions of chromosomes also often have a well-documented epigenetic plasticity. Subtelomeric regions in many organisms are heterochromatic. In several species, including *Drosophila*,[90,91] *S. cerevisiae*,[92,93] *S. pombe*,[94] *Candida glabrata*,[95] *Plasmodium falciparum*,[96] *Trypanosoma brucei*,[97] plants[98] and vertebrates,[99-101] genes inserted near telomeres are subject to telomere position effect (TPE) where they are able to switch at low but appreciable rates between transcriptionally silenced and transcriptionally active states. Where examined, this TPE is mediated in part through the presence or absence of specific modifications to histones present at subtelomeric regions.

Telomere function and TPE have important links connecting them. Proper telomere function is critical for TPE. For example, proteins interacting directly with telomeres, such as Rap1 and Ku in yeast, are required for TPE.[102,103] Also, telomere length can influence TPE. In both yeast and humans, short telomeres are associated with diminished TPE.[99,102] Similarly, the chromatin modifications associated with TPE can influence telomere function. In mammalian cells lacking histone methyltransfereases thought to modify subtelomeric chromatin, telomeres become abnormally elongated.[104] Additionally, in *S. cerevisiae*, Rif proteins involved in telomere length regulation compete with Sir proteins involved in TPE for the ability to bind to the telomeric Rap1 protein.[102]

TPE can readily be imagined to be able to act synergistically with the genetic plasticity of subtelomeric regions to allow populations of organisms to have considerable adaptive headroom to respond to environmental challenges. TPE might, for example, act as a buffering force to mitigate any reduced fitness from increased gene dosage caused by a nonreciprocal translocation of a subtelomeric region to another chromosome end. This could allow adequate time for mutational mechanisms to suppress the reduced fitness and perhaps also permit further advantageous divergence of the duplicated sequence.

The Relationship between Chromosome Ends and Centromeres

A final area of discussion about the influence of telomeres on chromosome evolution concerns centromeres. Telomeres and their adjoining subtelomeric regions often share functional, structural and/or evolutionary ties with centromeres. Not only are telomeres and centromeres essential components of chromosomes but they also must evolve together to remain in a fixed two-to-one ratio. Any splitting or fusion of chromosomes must ultimately lead to a corresponding change in the number of both telomeres and centromeres. Subtelomeric and pericentromeric sequences are also often physically nearby on a chromosome and may even effectively overlap in acrocentric and telocentric chromosomes. For example, the short p arms of mouse chromosomes, have only 1.8-11 kb of DNA separating telomeres from pericentric repeats.[105] The very fact that some organisms, such as mice, have only telocentric or acrocentric chromosomes clearly indicates that mutational or repair processes that can alter chromosome structure can favor outcomes where centromeres are adjacent to a telomere.

An additional tie between subtelomeric and centromeric regions is that both tend to be heterochromatic and composed of complex repeat families, some of which may even be shared in some organisms. Shared sequences would likely permit some level of ectopic homologous recombination to occur between subtelomeric and centromeric sequences. Indeed, the centromeric regions of *Schizosaccharomyces pombe* have been suggested to have evolved from head to head fusion of two sets of subtelomeric sequences.[106] Even the key function of centromeres, serving as attachment points for permitting microtubule-driven chromosome movement, can in some contexts be done by chromosome ends. The telomeres can lead chromosome movements in a microtubule-dependent fashion in the prophase of meiosis I.[26,107,108] Considerable more work will be needed to fully understand the multiple ways that telomeric and centromeric regions are interconnected.

Conclusion

Subtelomeric regions of chromosomes characteristically show a high level of genetic and epigenetic plasticity. A variety of reasons likely contribute to this. Perhaps the most fundamental is that, relative to more internal parts of chromosomes, subtelomeric regions are inherently more tolerant of nonconservative repair of DNA double strand breaks that produce terminal duplications and truncations. This in turn produces a genomic environment favorable for genes and sequences that are tolerant of and may benefit from, frequent recombination, changing copy number and the transcriptional irregularity of telomere position effect. Of huge importance to the genetic stability of subtelomeric sequences is the protective capping function of telomeres. Clearly the predominant role of telomeres in cells is to protect chromosome ends from eliciting DNA repair processes. However, it is hypothesized that a regulated relaxation of the protective capping function, termed adaptive telomere failure, could be an important mechanism helping to shape the genetic plasticity of subtelomeric regions in at least some organisms. Experiments are needed to test predictions made by this hypothesis.

References

1. de Lange T, Lundblad V, Blackburn EH eds. Telomeres, second edition. Cold Spring Harbor, New York: Cold Spring Harbor Laboratory Press; 2006.
2. Greider CW, Blackburn EH. A telomeric sequence in the RNA of Tetrahymena telomerase required for telomere repeat synthesis. Nature 1989; 337(6205):331-337.
3. Cech TR. Beginning to understand the end of the chromosome. Cell 2004; 116(2):273-279.
4. Autexier C, Lue NF. The Structure and Function of Telomerase Reverse Transcriptase. Annu Rev Biochem 2006; 75:493-517.
5. Bertuch AA, Lundblad V. The maintenance and masking of chromosome termini. Curr Opin Cell Biol 2006; 18(3):247-253.
6. Griffith JD, Comeau L, Rosenfield S et al. Mammalian telomeres end in a large duplex loop Cell 1999; 97(4):503-514.
7. Azzalin CM, Nergadze SG, Giulotto E. Human intrachromosomal telomeric-like repeats: sequence organization and mechanisms of origin. Chromosoma 2001; 110(2):75-82.
8. Bolzan AD, Bianchi MS. Telomeres, interstitial telomeric repeat sequences and chromosomal aberrations. Mutat Res 2006; 612(3):189-214.
9. Aylon Y, Kupiec M. DSB repair: the yeast paradigm. DNA Repair (Amst) 2004; 3(8-9):797-815.
10. Daley JM, Palmbos PL, Wu D et al. Nonhomologous end joining in yeast. Annu Rev Genet 2005; 39:431-451.
11. Hefferin ML, Tomkinson AE. Mechanism of DNA double-strand break repair by nonhomologous end joining. DNA Repair (Amst) 2005; 4(6):639-648.

12. Pryde FE, Gorham HC, Louis EJ. Chromosome ends: all the same under their caps. Curr Opin Genet Dev 1997; 7(6):822-828.
13. Mewborn SK, Lese Martin C, Ledbetter DH. The dynamic nature and evolutionary history of subtelomeric and pericentromeric regions. Cytogenet Genome Res 2005; 108(1-3):22-25.
14. Mefford HC, Trask BJ. The complex structure and dynamic evolution of human subtelomeres. Nat Rev Genet 2002; 3(2):91-102.
15. Craven RJ, Petes TD. Dependence of the regulation of telomere length on the type of subtelomeric repeat in the yeast Saccharomyces cerevisiae. Genetics 1999; 152(4):1531-1541.
16. Ray A, Runge KW. The yeast telomere length counting machinery is sensitive to sequences at the telomere-nontelomere junction. Mol Cell Biol 1999; 19(1):31-45.
17. Figueiredo LM, Freitas-Junior LH, Bottius E et al. A central role for Plasmodium falciparum subtelomeric regions in spatial positioning and telomere length regulation. EMBO J 2002; 21(4):815-824.
18. Jacob NK, Stout AR, Price CM. Modulation of telomere length dynamics by the subtelomeric region of tetrahymena telomeres. Mol Biol Cell 2004; 15(8):3719-3728.
19. Berthiau AS, Yankulov K, Bah A et al. Subtelomeric proteins negatively regulate telomere elongation in budding yeast. EMBO J 2006; 25(4):846-856.
20. Pryde FE, Louis EJ. Limitations of silencing at native yeast telomeres. EMBO J 1999; 18(9):2538-2550.
21. Fourel G, Revardel E, Koering CE et al. Cohabitation of insulators and silencing elements in yeast subtelomeric regions. EMBO J 1999; 18(9):2522-2537.
22. Cryderman DE, Morris EJ, Biessmann H et al. Silencing at Drosophila telomeres: nuclear organization and chromatin structure play critical roles. EMBO J 1999; 18(13):3724-3735.
23. Lundblad V, Blackburn EH. An alternative pathway for yeast telomere maintenance rescues est1- senescence. Cell 1993; 73(2):347-360.
24. McEachern MJ, Iyer S. Short telomeres in yeast are highly recombinogenic. Mol Cell 2001; 7(4):695-704.
25. McEachern MJ, Haber JE. Break-induced replication and recombinational telomere elongation in yeast. Annu Rev Biochem 2006; 75:111-135.
26. Chikashige Y, Ding DQ, Funabiki H et al. Telomere-led premeiotic chromosome movement in fission yeast. Science 1994; 264(5156):270-273.
27. Bass HW. Telomere dynamics unique to meiotic prophase: formation and significance of the bouquet. Cell Mol Life Sci 2003; 60(11):2319-2324.
28. Barton AB, Su Y, Lamb J et al. A function for subtelomeric DNA in Saccharomyces cerevisiae. Genetics 2003; 165(2):929-934.
29. Ross LO, Maxfield R, Dawson D. Exchanges are not equally able to enhance meiotic chromosome segregation in yeast. Proc Natl Acad Sci USA 1996; 93(10):4979-4983.
30. Klein S, Zenvirth D, Dror V et al. Patterns of meiotic double-strand breakage on native and artificial yeast chromosomes. Chromosoma 1996; 105(5):276-284.
31. Gerton JL, DeRisi J, Shroff R et al. Global mapping of meiotic recombination hotspots and coldspots in the yeast Saccharomyces cerevisiae. Proc Natl Acad Sci USA 2000; 97(21):11383-11390.
32. Kong A, Gudbjartsson DF, Sainz J et al. A high-resolution recombination map of the human genome. Nat Genet 2002; 31(3):241-247.
33. Matise TC, Sachidanandam R, Clark AG et al. A 3.9-centimorgan-resolution human single-nucleotide polymorphism linkage map and screening set. Am J Hum Genet 2003; 73(2):271-284.
34. Moxon ER, Rainey PB, Nowak MA et al. Adaptive evolution of highly mutable loci in pathogenic bacteria. Curr Biol 1994; 4(1):24-33.
35. Barry JD, Ginger ML, Burton P et al. Why are parasite contingency genes often associated with telomeres? Int J Parasitol 2003; 33(1):29-45.
36. Horn D, Barry JD. The central roles of telomeres and subtelomeres in antigenic variation in African trypanosomes. Chromosome Res 2005; 13(5):525-533.
37. Borst P. Antigenic variation and allelic exclusion. Cell 2002; 109(1):5-8.
38. Figueiredo L, Scherf A. Plasmodium telomeres and telomerase: the usual actors in an unusual scenario. Chromosome Res 2005; 13(5):517-524.
39. Keely SP, Renauld H, Wakefield AE et al. Gene arrays at Pneumocystis carinii telomeres. Genetics 2005; 170(4):1589-1600.
40. De Las Penas A, Pan SJ, Castano I et al. Virulence-related surface glycoproteins in the yeast pathogen Candida glabrata are encoded in subtelomeric clusters and subject to RAP1- and SIR-dependent transcriptional silencing. Genes Dev 2003; 17(18):2245-2258.
41. Riethman H, Ambrosini A, Paul S. Human subtelomere structure and variation. Chromosome Res 2005; 13(5):505-515.

42. Linardopoulou EV, Williams EM, Fan Y et al. Human subtelomeres are hot spots of interchromosomal recombination and segmental duplication. Nature 2005; 437(7055):94-100.
43. Knight SJ, Flint J. The use of subtelomeric probes to study mental retardation. Methods Cell Biol 2004; 75:799-831.
44. Cornforth MN, Eberle RL. Termini of human chromosomes display elevated rates of mitotic recombination. Mutagenesis 2001; 16(1):85-89.
45. Rudd MK, Friedman C, Parghi SS et al. Elevated rates of sister chromatid exchange at chromosome ends. PLoS Genet 2007; 3(2):e32.
46. Carlson M, Celenza JL, Eng FJ. Evolution of the dispersed SUC gene family of Saccharomyces by rearrangements of chromosome telomeres. Mol Cell Biol 1985; 5(11):2894-2902.
47. Teunissen AW, Steensma HY. Review: the dominant flocculation genes of Saccharomyces cerevisiae constitute a new subtelomeric gene family. Yeast 1995; 11(11):1001-1013.
48. Denayrolles M, de Villechenon EP, Lonvaud-Funel A et al. Incidence of SUC-RTM telomeric repeated genes in brewing and wild wine strains of Saccharomyces. Curr Genet 1997; 31(6):457-461.
49. Ness F, Aigle M. RTM1: a member of a new family of telomeric repeated genes in yeast. Genetics 1995; 140(3):945-956.
50. Kellis M, Patterson N, Endrizzi M et al. Sequencing and comparison of yeast species to identify genes and regulatory elements. Nature 2003; 423(6937):241-254.
51. Harrison P, Kumar A, Lan N et al. A small reservoir of disabled ORFs in the yeast genome and its implications for the dynamics of proteome evolution. J Mol Biol 2002; 316(3):409-419.
52. Lafontaine I, Fischer G, Talla E et al. Gene relics in the genome of the yeast Saccharomyces cerevisiae. Gene 2004; 335:1-17.
53. Greig D, Travisano M. The Prisoner's Dilemma and polymorphism in yeast SUC genes. Proc Biol Sci 2004; 271 Suppl 3:S25-26.
54. McClintock B. The stability of broken ends of chromosomes in Zea mays. Genetics 1941; 26:234-282.
55. McClintock B. The fusion of broken ends of chromosomes following nuclear fusion. Proc Natl Acad Sci USA 1942; 28:458-463.
56. Murnane JP. Telomeres and chromosome instability. DNA Repair (Amst) 2006; 5(9-10):1082-1092.
57. Kramer KM, Haber JE. New telomeres in yeast are initiated with a highly selected subset of TG1-3 repeats. Genes Dev 1993; 7(12A):2345-2356.
58. Schulz VP, Zakian VA. The Saccharomyces PIF1 DNA helicase inhibits telomere elongation and de novo telomere formation. Cell 1994; 76(1):145-155.
59. Mangahas JL, Alexander MK, Sandell LL et al. Repair of chromosome ends after telomere loss in Saccharomyces. Mol Biol Cell 2001; 12(12):4078-4089.
60. Pennaneach V, Putnam CD, Kolodner RD. Chromosome healing by de novo telomere addition in Saccharomyces cerevisiae. Mol Microbiol 2006; 59(5):1357-1368.
61. Ricchetti M, Dujon B, Fairhead C. Distance from the chromosome end determines the efficiency of double strand break repair in subtelomeres of haploid yeast. J Mol Biol 2003; 328(4):847-862.
62. Therizols P, Fairhead C, Cabal GG et al. Telomere tethering at the nuclear periphery is essential for efficient DNA double strand break repair in subtelomeric region. J Cell Biol 2006; 172(2):189-199.
63. Stewart SA, Weinberg RA. Telomeres: cancer to human aging. Annu Rev Cell Dev Biol 2006; 22:531-557.
64. Heller RC, Marians KJ. Replisome assembly and the direct restart of stalled replication forks. Nat Rev Mol Cell Biol 2006; 7(12):932-943.
65. McEachern MJ, Blackburn EH. Cap-prevented recombination between terminal telomeric repeat arrays (telomere CPR) maintains telomeres in Kluyveromyces lactis lacking telomerase. Genes Dev 1996; 10(14):1822-1834.
66. Topcu Z, Nickles K, Davis C et al. Abrupt disruption of capping and a single source for recombinationally elongated telomeres in Kluyveromyces lactis. Proc Natl Acad Sci USA 2005; 102:3348-3353.
67. Natarajan S, McEachern MJ. Recombinational telomere elongation promoted by DNA circles. Mol Cell Biol 2002; 22(13):4512-4521.
68. Lin CY, Chang HH, Wu KJ et al. Extrachromosomal telomeric circles contribute to Rad52-, Rad50- and polymerase delta-mediated telomere-telomere recombination in Saccharomyces cerevisiae. Eukaryot Cell 2005; 4(2):327-336.
69. Smogorzewska A, Karlseder J, Holtgreve-Grez H et al. DNA ligase IV-dependent NHEJ of deprotected mammalian telomeres in G1 and G2. Curr Biol 2002; 12(19):1635-1644.
70. Mieczkowski PA, Mieczkowska JO, Dominska M et al. Genetic regulation of telomere-telomere fusions in the yeast Saccharomyces cerevisae. Proc Natl Acad Sci USA 2003; 100(19):10854-10859.
71. Ferreira MG, Cooper JP. Two modes of DNA double-strand break repair are reciprocally regulated through the fission yeast cell cycle. Genes Dev 2004; 18(18):2249-2254.

72. Heacock M, Spangler E, Riha K et al. Molecular analysis of telomere fusions in Arabidopsis: multiple pathways for chromosome end-joining. EMBO J 2004; 23(11):2304-2313.
73. Celli GB, de Lange T. DNA processing is not required for ATM-mediated telomere damage response after TRF2 deletion. Nat Cell Biol 2005; 7(7):712-718.
74. Celli GB, Denchi EL, de Lange T. Ku70 stimulates fusion of dysfunctional telomeres yet protects chromosome ends from homologous recombination. Nat Cell Biol 2006; 8(8):885-890.
75. Carter SD, Iyer S, Xu J et al. The Role Of Nonhomologous End-Joining Components in Telomere Metabolism in Kluyveromyces lactis. Genetics 2007; 175(3):1035-1045.
76. Naito T, Matsuura A, Ishikawa F. Circular chromosome formation in a fission yeast mutant defective in two ATM homologues Nat Genet 1998; 20(2):203-206.
77. Nakamura TM, Cooper JP, Cech TR. Two modes of survival of fission yeast without telomerase Science 1998; 282(5388):493-496.
78. McEachern MJ, Iyer S, Fulton TB et al. Telomere fusions caused by mutating the terminal region of telomeric DNA. Proc Natl Acad Sci USA 2000; 97:11409-11414.
79. Garvik B, Carson M, Hartwell L. Single-stranded DNA arising at telomeres in cdc13 mutants may constitute a specific signal for the RAD9 checkpoint. Mol Cell Biol 1995; 15(11):6128-6138.
80. Teng SC, Zakian VA. Telomere-Telomere Recombination Is an Efficient Bypass Pathway for Telomere Maintenance in Saccharomyces cerevisiae. Mol Cell Biol 1999; 19(12):8083-8093.
81. Natarajan S, Groff-Vindman C, McEachern MJ. Factors influencing the recombinational expansion and spread of telomeric tandem arrays in Kluyveromyces lactis. Eukaryot Cell 2003; 2(5):1115-1127.
82. Hackett JA, Feldser DM, Greider CW. Telomere dysfunction increases mutation rate and genomic instability. Cell 2001; 106(3):275-286.
83. Myung K, Chen C, Kolodner RD. Multiple pathways cooperate in the suppression of genome instability in Saccharomyces cerevisiae. Nature 2001; 411(6841):1073-1076.
84. Liti G, Louis EJ. NEJ1 prevents NHEJ-dependent telomere fusions in yeast without telomerase. Mol Cell 2003; 11(5):1373-1378.
85. Meyer DH, Bailis AM. Telomere Dysfunction Drives Increased Mutation by Error-Prone Polymerases Rev1 and {zeta} in Saccharomyces cerevisiae. Genetics 2007; 175(3):1533-1537.
86. Dreesen O, Cross GA. Consequences of telomere shortening at an active VSG expression site in telomerase-deficient Trypanosoma brucei. Eukaryot Cell 2006; 5(12):2114-2119.
87. Dreesen O, Li B, Cross GA. Telomere structure and function in trypanosomes: a proposal. Nat Rev Microbiol 2007; 5(1):70-75.
88. Denamur E, Matic I. Evolution of mutation rates in bacteria. Mol Microbiol 2006; 60(4):820-827.
89. Shay JW, Wright WE. Senescence and immortalization: role of telomeres and telomerase. Carcinogenesis 2005; 26(5):867-874.
90. Hazelrigg T, Levis R, Rubin GM. Transformation of white locus DNA in Drosophila: dosage compensation, zeste interaction and position effects. Cell 1984; 36(2):469-481.
91. Levis R, Hazelrigg T, Rubin GM. Effects of genomic position on the expression of transduced copies of the white gene of Drosophila. Science 1985; 229(4713):558-561.
92. Gottschling DE, Aparicio OM, Billington BL et al. Position effect at S. cerevisiae telomeres: reversible repression of Pol II transcription. Cell 1990; 63(4):751-762.
93. Tham WH, Zakian VA. Transcriptional silencing at Saccharomyces telomeres: implications for other organisms. Oncogene 2002; 21(4):512-521.
94. Nimmo ER, Cranston G, Allshire RC. Telomere-associated chromosome breakage in fission yeast results in variegated expression of adjacent genes. EMBO J 1994; 13(16):3801-3811.
95. Castano I, Pan SJ, Zupancic M et al. Telomere length control and transcriptional regulation of subtelomeric adhesins in Candida glabrata. Mol Microbiol 2005; 55(4):1246-1258.
96. Scherf A, Hernandez-Rivas R, Buffet P et al. Antigenic variation in malaria: in situ switching, relaxed and mutually exclusive transcription of var genes during intra-erythrocytic development in Plasmodium falciparum. EMBO J 1998; 17(18):5418-5426.
97. Horn D, Cross GA. A developmentally regulated position effect at a telomeric locus in Trypanosoma brucei. Cell 1995; 83(4):555-561.
98. Matzke MA, Moscone EA, Park YD et al. Inheritance and expression of a transgene insert in an aneuploid tobacco line. Mol Gen Genet 1994; 245(4):471-485.
99. Baur JA, Zou Y, Shay JW et al. Telomere position effect in human cells. Science 2001; 292(5524):2075-2077.
100. Koering CE, Pollice A, Zibella MP et al. Human telomeric position effect is determined by chromosomal context and telomeric chromatin integrity. EMBO Rep 2002; 3(11):1055-1061.
101. Pedram M, Sprung CN, Gao Q et al. Telomere position effect and silencing of transgenes near telomeres in the mouse. Mol Cell Biol 2006; 26(5):1865-1878.

102. Kyrion G, Liu K, Liu C et al. RAP1 and telomere structure regulate telomere position effects in Saccharomyces cerevisiae. Genes Dev 1993; 7(7A):1146-1159.
103. Boulton SJ, Jackson SP. Components of the Ku-dependent nonhomologous end-joining pathway are involved in telomeric length maintenance and telomeric silencing. EMBO J 1998; 17(6):1819-1828.
104. Gonzalo S, Jaco I, Fraga MF et al. DNA methyltransferases control telomere length and telomere recombination in mammalian cells. Nat Cell Biol. Apr 2006; 8(4):416-424.
105. Kalitsis P, Griffiths B, Choo KH. Mouse telocentric sequences reveal a high rate of homogenization and possible role in Robertsonian translocation. Proc Natl Acad Sci USA 2006; 103(23):8786-8791.
106. Hansen KR, Ibarra PT, Thon G. Evolutionary-conserved telomere-linked helicase genes of fission yeast are repressed by silencing factors, RNAi components and the telomere-binding protein Taz1. Nucleic Acids Res 2006; 34(1):78-88.
107. Chikashige Y, Ding DQ, Imai Y et al. Meiotic nuclear reorganization: switching the position of centromeres and telomeres in the fission yeast Schizosaccharomyces pombe. EMBO J 1997; 16(1):193-202.
108. Scherthan H. Telomere attachment and clustering during meiosis. Cell Mol Life Sci 2007; 64(2):117-124.

Chapter 8

Evolution, Composition and Interrelated Functions of Telomeres and Subtelomeres:

Lessons from Plants

Jiří Fajkus,* Andrew R. Leitch, Michael Chester and Eva Sýkorová

Abstract

The terminal parts of chromosomes, telomeres and their subtelomeric neighbors are reviewed. Both have common and specific features that interact to generate the unique and essential biology that characterizes telomeres. The chromatin of both chromosome domains has a nucleosomal structure and their DNA component is often, but not always formed by tandem repeats. The relationships between telomere and subtelomere domains are complex and are discussed at sequence and folding levels. The complexity of the interaction arises from, and enables, the multiple functions of the telomere domain. For example, telomere length influences the expression of genes adjacent to the telomere via controlled heterochromatinization, while at the same time the conformation changes in the heterochromatin influences telomere length regulation. Here we discuss the process of heterochromatinization of terminal chromosome domains, including the role of RNA interference, telomere and subtelomere boundary domains and the evolution of telomere minisatellite and satellite repeats. Variability in telomere sequences and telomere-maintenance strategies is exemplified by telomere evolution in the plant order Asparagales, which is reviewed together with the latest molecular data on the telomerases that synthesize these variant repeats. We also highlight a surprising interconnectivity between telomere and nucleolar function.

Introduction

The ends of eukaryotic chromosomes are capped by a special structure called the telomere whose DNA component is formed by long arrays of tandemly repeated short minisatellite sequences, e.g., TTAGGG in vertebrates (first described in humans[1,2]), TTTAGGG in model plants (*Arabidopsis*[3]) and TTAGG in some insects (*Bombyx*[4]). The maintenance of these telomeres is performed by telomerase. The enzyme consists of a protein catalytic subunit (telomerase reverse transcriptase, TERT) and an RNA subunit (TR), which provides the template for telomere synthesis.[5] TR was first cloned and sequenced in *Tetrahymena thermophila*[6] and TERT was first cloned in another ciliate, *Euplotes aediculatus*.[7] TERTs and TRs from many model organisms have now been characterized, including from yeasts, protozoa, humans and other vertebrates (see e.g., ref. 8 for a review). Recently, insect TERTs have also been reported.[9,10] Plant TERTs have been characterized in the model species *Arabidopsis thaliana* and *Oryza sativa*,[11-14] and in six other

*Corresponding Author: Jiří Fajkus—Department of Functional Genomics and Proteomics, Masaryk University, Kamenice 5, CZ-62500 Brno, Czech Republic. Email: fajkus@sci.muni.cz

Origin and Evolution of Telomeres, edited by Jozef Nosek and Ľubomír Tomáška.

species (*Zea mays* and species in order Asparagales,[15] see below). However no RNA subunit has been described for plants.

The first exceptions to the view that telomeric repeats define major eukaryote lineages were in insects of the order Diptera[16-18] and plant species in the genus *Allium*, e.g., *Allium cepa* (Alliaceae).[19] Later Sahara et al[20] showed that the telomeric repeat typical of insects was missing in several insect orders and in a spider. Subsequently further insects and other arthropods[21,22] and two plant groups (the eudicot family Solanaceae[23] and the monocot order Asparagales[24-26]) were added to the growing list of organisms lacking the "expected" telomeric motifs.

The maintenance of chromosome ends requires specific telomere-binding proteins (TBP) and proteins involved in the formation of the telomere chromatin structure. The telomere at its proximal end interfaces with the subtelomere, which shares some similarities with the telomere and provides a buffer zone between the telomere and coding regions. The telomere's distal end is composed of double-stranded (ds) telomeric DNA and a single-stranded (ss) 3'-overhang, which recruits proteins of different ds/ssDNA binding and sequence recognition specificities. There has been considerable recent interest in resolving the dynamic interplay between these proteins and DNA/RNA that comprise the machinery involved in protecting telomeres against nucleolytic cleavage and terminal degradation.

Telomeres and DNA Folding

Telomeric chromatin displays some specific features that distinguish it from the bulk and subtelomeric chromatin. When telomeric chromatin is digested with micrococcal nuclease (MNase), short nucleosomal spacing (usually about 30-40 bp less than the bulk chromatin) and extensive subnucleosomal fragmentation of short mono- and oligonucleosome-size particles is observed.[27-30] This fragmentation of telomeric DNA probably arises because histone octamers can slide on short stretches of telomere sequence arrays lacking nucleosome positioning signals.[30,31] However longer stretches (≥4 nucleosomes) have stable nucleosome positioning and MNase digestions generate fragments with regular periodicity. Thus the telomere domain has different and specific packing properties as described in the columnar model of telomere nucleosome arrangement.[32] In this model DNA is continuously wound around stacked histone octamers. Data on the structure of reconstituted telomeric nucleosomes obtained using atomic force microscopy (AFM) imaging and theoretical modeling support observations of short internucleosomal distances and a role of histone-histone interactions in chromatin assembly.[33,34] Importantly, the occurrence of nucleosome structure at telomeres and their t-loops[35] does not exclude the binding of telomere-specific proteins.[36,37]

The unusual nucleosome organization at telomeres may have additional undiscovered importance as is suggested by the intriguing biology of telomeres in mammalian sperm nuclei. In these nuclei chromatin is remodeled with protamines into a highly condensed structure, but telomeres remain in a nucleohistone structure.[38] These authors speculate that telomeres are among the first male pronucleus structures to respond to oocyte signals at fertilization. In addition, telomeres in sperm tend to form dimers at the nuclear periphery. These dimers arise from the two ends of each chromosome being connected in a looped conformation.[39,40] Perhaps these unusual features of higher-order chromatin organization are required for chromosome decondensation during the early fertilization events leading to zygote formation.

Subtelomere Domains

Unlike telomeric domains that can have nucleosomes sliding on short stretches of DNA, subtelomeric domains have stable nucleosome positioning. The differences between these domains may be partly due to their DNA sequences. While telomeres are typically 6-8 bp minisatellite tandem repeats with little potential to define positioning of nucleosomes, subtelomeric DNA are typically satellite repeats of mono- and dinucleosomal size. In addition, DNA sequence heterogeneity is usually much higher at subtelomeres than at telomeres and frequently includes complex repeats, ribosomal DNA (rDNA) clusters, degenerated telomeric repeats (accumulated especially at the boundary between telomere and subtelomere) and single-copy sequences. Furthermore the influence of long arrays of satellite repeats probably dominates chromatin structure determination. Quite often members of the same satellite sequence family are distributed at subtelomeres of most or all chromosomes, as exemplified by the HRS60 family in tobacco,[41-43] TGRI family in tomato[44] and X43.1 in *Silene latifolia*.[45] But not all species have subtelomeric repetitive sequences, as e.g., *Arabidopsis thaliana*, which has unique subtelomeric sequences on each chromosome arm except the two that carry rDNA.[46]

The telomere-subtelomere boundary represents not only a junction between different DNA sequence clusters, but also boundaries between two distinct chromatin domains. This junction was analyzed in detail in several plant species. In *Secale cereale* (rye) two nucleosome periodicities were observed in subtelomeric satellite sequences pSc200 and pSc250. The shorter periodicity corresponded to telomeric nucleosomes and the longer periodicity to the phasing of bulk nucleosomes.[47] A similar result was obtained in the analysis of chromatin structure in *Silene latifolia* X43.1 repeats. These repeats showed nucleosome periodicities of 157 and 188 bp, corresponding to telomeric and bulk nucleosome phasing, respectively. The two periodicities could thus correspond to telomere-associated and proximal parts of the subtelomere, respectively. Moreover, another sequence (15Ssp), which is directly attached to telomere in *Silene* chromosomes, also shows 157 bp periodicity and nucleosome sliding, the features typical of telomeric chromatin.[48]

At meiosis, it remains unknown how chromosome pairing at telomeres is initiated and how homologous chromosome ends recognize each other. Given that the same telomere sequences are usually present at all, or most chromosome ends and a limited number of satellite sequences constitute a major fraction of most subtelomeres, a unique character could be achieved by the arrangements of a limited number of building blocks. In both rye and *Silene*, the lengths of telomere and subtelomere repeats, their mutual arrangement and the presence of specific spacer sequences between arrays results in a characteristic pattern at each chromosome end.[49,50] *Arabidopsis thaliana* has unique subtelomeric sequences that may directly determine the individual telomere identity. Furthermore the length of the telomere could be controlled by the subtelomere sequences. Certainly in crosses between ecotypes of *A. thaliana* with different telomere lengths, a new intermediate telomere-length is established.[51,52]

Subtelomeric repeats may act as buffer-zones to minimize the effects of telomere erosion in the absence of telomerase-mediated elongation. Due to their tandemly repetitive character, they can potentially change in copy number using several pathways, e.g., rolling-circle replication, gene conversion and unequal recombination. Like minisatellite telomeric repeats they are capable of forming lasso-shaped loops and extrachromosomal circles, structures considered as intermediates in mechanisms leading to rapid telomere lengthening or contraction (for a recent review, see ref. 53). Satellite repeat copy number expansion mechanisms function in some organisms and cell lines that use alternative lengthening of telomeres (ALT) pathways. Subtelomeric repeats have taken on the role of telomeres in yeast telomerase mutants[54] and perhaps in species of *Allium* (Alliaceae) and *Cestrum* (Solanaceae), which are both plant genera that lost G-rich telomeric minisatellites and telomerase activity at some point in their ancestry.[23,26,55,56] In species of the insect genus *Chironomus* satellite repeats act as telomeres,[57-59] perhaps these too had their origin as subtelomeric repeats in the ancestry of this genus. Satellite repeat amplification may also maintain the ends of linear mitochondrial chromosomes in the yeast *Candida parapsilosis*.[60,61]

Telomeric and Subtelomeric Heterochromatin

Both telomeres and subtelomeres could be described as heterochromatic chromosome domains with no protein-coding capacity. But the classical definition of "heterochromatin" is poor in that it considers the chromatin as transcriptionally silent. Now we know that transcription and the RNA interference (RNAi) machinery can play an important role in the formation of heterochromatin, particularly in subtelomeric domains. Thus paradoxically transcriptional silencing of one strand depends on continuous transcription of the other strand.[62]

Subtelomeres are a domain potentially influenced by the telomere position effect (TPE), but the universality and significance of TPE across eukaryotes remains unclear. TPE is best known in yeast and results in transcriptional silencing of genes near the telomere.[63] In *Saccharomyces cerevisiae*, the effect of TPE increases with telomere length,[64] but other "architectural" factors, like telomere folding, are also critical.[65] Recent studies in human cells have shown that among 450 endogenous subtelomeric genes, TPE was found only in interferon-stimulated gene 15 (ISG15),[66] although in a previous study by the same group, artificial constructs with the luciferase reporter gene showed a clear telomere-length-dependent silencing when inserted into telomere-adjacent positions.[67]

In yeast the establishment of heterochromatin at the telomere can be achieved via two pathways, the first uses telomere binding protein (Taz1) causing H3K9 methylation, which marks chromatin for heterochromatinization. The second pathway involves the RNA interference (RNAi) machinery, which is targeted at subtelomeric heterochromatin and can spread to the telomere domain.[68] The RNAi machinery couples DNA-dependent and RNA-dependent transcription. One DNA strand is transcribed and the transcript is rapidly processed by RNAi, resulting in the generation of siRNAs. These siRNAs then prime RNA-dependent transcription by RNA-dependent RNA polymerase (RdRp) to generate dsRNA. Double-stranded RNA is then processed by the RNAse Dicer to generate further siRNA molecules. RdRp then recruits histone methyltransferase (HMT), which methylates histone H3 at Lys9 (H3K9). In *S. cerevisiae*, this modification is a hallmark for binding of heterochromatin protein 1 (HP1), which is associated with a compact heterochromatin structure. Both HMT and HP1 proteins then promote spreading of heterochromatin since their chromodomains are capable of binding RNA as well as methylated histone tails.[69]

It is noteworthy that chromatin silencing mechanism using the RNAi machinery works more effectively on tandemly repeated sequences than on single-copy sequences. Single-copy sequences face the problem of a limited supply of siRNAs and, consequently, of finite number of reaction cycles of RdRp. The reason is that siRNA primers for each subsequent round of RdRp synthesis are generated from the RNA sequence downstream of the previous siRNA primer and no priming from upstream sites is possible. The situation is analogous to the end-replication problem of linear chromosomes.[62,70]

Evolutionary Divergence of Telomeric Sequences

The commonest type of eukaryote telomeres contain motifs based on $T_nA_mG_o$—minisatellite repeats. But other repetitive sequences are known to function as telomeres. In *Drosophila* (Diptera) HetA, TART and TAHRE are thought to preferentially retrotranspose to the telomere[17,71] and in *Chironomus* (also Diptera), telomeric satellites of 176 and 340-350 bp length are thought to form and maintain telomeric DNA using homologous recombination.[57-59] However, these examples were treated for a long time as exceptional. In the past decade it has become apparent that the minisatellite repeat at the functional telomere may not define major phylogenetic groups. This is the case of both plants (Asparagales,[23,24,72] Solanaceae[23]) and invertebrates.[20,22]

The first indication that the minisatellite sequence $(TTTAGGG)_n$ found at telomeres of *A. thaliana*[3] was not ubiquitous to all flowering plants was from Alliacaeae, a group of monocots in plant order Asparagales that includes the onions.[19,55] An extensive survey of Asparagales has shown a surprising diversity of telomere types. Early diverged families retain the *Arabidopsis*-type telomeric repeat, while more recently diverged families are associated with a change in telomerase activity and the synthesis of the human-type $(TTAGGG)_n$ telomeric repeat.[24,25,73] A derived group (genus *Allium*) from the clade synthesizing the human-type of telomere repeat appears to have lost telomeric minisatellites and telomerase activity altogether[25,26] (see Fig. 1). In Asparagales possessing human-type telomeres, analysis of the TRAP products revealed that telomerase has error-prone synthesis of minisatellite repeats. In Hyacinthaceae (Asparagales) with predominantly human-type repeats, variant telomeric motifs (of the *Arabidopsis*- and *Tetrahymena*-type) are also present at telomeres in mixed arrays.[74] In addition to a low telomerase fidelity, this might be the result of rearrangements in telomere-subtelomere junctions.

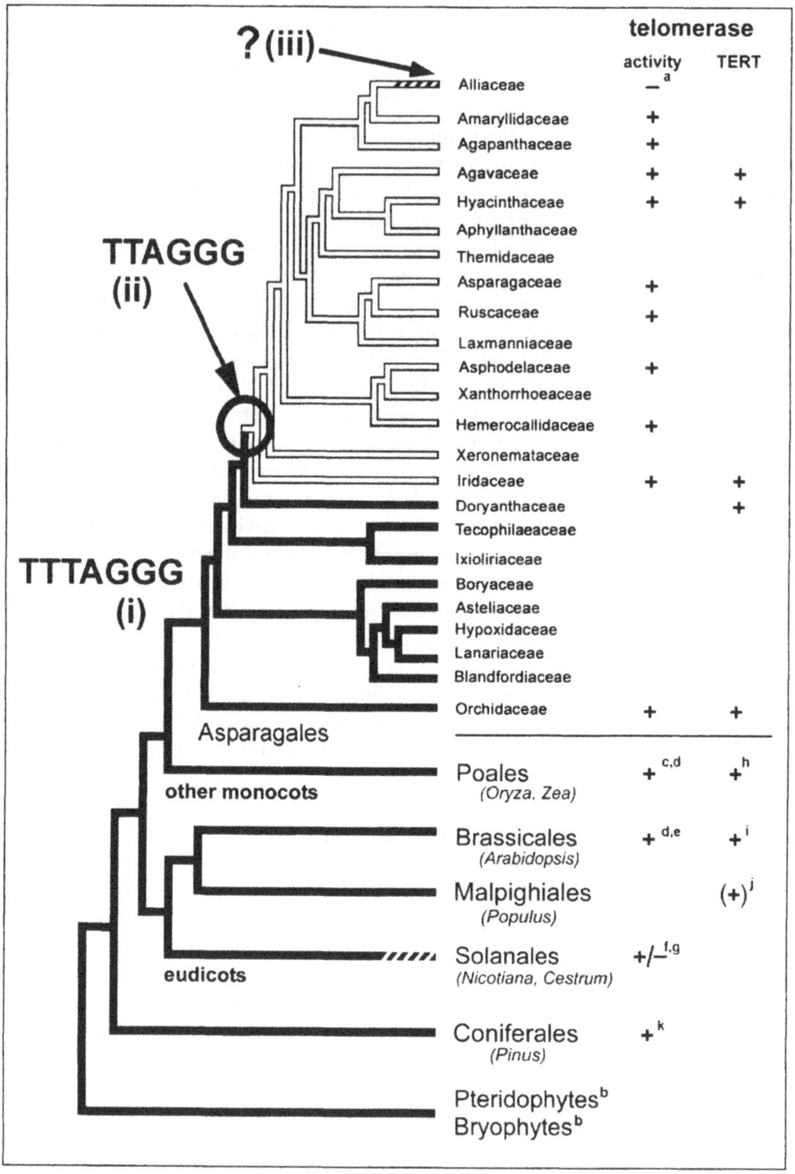

Figure 1. Summary of knowledge of telomeres and telomerases in plants. The phylogenetic scheme (on the left), data on telomerase activity in Asparagales (left column) and cloned TERT sequences (right column) are adapted from[25] and[15] unless otherwise stated. Asparagales are split into: group (i) with the *Arabidopsis*- type of telomeres (black line), group (ii) with human-type telomeres (white line) and group (iii) with uncharacterised telomeres (hatched line, a, *Allium cepa* also negative for telomerase activity). For comparison, the liverwort, fern [b,[115]] and some model plants are shown with relevant references. Note that the Solanales also include genera with unknown types of telomeres [hatched line,[23]]. References for telomerase activity are: c[12]; d,[116]; e,[117]; f, tobacco (+)[118]; g, *Cestrum* (–) (Sykorova, unpublished results); k, *Pinus longaeva*[119] and for TERT subunits: h, *Oryza sativa*[12,14] and *Zea mays*[15]; i, *Arabidopsis thaliana*[11,13]; j, *Populus trichocarpa*[120], search at http://genome.jgi-psf.org/Poptr1_1/Poptr1_1.home.html.

Incorrect telomeric repeats cause the loss of telomere cap integrity and mitotic defects in yeasts, *Tetrahymena* and human (refs. 75, 76; reviewed in Blackburn, 2005). Therefore changes in telomere motif in the ancestral Asparagales may have had consequences for telomere function. Analysis of putative telomere binding proteins in *Iris tectorum* and *Scilla peruviana* (both with human-type telomere motifs, Iris : Iridaceae) has shown ssDNA binding proteins have affinity to both *Arabidopsis*-type and human-type telomere motifs.[74] Protein extracts of *A. cepa* (lacking both *Arabidopsis* and human-type of telomeres) formed nucleoprotein complexes with *Arabidopsis*- and/or human-type of telomeric sequences.[78] These data indicate conservation of putative telomere binding proteins in Asparagales even though the telomere sequence itself is much diverged (especially in *A. cepa*) from the ancestral TTTAGGG motif. In *Allium*, the sequence at the telomere may have been replaced by one or more of candidate sequences, such as 375-bp satellite sequence (ACSAT), first described by Barnes et al,[79] rDNA, Ty1-copia group retrotransposons or En/Spm-like sequences.[55,80,81] Also the possibility of replacement by another minisatellite sequence cannot be completely excluded. It is also possible that the chromosome termini are not all the same in *Allium:* certainly the distribution of rDNA at, or near, the telomere can be highly variable between and within species, even at the level of an individual plant.[82] In contrast to Asparagales, where change of telomere motif can easily be explained by patterns of species divergence, in insects there are repeated losses of the TTAGG motif in different beetle and other insect groups.[21,22,83]

Evolution of Plant Telomerases

It is interesting to question how telomerase evolved to synthesize the human-type of telomeric repeat in the Asparagales and what was the consequence of the new repeat. There are two probable scenarios to explain the synthesis of a new repeat. The first is that a point mutation occurred in the template region of TR and that this was subsequently stabilized by adaptations of the protein machinery. This explanation is supported by the high tolerance to mutations in TR observed in many in vitro studies.[84] Such a hypothesis cannot be tested on Asparagales because no plant TR has been cloned. A second explanation for altered telomerase activity is that a structural change in TERT resulted in altered template usage, perhaps similar to the error-prone activity of telomerase in *Paramecium tetraurelia, Sterkiella nova* and *Euplotes crassus.*[85-87]

In order to determine if mutations in TERT were responsible for the switch to a human-type of minisatellite repeat in the divergence of Asparagales, several TERT genes were cloned. We searched for mutations in conserved motifs, including the reverse transcriptase (1, 2, A-E) and telomerase specific (CP, T, T2, QFP) domains.[15] There was lower similarity in the N-terminal part of protein and high similarities among all plant TERTs in the C-terminal extension (CTE) region and in QFP motif (Fig. 2). The size of cloned Asparagales TERTs ranged from 1227 to 1323 amino acids, the differences were accounted for by the length of sequence between conserved motifs in the N-terminal part of the protein and by the presence of additional sequence between the B' and C motifs. All investigated plant TERTs are organized into 12 exons. The TERTs synthesizing human-type of telomeric repeats have 8 amino acid changes inside and 14 outside, the conserved reverse transcriptase motifs, some of which might affect telomerase action (Fig. 2). Another example of the influence of mutations in catalytic protein subunit on telomerase fidelity has been found recently in *Saccharomyces cerevisiae* (Friedman K.L., *pers. commun.*).

In *Oryza sativa*, the regulation of telomerase via phosphorylation of the TERT subunit was suggested and putative target sites predicted,[14] but no such sites were detected in *Arabidopsis*.[13] Comparison of these data with other plant TERT sequences revealed a small conserved region in exon 1 in the N-terminal part of the gene. This region was proposed to be a site of serine phosphorylation in OsTERT (from *Oryza sativa*). But the serine residue is specific for the *O. sativa* sequence and is missing in *Zea mays* (also Poales) and all other plants investigated. A second potential serine phosphorylation site occurs in a conserved basic region of TERT. This region was predicted by Prosite and/or PSORT to be a nuclear localization-like signal (NLS) and can be found in all plant and some vertebrate (including human) TERTs.

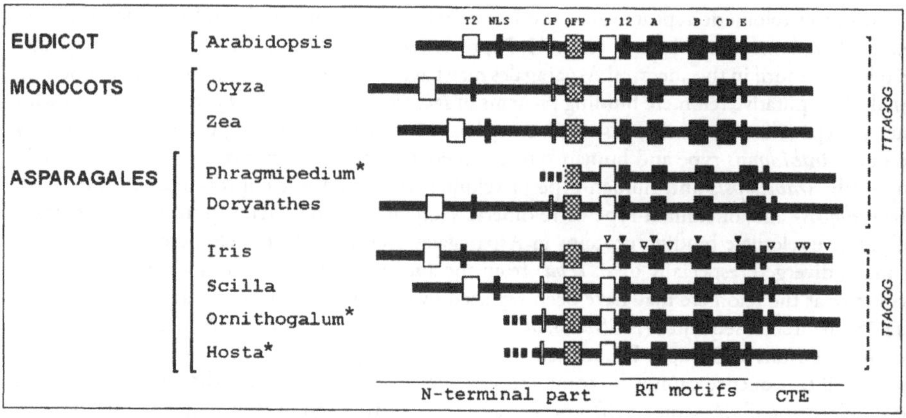

Figure 2. Organization of plant TERTs. The conserved motifs (T2, QFP, CP, 1, 2, A-E) are as described previously,[121-123] except for the nuclear localization-like signal (NLS) predicted by Prosite and/or PSORT.[15] Conserved amino acid changes in Asparagales TERTs synthesizing human-type repeats are indicated, closed triangles—inside reverse transcriptase motifs, open triangles—in T motif and outside RT motifs. Partially sequenced TERTs are marked by asterisks; the species taxonomy is shown on left; brackets on right group plant telomerases synthesizing *Arabidopsis*-type telomeres (TTTAGGG, *Arabidopsis thaliana, Oryza sativa, Zea mays, Phragmipedium longifolium*) and telomerases synthesizing human-type telomeres (TTAGGG, *Iris tectorum, Scilla peruviana, Ornithogalum virens, Hosta rectifolia*).

Telomeric repeat amplification protocol (TRAP) revealed that Asparagales telomerases synthesizing the human-type of telomeric repeat have two modes of elongation of substrate primers[15] reflecting differences in enzymatic properties. For instance, ***Bulbine glauca*** telomerase produced a short by-product in TRAP assays (probably resulting from G-slippage in the first repeat synthesized), while in ***Scilla peruviana***, the shorter by-product was barely detected (probably due to a stronger preference for the substrate-template annealing site). The differences in TERT activities were independent of species relationships in Asparagales phylogenetic schemes and are likely to be dependent on the secondary structure of the telomerase complex, which probably determines template usage or assembly of the telomerase/substrate complex. Different template usage behavior may account for error-prone synthesis of telomerase activity in species of *Ornithogalum* and *Bulbine*[25] and accurate synthesis of repeats reported in *Othocallis*.[73] Unfortunately, determining differences in the DNA- binding sites of TERTs synthesizing *Arabidopsis*- or human-type of sequence is difficult to address. Jacobs et al[88] identified three residues in the N-terminal region of *Tetrahymena thermophila* TERT that are partially responsible for DNA-recognition and substrate binding. These residues lie in the linker region near the T2 motif, one of which (Gln168) is conserved in all organisms including plant TERTs, but the other two (Phe178, Trp187) differ even between ciliates. These differences illustrate the difficulty in identifying candidate sites involved in DNA-recognition and substrate binding from protein sequence alone and no such regions have been determined in TERT from species of Asparagales.

In *Dictyostelium discoideum* the telomere is thought to be $AG_{1\text{-}8}$, with some evidence of a higher order repeat $(AG_6AG_6AGAG_6AG_2AG_3)_n$.[89] There are no reports for the presence of TERT or telomerase activity in this species, although our BLAST search revealed sequence similarities between a hypothetical protein (Genbank accession number XP_628872.1) and reverse transcriptase and T motifs of other TERTs. This hypothetical protein sequence has long stretches of in-frame triplets (AAT, CAA) giving homogeneous repeats of asparagine and glutamine. The hypothetical protein may not be a functional TERT, but a relic TERT and the mechanism of telomere elongation may be telomerase-independent. The phylogeny of social amoebas is currently under revision,[90] however *D. discoideum* and species from a sister genus (*Entamoeba histolytica*) have a published genome sequence.[91,92] The telomeric sequence of *E. histolytica* is not known and our BLAST revealed no candidate TERTs.

Does rDNA Have Functional Significance in Telomere Biology?

Replication of extrachromosomal t-circles occurs in the elongation of telomeres in human ALT cell lines and a variety of organisms including fungi, plants and animals.[61] Indeed it has been hypothesized that rolling circle replication may have been the original minisatellite telomere maintenance system of eukaryotes.[53] Extrachromosomal circles of all, or part of the rDNA unit and of telomeric sequence are known in several systems. In comparison extrachromosomal rDNA circles (ERCs) are a feature of aging yeast cells,[64] they are elevated 46-78 fold in H3K9 methylation mutants (mediated via RNAi pathway) of *Drosophila*[93] and form in cells of Bloom's syndrome patients,[94] who have increased incidence of cancer, premature ageing and unusually short telomeres. Perhaps there is a link in the biology that generates extrachromosomal circles of rDNA and telomere sequences. Certainly, recombination mechanisms are common to both.

Despite the potential for rDNA repeats to expand using recombination, there is no direct evidence that rDNA has a telomeric role. However in *Allium cepa* the number of distal rDNA loci appear remarkably unstable, even in the same plant appearing as if the locus was "jumping".[82] Potentially such processes could act to maintain telomeres and reverse the progressive degradation of DNA at each cell cycle.[95] However fast expansion of rDNA copy number has also been observed in allopolyploid plants,[96,97] and in these cases there is no indication that the expansion has a telomeric role.

rDNA Physically Associated with Telomeric DNA

In addition to the possibility that rDNA can have a telomeric role, fluorescent *in situ* hybridization (FISH) of rDNA probes to metaphase chromosomes frequently reveal rDNA located at subtelomeric positions e.g., in paeonies.[98] However the resolution is too low to know exactly how close rDNA is to the telomere, this is only possible by sequence analysis. Using that method, a very close association has been observed in a few organisms. For example the unicellular eukaryote *Giardia lamblia* has an abrupt transition between the minisatellite telomere and rDNA sequences,[99] the social amoebae *D. discoideum* probably has subtelomeric rDNA on all of its chromosome arms,[91] the nucleomorph (a relic nucleus of an ancient endosymbiosis) of the cryptomonad *Guillardia theta* has a repeat of the 23-mer $(AG)_7AAG_6A$ at the telomere which subtends a single copy of the rDNA unit on each of its three chromosomes[100] and *A. thaliana* has rDNA only a few hundred base pairs proximal to the telomere sequence motifs.[46]

The physical association of the two sequences can influence each other's respective biology. It has been suggested that telomere folding may influence the accessibility of telomerase and other proteins associated with telomere elongation.[101] Indeed length regulation of T_2G_4 telomeres in *Tetrahymena* is disrupted from the typical 250-350 bp arrays of the minisatellite repeat when subtelomeric rDNA structure is altered. The absence of minisatellite telomere sequences in TERT minus mutant cell lines of *A. thaliana* results in many cytological abnormalities, including aneuploidy, polyploidy, chromatin bridges and DNA C-value variation. In addition, these cells can show much heterogeneity in rDNA content,[102] which may be derived from a particularly high incidence of bridge formation at rDNA loci that are improperly capped by minisatellites. Given that there is a wide occurence of species with rDNA at or near subtelomeric domains, there might be functional significance in the close proximity, for example TERT can occur in the nucleolus in S-phase and its activity is at the telomere.[103]

Proteins, Telomeres and Nucleoli

Chromosomes can be organized in a Rabl configuration[104] with telomeres and centromeres clustered at opposite poles of the interphase nucleus. This organization is thought to reflect the arrangement of chromosomes established at the preceding anaphase and is most apparent in dividing cell types of organisms with large genomes.[105] In human fibroblasts the telomeres associate with the nuclear envelope,[106] but in T-lymphocytes they occupy a central domain of the nucleus and in these cells the centromeres re-distribute in a cell-cycle dependent manner.[107] In neurons and glial cells of the central nervous system, a probe against a sequence near the telomere (30-40 kb) on the Y chromosome reveals a close association with the nucleolus.[108] These data together reveal that the organization of chromatin at interphase in human cells is dynamic and changes through the cell cycle and in different cell types. Furthermore the redistribution of chromatin can bring telomeres in association with the nucleolus.

There is increasing realization that the nucleolus contains many proteins that are not directly involved in ribosome biogenesis, suggestive of multiple functions of this nuclear domain. Recently a reverse transcriptase associated with telomere elongation in *Chironomous* has been localized to nucleoli.[109] Likewise there is an association of hTERT with the nucleolus of a fraction of HeLa cells, perhaps those that occur at late S phase.[110] Possibly the nucleolus is the site of assembly of ribonuclear protein complexes (as for ribosomes themselves), or perhaps they play a role in the regulation of telomerase/telomere elongating mechanisms. One protein involved in the regulation of telomere length is TRF2. This has also been located in the nucleolus in G0 and S phase human cells.[111] Inhibition of RNA polymerase I transcription (which transcribes 45S rDNA) with actinomycin D delays the dispersal of TRF2 after S phase and causes chromosome fusions. Such fusions are also observed in TRF2 mutants.[112]

There is an increasing list of proteins involved in telomere maintenance that also have a role in DNA damage repair (e.g., Mre11/Rad50/Xrs2 complex, Ku), strongly implicating the recombination machinery in the activity of telomeres.[113] The recombination machinery also acts within rDNA arrays to drive copy number expansion in *S. cerevisiae* mutants greatly deficient in rDNA.[114] Similar mechanisms to rapidly increase copy numbers probably explain how rDNA arrays can expand so rapidly in synthetic polyploidy plants.[96,97] Potentially the nucleolus plays a role in both rDNA recombination (leading to homogenization) and the elongation of telomeres.

Conclusions

Understanding of telomere and telomerase variability has increased fast in the past few years. But there remains much to learn. In many groups of organisms we know their telomere sequences, but not whether they have telomerase genes, or telomerase activity. In other cases we know the TERT gene sequence without knowing the telomere. In addition to ALT mechanisms, telomerases remain the primary target for fundamental and applied research. More research is needed into the complex protein and nucleic acid interactions that govern telomerase assembly, binding, activity, fidelity, processivity, nucleolar interactions and cellular localization. We also need more research into the cell biology and genetic characterization of telomeres and their functional connection with subtelomere chromatin structures. Taking an evolutionary perspective to all these fields of research coordinates information between fields and gives valuable predictive power in the understanding of telomere biology. We know there is variability amongst telomeric sequences, flexibility of telomere-binding proteins and alternative mechanisms for telomere maintenance, yet we still do not have an integral and consistent view of many fundamental aspects of telomere and telomerase function.

Acknowledgements

This work was supported by the grants from GACR (521/05/0055), GA ASCR (IAA600040505), MSMT (LC06004) and the institutional funding (MSM0021622415, AVOZ50040507).

References

1. Cheng JF, Smith CL, Cantor CR. Isolation and characterization of a human telomere. Nucleic Acids Res 1989; 17(15):6109-6127.
2. Meyne J, Ratliff RL, Moyzis RK. Conservation of the human telomere sequence (TTAGGG)n among vertebrates. Proc Natl Acad Sci USA 1989; 86(18):7049-7053.

3. Richards EJ, Ausubel FM. Isolation of a higher eukaryotic telomere from Arabidopsis thaliana. Cell 1988; 53(1):127-136.
4. Okazaki S, Tsuchida K, Maekawa H et al. Identification of a pentanucleotide telomeric sequence, (TTAGG)n, in the silkworm Bombyx mori and in other insects. Mol Cell Biol 1993; 13(3):1424-1432.
5. Greider CW, Blackburn EH. The telomere terminal transferase of Tetrahymena is a ribonucleoprotein enzyme with two kinds of primer specificity. Cell 1987; 51(6):887-898.
6. Greider CW, Blackburn EH. A telomeric sequence in the RNA of Tetrahymena telomerase required for telomere repeat synthesis. Nature 1989; 337(6205):331-337.
7. Lingner J, Hughes TR, Shevchenko A et al. Reverse transcriptase motifs in the catalytic subunit of telomerase. Science 1997; 276(5312):561-567.
8. Krupp G, Parwaresch R. Telomerases, Telomeres and Cancer. Vol 22. 1. ed. New York: Kluwer Academic/ Plenum Publishers; 2002.
9. Osanai M, Kojima KK, Futahashi R et al. Identification and characterization of the telomerase reverse transcriptase of Bombyx mori (silkworm) and Tribolium castaneum (flour beetle). Gene 2006; 376(2):281-289.
10. Robertson HM, Gordon KH. Canonical TTAGG-repeat telomeres and telomerase in the honey bee, Apis mellifera. Genome Res 2006; 16(11):1345-1351.
11. Fitzgerald MS, Riha K, Gao F et al. Disruption of the telomerase catalytic subunit gene from Arabidopsis inactivates telomerase and leads to a slow loss of telomeric DNA. Proc Natl Acad Sci USA 1999; 96(26):14813-14818.
12. Heller-Uszynska K, Schnippenkoetter W, Kilian A. Cloning and characterization of rice (Oryza sativa L) telomerase reverse transcriptase, which reveals complex splicing patterns. Plant J 2002; 31(1):75-86.
13. Oguchi K, Liu H, Tamura K et al. Molecular cloning and characterization of AtTERT, a telomerase reverse transcriptase homolog in Arabidopsis thaliana. FEBS Lett 1999; 457(3):465-469.
14. Oguchi K, Tamura K, Takahashi H. Characterization of Oryza sativa telomerase reverse transcriptase and possible role of its phosphorylation in the control of telomerase activity. Gene 2004; 342(1):57-66.
15. Sykorova E, Leitch AR, Fajkus J. Asparagales telomerases which synthesize the human type of telomeres. Plant Mol Biol 2006; 60(5):633-646.
16. Biessmann H, Donath J, Walter MF. Molecular characterization of the Anopheles gambiae 2L telomeric region via an integrated transgene. Insect Mol Biol 1996; 5(1):11-20.
17. Biessmann H, Mason JM, Ferry K et al. Addition of telomere-associated HeT DNA sequences "heals" broken chromosome ends in Drosophila. Cell 1990; 61(4):663-673.
18. Nielsen L, Edstrom JE. Complex telomere-associated repeat units in members of the genus Chironomus evolve from sequences similar to simple telomeric repeats. Mol Cell Biol 1993; 13(3):1583-1589.
19. Fuchs J, Brandes A, Schubert I. Telomere sequence localization and karyotype evolution in higher plants. Pl Syst Evol 1995; 196:227-241.
20. Sahara K, Marec F, Traut W. TTAGG telomeric repeats in chromosomes of some insects and other arthropods. Chromosome Res 1999; 7(6):449-460.
21. Frydrychova R, Marec F. Repeated losses of TTAGG telomere repeats in evolution of beetles (Coleoptera). Genetica 2002; 115(2):179-187.
22. Vitkova M, Kral J, Traut W et al. The evolutionary origin of insect telomeric repeats, (TTAGG)n. Chromosome Res 2005; 13(2):145-156.
23. Sykorova E, Lim KY, Chase MW et al. The absence of Arabidopsis-type telomeres in Cestrum and closely related genera Vestia and Sessea (Solanaceae): first evidence from eudicots. Plant J 2003; 34(3):283-291.
24. Adams SP, Hartman TP, Lim KY et al. Loss and recovery of Arabidopsis-type telomere repeat sequences 5′-(TTTAGGG)(n)-3′ in the evolution of a major radiation of flowering plants. Proc Biol Sci 2001; 268(1476):1541-1546.
25. Sykorova E, Lim KY, Kunicka Z et al. Telomere variability in the monocotyledonous plant order Asparagales. Proc R Soc Lond B Biol Sci 2003; 270(1527):1893-1904.
26. Sykorova E, Fajkus J, Meznikova M et al. Minisatellite telomeres occur in the family Alliaceae but are lost in Allium. Am J Bot 2006; 93(6):814-823.
27. Lejnine S, Makarov VL, Langmore JP. Conserved nucleoprotein structure at the ends of vertebrate and invertebrate chromosomes. Proc Natl Acad Sci USA 1995; 92(6):2393-2397.
28. Makarov VL, Lejnine S, Bedoyan J et al. Nucleosomal Organization of Telomere-Specific Chromatin in Rat. Cell 1993; 73(4):775-787.
29. Tommerup H, Dousmanis A, Delange T. Unusual chromatin in human telomeres. Mol Cell Biol 1994; 14(9):5777-5785.
30. Fajkus J, Kovarik A, Kralovics R et al. Organization of telomeric and subtelomeric chromatin in the higher-plant nicotiana tabacum. Mol Gen Genet 1995; 247(5):633-638.
31. Rossetti L, Cacchione S, Fua M et al. Nucleosome assembly on telomeric sequences. Biochemistry 1998; 37(19):6727-6737.

32. Fajkus J, Trifonov EN. Columnar packing of telomeric nucleosomes. Biochem Biophys Res Commun 2001; 280(4):961-963.
33. Mechelli R, Anselmi C, Cacchione S et al. Organization of telomeric nucleosomes: atomic force microscopy imaging and theoretical modeling. FEBS Lett 2004; 566(1-3):131-135.
34. Pisano S, Pascucci E, Cacchione S et al. AFM imaging and theoretical modeling studies of sequence-dependent nucleosome positioning. Biophys Chem 2006; 124(2):81-89.
35. Nikitina T, Woodcock CL. Closed chromatin loops at the ends of chromosomes. J Cell Biol 2004; 166(2):161-165.
36. Rossetti L, Cacchione S, De Menna A et al. Specific interactions of the telomeric protein Rap1p with nucleosomal binding sites. J Mol Biol 2001; 306(5):903-913.
37. Galati A, Rossetti L, Pisano S et al. The human telomeric protein TRF1 specifically recognizes nucleosomal binding sites and alters nucleosome structure. J Mol Biol 2006; 360(2):377-385.
38. Zalenskaya IA, Bradbury EM, Zalensky AO. Chromatin structure of telomere domain in human sperm. Biochem Biophys Res Commun 2000; 279(1):213-218.
39. Zalensky AO, Tomilin NV, Zalenskaya IA et al. Telomere-telomere interactions and candidate telomere binding protein(s) in mammalian sperm cells. Exp Cell Res 1997; 232(1):29-41.
40. Solov'eva L, Svetlova M, Bodinski D et al. Nature of telomere dimers and chromosome looping in human spermatozoa. Chromosome Res 2004; 12(8):817-823.
41. Koukalova B, Reich J, Matyasek R et al. A bamhi family of highly repeated DNA-sequences of nicotiana tabacum. Theor Appl Genet 1989; 78(1):77-80.
42. Fajkus J, Kralovics R, Kovarik A et al. The telomeric sequence is directly attached to the Hrs60 subtelomeric tandem repeat in tobacco chromosomes. FEBS Lett 1995; 364(1):33-35.
43. Lim KY, Kovarik A, Matyasek R et al. Comparative genomics and repetitive sequence divergence in the species of diploid nicotiana section alatae. Plant J 2006; 48(6):907-919.
44. Ganal MW, Lapitan NLV, Tanksley SD. Macrostructure of the tomato telomeres. Plant Cell 1991; 3(1):87-94.
45. Buzek J, Koutnikova H, Houben A et al. Isolation and characterization of X chromosome-derived DNA sequences from a dioecious plant Melandrium album. Chromosome Res 1997; 5(1):57-65.
46. Copenhaver GP, Pikaard CS. RFLP and physical mapping with an rDNA-specific endonuclease reveals that nucleolus organizer regions of Arabidopsis thaliana adjoin the telomeres on chromosomes 2 and 4. Plant J 1996; 9(2):259-272.
47. Vershinin AV, Heslop-Harrison JS. Comparative analysis of the nucleosomal structure of rye, wheat and their relatives. Plant Mol Biol 1998; 36(1):149-161.
48. Sykorova E, Fajkus J, Ito M et al. Transition between two forms of heterochromatin at plant subtelomeres. Chromosome Res 2001; 9(4):309-323.
49. Sykorova E, Cartagena J, Horakova M et al. Characterization of telomere-subtelomere junctions in Silene latifolia. Mol Genet Genom 2003; 269(1):13-20.
50. Alkhimova OG, Mazurok NA, Potapova TA et al. Diverse patterns of the tandem repeats organization in rye chromosomes. Chromosoma 2004; 113(1):42-52.
51. Shakirov EV, Shippen DE. Length regulation and dynamics of individual telomere tracts in wild-type Arabidopsis. Plant Cell 2004; 16(8):1959-1967.
52. Maillet G, White CI, Gallego ME. Telomere-length regulation in inter-ecotype crosses of Arabidopsis. Plant Mol Biol 2006; 62(6):859-866.
53. Nosek J, Kosa P, Tomaska L. On the origin of telomeres: a glimpse at the pre-telomerase world. Bioessays 2006; 28(2):182-190.
54. Chen Q, Ijpma A, Greider CW. Two survivor pathways that allow growth in the absence of telomerase are generated by distinct telomere recombination events. Mol Cell Biol 2001; 21(5):1819-1827.
55. Pich U, Fuchs J, Schubert I. How do Alliaceae stabilize their chromosome ends in the absence of TTTAGGG sequences? Chromosome Res 1996; 4(3):207-213.
56. Sykorova E, Lim KY, Fajkus J et al. The signature of the cestrum genome suggests an evolutionary response to the loss of (TTTAGGG)n telomeres. Chromosoma 2003; 112(4):164-172.
57. Carmona MJ, Morcillo G, Galler R et al. Cloning and molecular characterization of a telomeric sequence from a temperature-induced balbiani ring. Chromosoma 1985; 92(2):108-115.
58. Lopez CC, Rodriguez E, Diez JL et al. Histochemical localization of reverse transcriptase in polytene chromosomes of chironomids. Chromosoma 1999; 108(5):302-307.
59. Rosen M, Edstrom J. DNA structures common for chironomid telomeres terminating with complex repeats. Insect Mol Biol 2000; 9(3):341-347.
60. Tomaska L, Nosek J, Makhov AM et al. Extragenomic double-stranded DNA circles in yeast with linear mitochondrial genomes: potential involvement in telomere maintenance. Nucleic Acids Res 2000; 28(22):4479-4487.

61. Nosek J, Rycovska A, Makhov AM et al. Amplification of telomeric arrays via rolling-circle mechanism. J Biol Chem 2005; 280(11):10840-10845.
62. Martienssen RA. Maintenance of heterochromatin by RNA interference of tandem repeats. Nature Genet 2003; 35(3):213-214.
63. Sandell LL, Zakian VA. Telomeric position effect in yeast. Trends Cell Biol 1992; 2(1):10-14.
64. Park Y, Lustig AJ. Telomere structure regulates the heritability of repressed subtelomeric chromatin in Saccharomyces cerevisiae. Genetics 2000; 154(2):587-598.
65. de Bruin D, Kantrow SM, Liberatore RA et al. Telomere folding is required for the stable maintenance of telomere position effects in yeast. Mol Cell Biol 2000; 20(21):7991-8000.
66. Shay JW. Molecular Mechanisms Initiating Cellular Senescence. Paper presented at: Telomeres and Genome Stability; 2006; Villars-sur-Ollon, Switzerland.
67. Baur JA, Zou Y, Shay JW et al. Telomere position effect in human cells. Science 2001; 292(5524):2075-2077.
68. Kanoh J, Sadaie M, Urano T et al. Telomere binding protein Taz1 establishes Swi6 heterochromatin independently of RNAi at telomeres. Curr Biol 2005; 15(20):1808-1819.
69. Bailis JM, Forsburg SL. RNAi hushes heterochromatin. Genome Biol 2002; 3(12): reviews 1035.1–1035.4.
70. Watson JD. Origin of concatemeric T7 DNA. Nat New Biol 1972; 239(94):197-201.
71. George JA, DeBaryshe PG, Traverse KL et al. Genomic organization of the Drosophila telomere retrotransposable elements. Genome Res 2006; 16(10):1231-1240.
72. Weiss H, Scherthan H. Aloe spp—plants with vertebrate-like telomeric sequences. Chromosome Res 2002; 10(2):155-164.
73. Weiss-Schneeweiss H, Říha K, Jang CG et al. Chromosome termini of the monocot plant Othocallis siberica are maintained by telomerase, which specifically synthesises vertebrate-type telomere sequences. Plant J 2004; 37(4):484-493.
74. Rotkova G, Sklenickova M, Dvorackova M et al. An evolutionary change in telomere sequence motif within the plant section Asparagales had significance for telomere nucleoprotein complexes. Cytogenet Genome Res 2004; 107(1-2):132-138.
75. Petcherskaia M, McGuire JM, Pherson JM et al. Loss of cap structure causes mitotic defect in Tetrahymena thermophila telomerase mutants. Chromosoma 2003; 111(7):429-437.
76. Lin J, Smith DL, Blackburn EH. Mutant telomere sequences lead to impaired chromosome separation and a unique checkpoint response. Mol Biol Cell 2004; 15(4):1623-1634.
77. Blackburn EH. Telomeres and telomerase: their mechanisms of action and the effects of altering their functions. FEBS Lett 2005; 579(4):859-862.
78. Fajkus J, Sykorova E, Leitch AR. Telomeres in evolution and evolution of telomeres. Chromosome Res 2005; 13(5):469-479.
79. Barnes SR, James AM, Jamieson G. The organisation, nucleotide sequence and chromosomal distribution of satellite DNA from Allium cepa. Chromosoma 1985; 92:185-192.
80. Pich U, Fritsch RM, Schubert I. Closely related Allium species (Alliaceae) share a very similar satellite sequence. Pl Syst Evol 1996; 202:255-264.
81. Pearce SR, Pich U, Harrison G et al. The Ty1-copia group retrotransposons of Allium cepa are distributed throughout the chromosomes but are enriched in the terminal heterochromatin. Chromosome Res 1996; 4(5):357-364.
82. Schubert I, Wobus U. Insitu hybridization confirms jumping nucleolus organizing regions in allium. Chromosoma 1985; 92(2):143-148.
83. Frydrychova R, Grossmann P, Trubac P et al. Phylogenetic distribution of TTAGG telomeric repeats in insects. Genome 2004; 47(1):163-178.
84. Ware TL, Wang H, Blackburn EH. Three telomerases with completely non-telomeric template replacements are catalytically active. EMBO J 2000; 19(12):3119-3131.
85. McCormick-Graham M, Romero DP. A single telomerase RNA is sufficient for the synthesis of variable telomeric DNA repeats in ciliates of the genus Paramecium. Mol Cell Biol 1996; 16(4):1871-1879.
86. Ye AJ, Haynes WJ, Romero DP. Expression of mutated Paramecium telomerase RNAs in vivo leads to templating errors that resemble those made by retroviral reverse transcriptase. Mol Cell Biol 1999; 19(4):2887-2894.
87. Melek M, Davis BT, Shippen DE. Oligonucleotides complementary to the Oxytricha nova telomerase RNA delineate the template domain and uncover a novel mode of primer utilization. Mol Cell Biol 1994; 14(12):7827-7838.
88. Jacobs SA, Podell ER, Cech TR. Crystal structure of the essential N-terminal domain of telomerase reverse transcriptase. Nat Struct Mol Biol 2006; 13(3):218-225.
89. Emery HS, Weiner AM. An irregular satellite sequence is found at the termini of the linear extrachromosomal rDNA in Dictyostelium discoideum. Cell 1981; 26(3 Pt 1):411-419.

90. Schaap P, Winckler T, Nelson M et al. Molecular phylogeny and evolution of morphology in the social amoebas. Science 2006; 314(5799):661-663.
91. Eichinger L, Pachebat JA, Glockner G et al. The genome of the social amoeba Dictyostelium discoideum. Nature 2005; 435(7038):43-57.
92. Loftus B, Anderson I, Davies R et al. The genome of the protist parasite Entamoeba histolytica. Nature 2005; 433(7028):865-868.
93. Peng JC, Karpen GH. H3K9 methylation and RNA interference regulate nucleolar organization and repeated DNA stability. Nat Cell Biol 2007; 9(1):25-35.
94. Heo SJ, Tatebayashi K, Ohsugi I et al. Bloom's syndrome gene suppresses premature ageing caused by Sgs1 deficiency in yeast. Genes Cells 1999; 4(11):619-625.
95. Pich U, Schubert I. Terminal heterochromatin and alternative telomeric sequences in Allium cepa. Chromosome Res 1998; 6(4):315-321.
96. Pontes O, Neves N, Silva M et al. Chromosomal locus rearrangements are a rapid response to formation of the allotetraploid Arabidopsis suecica genome. Proc Natl Acad Sci USA 2004; 101(52):18240-18245.
97. Skalicka K, Lim KY, Matyasek R et al. Rapid evolution of parental rDNA in a synthetic tobacco allotetraploid line. Am J Bot 2003; 90(7):988-996.
98. Zhang DM, Sang T. Physical mapping of ribosomal RNA genes in peonies (paeonia, paeoniaceae) by fluorescent in situ hybridization: implications for phylogeny and concerted evolution. Am J Bot 1999; 86(5):735-740.
99. Adam RD, Nash TE, Wellems TE. Telomeric location of Giardia rDNA genes. Mol Cell Biol 1991; 11(6):3326-3330.
100. Zauner S, Fraunholz M, Wastl J et al. Chloroplast protein and centrosomal genes, a tRNA intron and odd telomeres in an unusually compact eukaryotic genome, the cryptomonad nucleomorph. Proc Natl Acad Sci USA 2000; 97(1):200-205.
101. Jacob NK, Stout AR, Price CM. Modulation of telomere length dynamics by the subtelomeric region of Tetrahymena telomeres. Mol Biol Cell 2004; 15(8):3719-3728.
102. Watson JM, Bulankova P, Riha K et al. Telomerase-independent cell survival in Arabidopsis thaliana. Plant J 2005; 43(5):662-674.
103. Hug N, Lingner J. Telomere length homeostasis. Chromosoma 2006; 115(6):413-425.
104. Rabl C. Über Zelltheilung. Morpholgische Jahrubucke 1885; 10:214-330.
105. Leitch AR. Higher levels of organization in the interphase nucleus of cycling and differentiated cells. Microbiol Mol Biol Rev 2000; 64(1):138-152.
106. Lichter P, Cremer T, Borden J et al. Delineation of individual human chromosomes in metaphase and interphase cells by in situ suppression hybridization using recombinant DNA libraries. Hum Genet 1988; 80(3):224-234.
107. Ferguson M, Ward DC. Cell cycle dependent chromosomal movement in pre-mitotic human T-lymphocyte nuclei. Chromosoma 1992; 101(9):557-565.
108. Manuelidis L, Borden J. Reproducible compartmentalization of individual chromosome domains in human CNS cells revealed by in situ hybridization and three-dimensional reconstruction. Chromosoma 1988; 96(6):397-410.
109. Diez JL, Vilarino VR, Medina FJ et al. Nucleolar localization of a reverse transcriptase related to telomere maintenance in Chironomus (Diptera). Histochem Cell Biol 2006; 126(4):445-452.
110. Yang Y, Chen Y, Zhang C et al. Nucleolar localization of hTERT protein is associated with telomerase function. Exp Cell Res 2002; 277(2):201-209.
111. Zhang S, Hemmerich P, Grosse F. Nucleolar localization of the human telomeric repeat binding factor 2 (TRF2). J Cell Sci 2004; 117(Pt 17):3935-3945.
112. Smogorzewska A, de Lange T. Different telomere damage signaling pathways in human and mouse cells. EMBO J 2002; 21(16):4338-4348.
113. Slijepcevic P. The role of DNA damage response proteins at telomeres—an "integrative" model. DNA Repair (Amst) 2006; 5(11):1299-1306.
114. Kobayashi T, Ganley AR. Recombination regulation by transcription-induced cohesin dissociation in rDNA repeats. Science 2005; 309(5740):1581-1584.
115. Suzuki K. Characterization of telomere DNA among five species of pteridophytes and bryophytes. J Bryol 2004; 26:175-180.
116. Fitzgerald MS et al. Characterization and developmental patterns of telomerase expression in plants. Proc Natl Acad Sci USA 1996; 93(25):14422-14427.
117. Fitzgerald MS, Shakirov EV, Hood EE et al. Different modes of de novo telomere formation by plant telomerases. Plant J 2001; 26(1):77-87.
118. Fajkus J, Kovarik A, Kralovics R. Telomerase activity in plant cells. FEBS Lett 1996; 391(3):307-309.
119. Flanary BE, Kletetschka G. Analysis of telomere length and telomerase activity in tree species of various life-spans and with age in the bristlecone pine Pinus longaeva. Biogerontology 2005; 6(2):101-111.

120. Tuskan GA, Difazio S, Jansson S et al. The genome of black cottonwood, Populus trichocarpa (Torr. & Gray). Science 2006; 313(5793):1596-1604.
121. Nakamura TM, Morin GB, Chapman KB et al. Telomerase catalytic subunit homologs from fission yeast and human. Science 1997; 277(5328):955-959.
122. Malik HS, Burke WD, Eickbush TH. Putative telomerase catalytic subunits from Giardia lamblia and Caenorhabditis elegans. Gene 2000; 251(2):101-108.
123. Xia J, Peng Y, Mian IS et al. Identification of functionally important domains in the N-terminal region of telomerase reverse transcriptase. Mol Cell Biol 2000; 20(14):5196-5207.

Chapter 9

Telomere Position Effect and the Evolution of the Genome

Frederique Magdinier, Alexandre Ottaviani and Eric Gilson*

Abstract

It is now well known that chromatin structure and subnuclear environment play universal roles in the regulation of gene transcription and any type of DNA transactions, including the 3Rs (replication, recombination and repair). Most telomeres and subtelomeres from *Saccharomyces cerevisiae* to *Homo sapiens* repress neighboring genes by a mechanism named Telomere Position Effect (TPE). Such a widespread conservation among eukaryotes suggests a fundamental role of TPE for genome function.

Introduction

The organization of genomic DNA has greatly evolved from unicellular to complex organisms. Despite the increasing complexity of the information carried by the DNA sequence throughout evolution and subsequently, the chromatin architecture that controls the folding of the genome within the nucleus, shared mechanisms orchestrate the epigenetic regulation of chromatin compaction and its influence on gene expression. Eukaryotic genomes are organized into two distinct and interconvertible states, euchromatin and heterochromatin. Each chromatin state can be defined by the compaction and sensitivity to enzymatic digestion of the chromatin fiber, its histone code and the covalent modification of the underlying DNA.[1]

Heterochromatin was originally defined as a portion of the genome deeply stained from metaphase to interphase associated with the pericentric regions, telomeres and some interstitial domains. In higher eukaryotes, constitutive heterochromatin is enriched in methylated DNA, histone H3K9 methylation and HP1 binding and can spread over genomic regions inducing thereby the silencing of other sequences. A classical example of silencing is known as position effect variegation (PEV) and occurs when a gene is juxtaposed to heterochromatin.[2,3]

The nature of telomeric chromatin differs from global constitutive heterochromatin due to the specificity of its DNA sequence, the particular spacing of the nucleosomes and the binding of specific factors.[4-6] However, despite these structural differences, most telomeres from *Saccharomyces cerevisiae* to *Homo sapiens* repress neighboring genes by a mechanism named Telomere position effect (TPE). Such a widespread conservation of telomeric silencing among eukaryotes suggests that it is fundamental for telomere function and consequently telomere maintenance.

TPE and Chromatin Architecture

TPE was first discovered in *Drosophila melanogaster*[7-9] but has been extensively investigated in baker's yeast.[10,11] Unlike *Drosophila*, telomeres in *Saccharomyces cerevisiae* are constituted of stretches

*Corresponding Author: Eric Gilson—Laboratoire de Biologie Moléculaire de la Cellule, CNRS UMR5239, Ecole Normale Supérieure de Lyon, UCBL1, IFR128, 46 allée d'Italie, 69364 Lyon Cedex 07, France. Email: Eric.Gilson@ens-lyon.fr

Origin and Evolution of Telomeres, edited by Jozef Nosek and Ľubomír Tomáška.

of highly repetitive telomerase-added repeats and resemble most of the eukaryotic telomeres therefore constituting a powerful genetic system for the understanding of TPE.

Experiments in *S. cerevisiae* allowed the identification of more than 50 proteins that can modulate this epigenetic mechanism. However, only a few of them are strictly necessary for telomeric silencing: the Sir-complex (Sir2p, Sir3p and Sir4p for Silent Information Regulators),[12] the Ku heterodimer components (yKu70p and yKu80p),[13,14] and the C-terminal domain of Rap1p.[15] In the current model for TPE establishment, Rap1p and Ku are responsible for the initiation/nucleation step and linear spreading. They recruit the Sir-proteins at the telomere and thereby initiate the formation of heterochromatin that will propagate toward the centromere through the interaction of the Sir-complex with histone tails.[16-19] This spreading requires the Sir2p NAD-dependent histone deacetylase activity that permits Sir3p and Sir4p binding[20,21] and distributes along the DNA sequence by sequential deacetylation of the histone tails and binding (Fig. 1A).

The posttranslational modifications of the core histone (H2A, H2B, H3 and H4) provide heritable information that orchestrates transcription, replication and chromatin organization through cell divisions. TPE is induced by such modifications and the first clues were given by mutations in histone H4 N-terminal residues that suppress TPE in baker's yeast.[12] Furthermore, deletions of N-terminal tails of histones H3 and H2A also impede telomeric silencing[22-24] by altering Sir complex recruitment and histone acetylation that plays a preponderant role in TPE modulation.[19,25] Moreover, TPE induced by heterochromatin spreading is dependent upon Sir2p HDAC activity[21] and can be counteracted by Sas2p-dependent acetylation of H4-K16.[26] The same type of anti-silencing effect has recently been reported for the H2A variant, H2A.Z, enriched at subtelomeric regions[27,28] and for the acetyl-transferase, Nu4A.[29] In conclusion, acetylation seems to act as an important boundary against the propagation of telomeric silencing. To the present knowledge in *S. cerevisiae*, histone methylation only affects some lysines on H3: Lys4, Lys36 and Lys79[30] that are methylated by specific methyltransferases[31] and can also be demethylated[32] except for Lys79, located within the histone core. Methylation of these residues displays anti-silencing properties mainly by preventing Sir complex association to histone H3 tail.[33] Noteworthy, deletion mutants of *Set1p* and *Dot1p* exhibit reduction in telomeric silencing level due to the limiting amount of Sir3p that is displaced toward euchromatic regions.

The effect of a particular modification might be modulated by another adjacent modification and the combinatory use of different information gives enormous potential for the variability of the biological response. For example H2B ubiquitination is necessary for H3-K4 and H3-K79 methylation, which prevents Sir4p binding, whereas Sir4p recruits Ubp10p/Dot4p that deubiquitinate H2B and Sir2p that deacetylates H4-K16, both favoring establishment of telomeric silencing.[34]

In other organisms, most of the few factors mediating TPE identified so far are functional homologs of *S. cerevisiae* proteins. In *Schizosaccharomyces pombe*, the telomeric repeats-binding protein Taz1p recruits spRap1 (homologous to Rap1p in *S. cerevisiae*)[35,36] at the telomere and both are required for TPE[37,38] (Fig. 1B). Nevertheless in this model, no link could be established between Ku and TPE. Although they have no homolog in *S. cerevisiae*, HP1 proteins are involved in TPE in *Drosophila*[39] and fission yeast[36,40] where they could play similar roles of the budding yeast Sir3p and Sir4p in the spreading of heterochromatin. In *S. pombe*, the chromatin changes associated with the TPE machinery are more closely related to that of *Drosophila* and mammals. Firstly, it requires the methylation of H3K9 residues by the Clr4 methyltransferase[41] allowing the binding of Swi6 (the ortholog of HP1). Secondly, specialized repeats (*dg* and *dh*) present at subtelomeres and heterochromatin *loci* contribute to telomeric silencing through the formation of a RNAi-induced transcriptional silencing complex (RITS).[42] Recently, a new complex termed SHREC and composed of Clr1, Clr2, the Clr3 histone deacetylase and the SNF2 chromatin-remodeling factor homolog Mit1 has been identified.[43] At telomeres, SHREC interacts with Ccq1 that along with Taz1 and Swi6/HP1 act in parallel to the RNAi pathway to restrict polymerase II and stabilize heterochromatin (Fig. 1B).

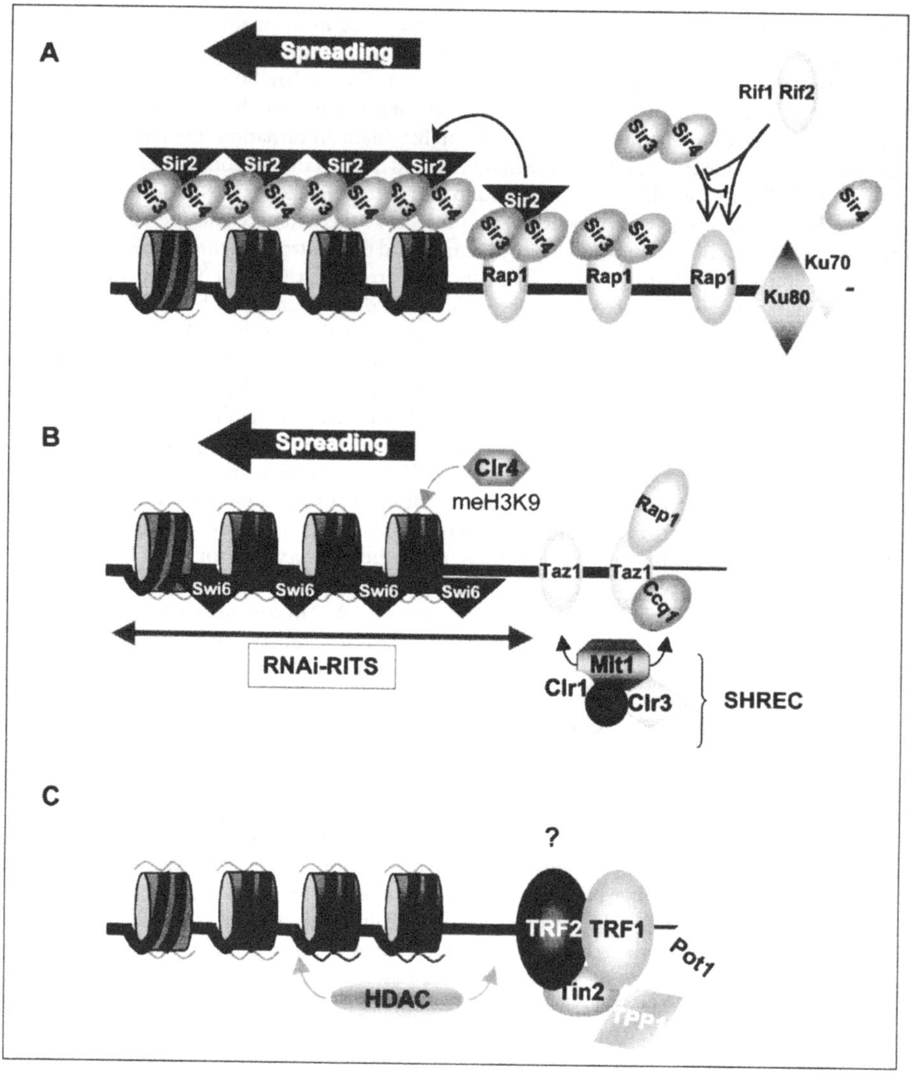

Figure 1. Implication of the telomeres and subtelomeres in the regulation of telomere position effect. A) In *Saccharomyces cerevisiae*, telomere position effect is regulated by proteins bound to the telomere and the subtelomeres that allow the spreading of silent chromatin toward the centromeres. At natural telomeres, the Sir complex is recruited at telomeres by Rap1 and the Sir2 histone deacetylase sequentially removes the acetyl residues along the subtelomeres. B) In *Schizosaccharomyces pombe*, the telomeric protein Taz1 and methylation of histone H3 K9 residues by the Clr4 histone methyltransferase recruit Swi6 to telomeric sequences and spread silencing toward the centromere to cover the subtelomeric region over 45-75 kb in cooperation with the RNAi-RITS machinery. The SHREC complex containing the Clr3 histone deacetylase and the Mit1 chromatin remodeling factor associates with Ccq1 and Swi6 and cooperates with the Taz1 and RITS pathway to facilitate chromatin condensation and telomeric silencing. C) In Human cells, the telomere position effect may involve the cooperation of telomere binding proteins such as TRF1 and classical chromatin remodeling factors such as class I or II histone deacetylases. The identity of the subtelomeric regions might influence TPE and explain different pathologies associated with the rearrangement of these regions.

Telomere Position Effect and Nuclear Periphery: The Reservoir Model

In budding yeast, the 32 telomeres are clustered into 4-6 *foci*, primarily associated with the nuclear envelope.[44] This peripheral localization of telomeres is dependent on redundant pathways.[45] One acts through Ku and the second through Sir4.[46,47] Neither Ku nor Sir4 possess a transmembrane domain and their recruitment to the nuclear envelope depends upon other factors. Sir4 is tethered to the nuclear periphery through interaction with Esc1, which is localized at the edge of the nucleus between nuclear pores independently of silent chromatin.[46] The anchoring of Ku also depends upon the presence of Esc1 but only during S phase.[46] The factors responsible for the peripheral localization of Ku during G1 have not yet been characterized.

Mating-type silencers require telomere proximity to be fully functional[48,49] and tethering of a silencer-flanked reporter to the nuclear periphery facilitates its repression.[50] However, relocation to this peripheral nuclear compartment probably does not cause repression per se[51] and silencing can be maintained without perinuclear anchoring.[52] All of these data, together with microscopy analyses, converge toward a reservoir model where telomere clusters act as a subnuclear compartment concentrating key silencing factors like the Sir proteins[48] (Fig. 2A). Although the perinuclear location is not strictly required to maintain the silent state, silencers and telomeres would need to be somehow associated to this compartment to be in a local microenvironment containing enough silencing factors to shut-down gene expression. In agreement with this model, close interactions between *HML* silencers and telomeres have been observed by using a methyltransferase targeting assay.[53]

Consistent with previous work showing that insulator activity in various species correlates with a particular spatial arrangement in the nucleus,[54-57] the boundary elements named STAR (for SubTelomeric Antisilencing Regions) antagonize the peripheral localization of telomeres, suggesting a mechanism by which they block TPE spreading.[58] Therefore, the individual organization of native subtelomeres is likely to directly influence telomere subnuclear localization and TPE capacities.

Nuclear periphery can also confer an optimal environment for transcriptional activation through the binding of genes to the nuclear pore complex.[59] Promoting a stronger association with the nuclear envelope improves silencing in non-inductive conditions but increases expression in inductive conditions, accounting for an ambivalent role of nuclear rim association.[60] Since telomeres are mainly localized between pores,[46] one can predict that silent telomeres and activated genes distribute to different subdomains at the nuclear envelope (Fig. 2A).

Modulation by the Subtelomere: Importance of the Telomere Identity

The strength of TPE varies widely between chromosome ends.[61] Indeed, in diploid cells, each of the 32 yeast chromosomes has a different composition in subtelomeric elements that can modulate TPE and polymorphisms in these regions are also found in the different strains.[62,63] In *S.cerevisiae*, these modulators include two types of subtelomeric repeats: X and Y' (Fig. 3).

The X element contains a "core X" element consisting in an ARS consensus sequence (ACS) and an Abf1p site present at all yeast telomeres.[63] This core X sequence behaves as a protosilencer, i.e., it does not act as a silencer by itself but reinforces silencing when located in the proximity of a master silencer.[64,65] At some chromosome ends, the core X element can be associated to the STAR boundary elements consisting of binding sites for Tbf1p and Reb1p.[66]

The Y' element is found on 50 to 70% of yeast telomeres, in 1 to 4 copies inserted between the X element and the telomere.[67] It contains two open reading frames (ORF), an ARS that can also bind the ORC complex and a STAR element immediately abutting the telomeric repeats, which similarly to the X STAR sequences consists of Tbf1 binding sites.[66]

Different combinations of STAR and protosilencer at native telomeres are likely to contribute to their respective behaviors with regard to TPE. In general terms, X-only telomeres usually exhibit a strong silencing[61] even if two of them do not seem to propagate silencing or bind Rap1p.[68] Mutations in the "core X" that impair Abf1p binding or the X-ACS sequence recruitment of ORC strongly reduce telomeric silencing at XI-L chromosome.[61] Sir1p, that does not seem to have any effect on truncated telomeres,[12] participates in silencing at the XI-L telomere[61,66] possibly through

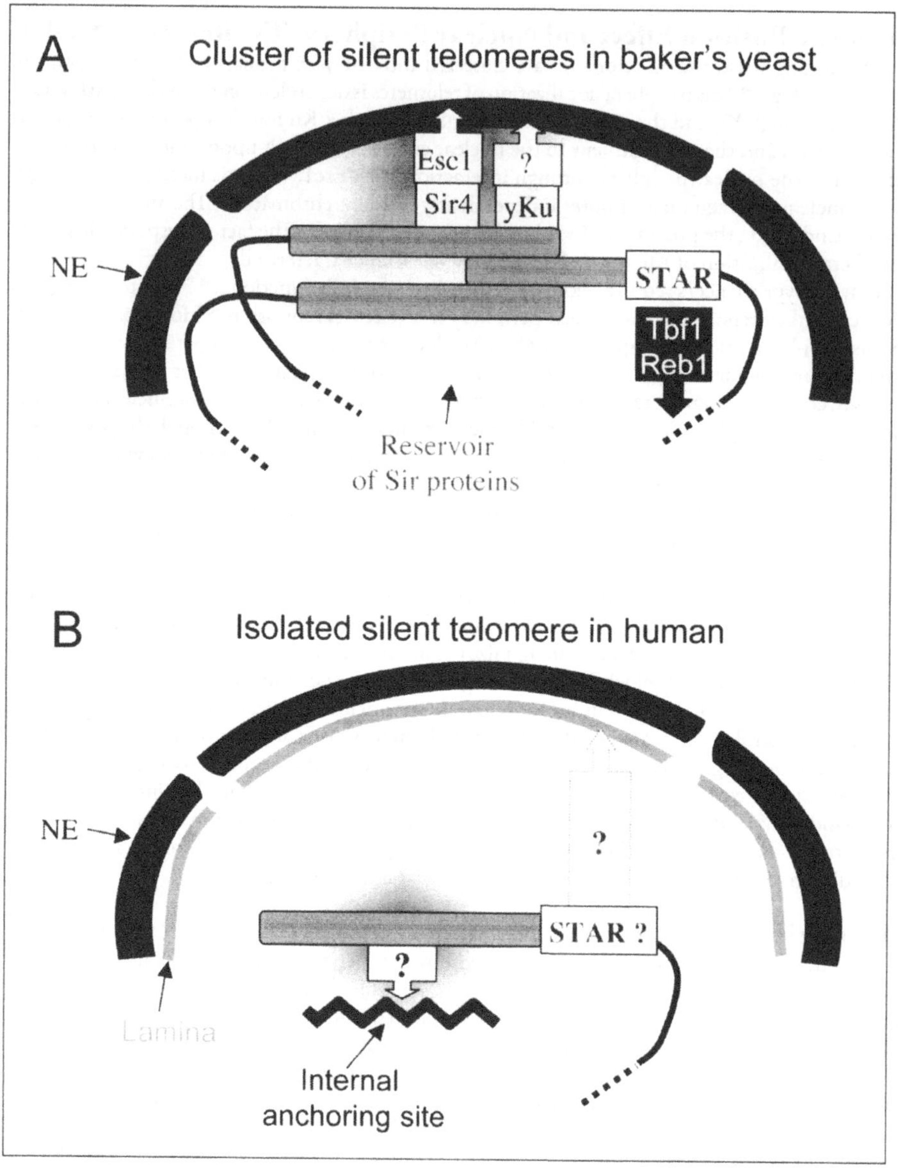

Figure 2. Telomeric position and the subnuclear organization in yeast and human cells. A) In baker's yeast, the nuclear periphery acts as a subnuclear compartment where telomeres are clustered in order to recruit and bind key heterochromatic factors. The Sir4 and yKu proteins are redundant anchors for telomeres at the nuclear envelope (NE). Sir4 binds the NE protein Esc1 while the partner of Ku for anchoring is unknown. The telomere-NE association is antagonized by subtelomeric insulator elements called STAR, linking subnuclear localization and transcriptional insulation. B) In human cells, most of the telomeres are localized within the nucleoplasm, probably anchored at defined subnuclear sites. These sites might serve to nucleate heterochromatin at telomeres. Some telomeres, like the 4q, is found associated to the periphery. The peripheral localization is caused by specific subtelomeric elements, which might be the functional homologs of yeast STAR.

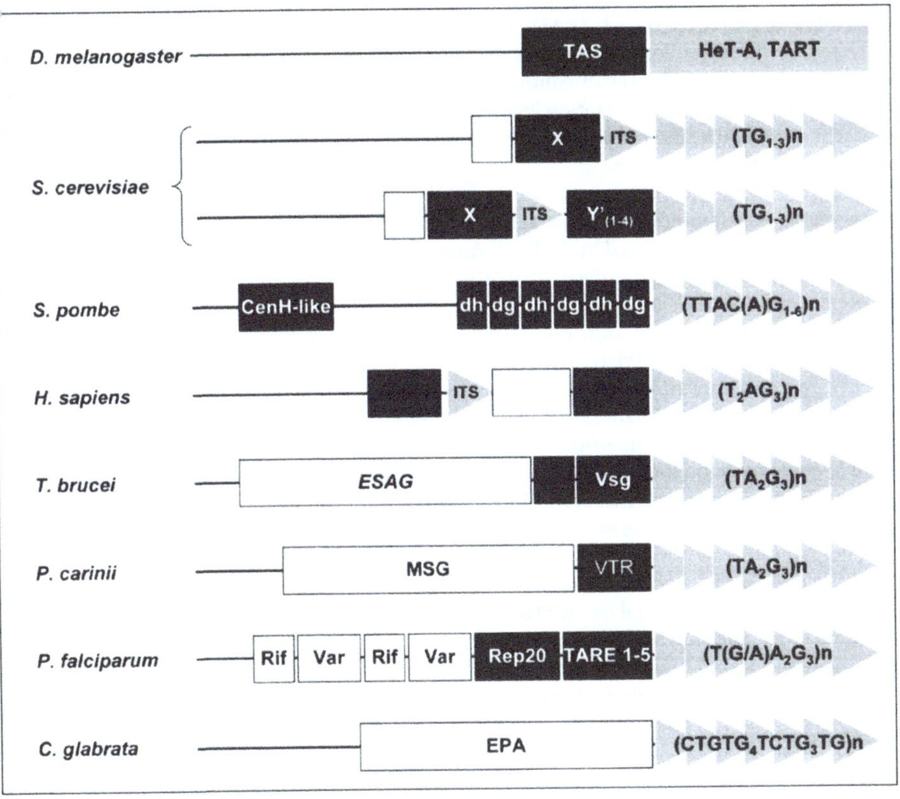

Figure 3. Details of telomere regions of various eukaryotic organisms exhibiting telomere position effect. In eukaryotes, the simple repeats that constitute the telomeric DNA by the telomerase enzyme are represented by triangles and the specific sequence is indicated for every organism. The end of chromosomes of *D. melanogaster* is not synthesized by a telomerase enzyme but is formed through the retrotransposition of the nonLTR retroptransposons *HeT-A* and *TART* (grey box). The subtelomeres are patchworks of genes (white boxes) interspersed within repeated elements (black boxes). In *Saccharomyces cerevisiae* and human, large polymorphic blocks of repeated sequences are distributed between the different chromosomes and subtelomeres contain genes. VTR, Variable Tandem Repeat. TAS, Telomere Associated Repeat. ITS, Intersititial Telomeric Repeats. ESAG, *Trypanosoma brucei* Expression Site Associated Gene; Vsg, variant coat protein. MSG, *Pseudomonas carinii* surface glycoprotein gene. Var, Rif, *Plasmodium falciparum* subtelomeric genes; *Rep20, TARE 1-5*, subtelomeric repeats. EPA, *Candida glabrata* adhesin gene. Dh, dg, *Schizosaccharomyces pombe* subtelomeric repeats.

interaction with the ORC complex.[69,70] The Y' element possesses anti-silencing properties linked to the boundary activity of the STAR sequence and limits the extend of spreading toward the subtelomeric regions.[66] On truncated VII-L, the level of telomeric silencing decreases exponentially and continuously with distance, but on some native telomeres, combination of X and Y' elements results in variable levels of gene silencing with regions protected from TPE by the STAR boundaries and resumed silencing at a distance from the telomere by protosilencers suggesting that telomere position effect can be a discontinuous mechanism relayed from place to place by silencers and protosilencers.[61,66] Such a discontinuity argues in favor of a coalescence model in which silencers and protosilencers form interactions leading to an apparent spreading of silent chromatin that might be further facilitated by nuclear positioning.

Biological Functions of Yeast TPE

Determining the biological relevance of TPE is difficult due to its link to the composition of individual telomeres and the discontinuity in regions of genes exposed to telomeric silencing. Genome-wide studies of transcription levels in *S. cerevisiae* could determine that about 267 genes located less than 20 kb from the telomeres display a mean expression level that is roughly 20% of nontelomeric genes expression.[71]

Recently, a genome comparison of closely related Hemiascomycetes species revealed that subtelomeric gene families are in general specific for the different species.[72] However, the expression of these genes is influenced by TPE.[73,74] A subtelomeric enrichment of genes related to stress response and metabolism in non-optimal growth conditions appears to be a conserved feature in several yeast species.[75] Most of these genes are silenced under optimal growth conditions. In budding yeast, different types of stress-like nutrient starvation, heat shock or chemical treatment can induce a hyperphosphorylation of Sir3p and a consequent decrease in TPE at the truncated VII-L.[76] This also leads to an increase in the expression of natural subtelomeric genes such as the *PAU* genes involved in cell wall constitution and drug resistence.[77] Interestingly, genes of cell wall proteins are also submitted to Sir-dependent TPE in *Candida glabrata*.[78]

In *S. cerevisiae*, subtelomeric HAST domains also contain clusters of genes involved in gluconeogenesis or stress response that are normally silenced.[75] For example, the *FLO* genes, involved in cellular adherence are silenced in a Sir-independent way which is however dependent upon Sir2p homologs, Hst1p and Hst2p.[79]

In *S. pombe*, many genes involved in response to nitrogen starvation are also clustered in subtelomeric regions and silenced by Hda1p ortholog, Clr3.[80]

Besides their effect on transcriptional regulation, telomeres also exert position effect on others DNA transactions. In budding yeast, the telomeres replicate at the late stage of the S-phase[81,82] mainly because they inactivate or delay most of the origins present in the subtelomeric X and Y' elements.[10,83-86] Reminiscent of TPE, short telomeres can replicate earlier than long telomeres,[87] Sir3 inactivation leads to a premature activation of Y' origins[84] and tethering of Sir proteins near an origin can reset replication timing from early to late S-phase.[88] These results are in favor of a model in which the Sir-mediated silent chromatin emanating from telomeres blocks replication initiation in the subtelomeric regions. However, the telomere position effect on replication timing extends over a distance (~35 kb) that is beyond the 6-8 kb seen for the Sir-dependent gene repression.[61,71] This suggests that Sir-independent chromatin-mediated mechanisms can also contribute to the late activation of telomere-associated origins.

Subtelomeric domains are cold-spots for meiotic recombination in a variety of organisms and are expected to favor unequal crossovers and promote homolog disjunction suggesting that the anti-recombination properties of these regions have been actively selected during evolution because of detrimental effects on chromosome stability. For instance, the crossovers are rare in the telomeric regions of grasshopper chromosomes[89] and the double-strand breaks that initiate meiotic recombination in yeast are quasi-absent from sequences within 50 kb from telomeres in a Sir4-independent manner.[90] However, it is worth noting that meiotic recombination occurs at an elevated rate near some human telomeres and can have both advantageous and pathological consequences in human biology.[91] In mitotic yeast cells, telomere proximity represses the homologous recombination between two internal stretches of telomeric DNA but not between two nontelomeric sequences.[92] Interestingly, similarly to the TPE on meiotic recombination, this anti-recombination effect does not rely on a Sir-dependent mechanism, revealing that Sir and nonSir dependent mechanisms are responsible of various classes of TPE. In addition, the anchoring of the telomeres to the nuclear pores that contributes to the regulation of TPE, is essential for repair of double strand breaks at subtelomeric zones, likely by protecting the chromatin structure.[93]

TPE Is Conserved at the Unusual Telomeres of *Drosophila*

Unlike many organisms, *Drosophila* species lack telomerase but maintain their telomeres by the transposition of the retrotransposons *HeTA* and *TART* to chromosome ends[94] (Fig. 3). Proximal to the terminal transposon array, *Drosophila* telomeres carry several kilobases of conserved complex satellites termed Telomere Associated Sequences (TAS) (reviewed in ref. 95). Despite these structural differences, *Drosophila melanogaster* exhibits telomeric silencing as observed when reporter genes are inserted at a telomeric position.[7] All the telomeric transgenes subjected to variegation lie adjacent to TAS suggesting that these sequences also contribute to telomeric silencing as for subtelomeric elements in other organisms.[96] Interestingly, genetic modifiers of position effect variegation (PEV) display little or no effect on TPE suggesting the existence of specialized mechanisms dependent either upon? the Polycomb group genes[97] or HP1.[39,98-100] HP1 is present at all telomeres in *Drosophila* cells and its absence in mutant *Su(var)2-5* cells causes multiple telomere-telomere fusions.[101] Interestingly, chromosome protection and telomeric transcriptional repression are separable mechanisms associated with two types of HP1 binding.[102] HP1 caps the telomere by direct binding while it contributes to telomeric silencing by interacting with trimethylation of Lysine 9 at the histone H3 tail.[102,103] Mutation in the gene for HP1 also increases the abundance of *HeT-A* and *TART* RNA and their frequency of transposition to broken telomeric ends.[104] In contrast, mutations of *Ku80* or *Ku70* strongly increase transposition but do not affect expression of *HeT-A*[105] (Fig. 3). The remarkable capacity of yeast and fly telomeres to uncouple telomere protection from TPE functions might provide a unique ability of rapid adaptation through changes in the subtelomeric transcriptional program and gene shuffling without altering the integrity of the rest of chromosomes.

In addition, *atm* mutations may suppress TPE, by affecting normal telomere chromatin structure and localization of HP1 to telomeres.[106] Mutants of *gpp*, the fly ortholog of the *Saccharomyces cerevisiae Dot1* gene, dominantly suppress silencing by telomeric, but not centromeric heterochromatin by affecting H3K79 methylation suggesting that as observed in yeast, TPE in flies is associated with modifications of the histone code.[107]

In ovaries and oocytes, *HeT-A* and *TART* are partially regulated by the RNAi machinery. Indeed, mutation in the RNA helicase gene, *Spindle-E* (*spnE*) encoding a DEAD-box helicase and the argonaute gene *aubergine* (*aub*) necessary for the assembly of the RNA-induced silencing complex (RISC), increases the rate of *HeT-A* and *TART* transcription and is accompanied by more frequent transpositions to chromosome ends.[108]

Thus, despite their unique and distinct composition and maintenance process, *Drosophila* telomeres share many of the features of telomeres in other species with regard to the silencing at telomeres and subtelomeres.

Telomere Position Effect in Higher Eukaryotes

The heterochromatin nature of mammalian telomeres and their capacity to induce position effect have been controversial for many years. The first example of telomere position effect in vivo came from the analysis of replication timing shifted from early S phase to later time of human chromosome 22 carrying chromosomal abnormality.[109] This delayed replication is not associated with differences in DNA methylation, condensation of the chromatin or silencing of some subtelomeric genes located 50 kb from the telomere suggesting that the large distance between the telomere and the genes may protect from the spreading of telomeric silencing.[109] However, other studies imply that human telomeres neither modulate the expression of nearby genes nor affect the homeostasis of telomeres.[110,111]

Compelling evidence for transcriptional silencing in the vicinity of human telomeres was provided experimentally by using transgenes inserted adjacent to telomeres, similar to the approach used with yeast after telomere fragmentation.[112,113] By a telomere seeding procedure, natural telomeric regions have been replaced by artificial ones containing a reporter gene. Using this method, reporter genes in the vicinity of telomeric repeats were found to be expressed on average ten-fold lower than reporter at nontelomeric sites. Overexpression of the human telomerase reverse transcriptase (*hTERT*) in the telomeric clones resulted in telomere extension and decrease in transgene expression[112] while overexpression of *TRF1*, involved in telomere length regulation, lead to the re-expression of the transgene[113] indicating the involvement of both the telomere length and architecture in TPE as observed in yeast. In addition, the treatment of cells with Trichostatin A, an inhibitor of class I and II histone deacetylases antagonizes TPE. In human cells, TPE is not sensitive to DNA methylation[113] while hypermethylation of the transgene appears as a secondary effect in TPE in mouse ES cells.[114]

In mammals, all three HP1 paralogs are found at telomeres and loss of histone H3 methyltransferases leads to reduced levels of HP1 proteins at telomeres.[113,115,116] In human cells, there is a correlation between HP1 delocalization and TPE alleviation by TSA treatment.[113] Taken together, these data suggest that in mammals, like in other simpler eukaryotic organisms, classical heterochromatin factors cooperate with telomere-associated proteins in the remodeling of the telomeric and subtelomeric regions and the propagation of the silencing at chromosome ends (reviewed in ref. 5). By comparison to position effect variegation, TPE might thus be an alternative and specialized silencing process acting for instance through the interaction between the chromatin remodeling factor SALL1 and TRF1[117] or the telomeric shelterin component TIN2 and HP1[118] (Fig. 1C). In mouse cells, the progressive silencing of telomeric transgenes observed from undifferentiated cells to adult tissues suggest that TPE may act as a switch controlling the expression of genes involved in embryonic development and cell differentiation.[119]

In human cells, most of the telomeres are localized internally within the nucleoplasm, except in certain cell type such as sperm cells.[120,121] Interestingly, the 4q telomeres appear an exception since it is preferentially found at the periphery of the nucleus in various cell types in a A-type lamin dependent manner.[122,123] These studies revealed that certain subtelomeric regions might help the targeting to the nuclear envelope. Indeed, a single D4Z4 subtelomeric repeat, similar to those present at the 4q telomere, is sufficient to tether any chromosome end to the nuclear periphery (Ottaviani, Rival-Gervier, Foerster, Boussouar, Bauwens, Gilson & Magdinier, in preparation). The human telomeres are firmly associated to nuclear scaffold preparations,[124] suggesting the existence of internal anchoring sites specific for telomeres. These sites might serve as nucleation platforms for telomeric heterochromatin. In agreement with this hypothesis, both yeast and human telomeres might require the association to a particular subnuclear structure in order to become silent: the nuclear envelop in yeast and internal anchoring sites in humans. It is tempting to speculate that, in human cells, the D4Z4 subtelomeric repeat behaves as yeast STAR elements by tethering the proximal telomere to the nuclear periphery. In summary, such divergent species might use different means to link nuclear organization to chromosome-specific regulations of TPE (Fig. 2).

Telomeric Silencing and Parasitic Infection

Antigenic variation is a highly efficient survival strategy employed by various pathogens to bypass the eradication by the immune response of the host. Sets of genes responsible for such challenges called contingency genes are subject to spontaneous mutations resulting in pathogen diversity.[125] In the widely different species of parasites, variant gene families are within or just upstream of patchworks of subtelomeric repeats adjacent to telomeres and can be regulated by TPE. Such regulation has been described for *Trypanosoma brucei* and *cruzi*, *Plasmodium falciparum* and related species, *Pneumocystis carinii*, *Candida glabrata* (Fig. 3) and *Leishmania major*.

Antigenic variation can be achieved by at least two distinct mechanisms facilitated by telomeres. In *T. brucei* and *P. carinii*, exclusive expression of a single *vsg* or *MSG* gene respectively occurs from a unique 'expression site' into which silent genes are sequentially rearranged near to a telomere.[126,127]

By contrast *P. falciparum* and *C. glabrata* control their variant gene families by default subtelomeric silencing similar to what was described in yeast. For instance, the malaria parasite *Plasmodium falciparum* undergoes allelic variation through switching expression of variant surface proteins pfEMP1 encoded by the *var* gene family. Evidence for the epigenetic regulation at *P. falciparum*, was obtained experimentally by the insertion of a human marker gene into the end of chromosome 3.[128] The *P. falciparum* genome encodes a SIR2 homolog (*pfSir2*) that is involved in this silencing and the knock-out of *pfSir2* simultaneously derepresses many subtelomeric genes, including *var* genes and *rifin* genes. Subtelomeric genes are placed 40 kb from the telomere itself and Sir2 binding and concomitant TPE can spread to such a distance. Furthermore, *var* gene activation correlates with the repositioning into a location that may be permissive for transcription and changes in histone acetylation.[129-131] Reminiscent of *S. cerevisiae*, silent *var* genes on clustered telomeres are heterochromatinized and localized at the nuclear periphery in the vicinity of silencing factors. Upon activation, one telomere might move toward a permissive area of the nucleoplasm and boundary elements prevent transcription of the adjacent genes. The *var* gene that was expressed moves out of the active area and relocalizes with the other silent telomeres. The involvement of boundary elements remains hypothetical but likely play a role in the switching of the *var* genes upon parasitic infection.

Conclusions

Through the description of the numerous variations of telomeric silencing existing in a wide range of eukaryotic organisms, it appears that TPE on natural chromosomes depends on telomere length, on the structure of the telomeric chromatin, on the composition of the subtelomeric regions and on the spatial organization of chromosome ends. Consequently, a number of factors that can influence directly or indirectly telomere structure will likely affect TPE by changing telomere conformation and maintenance and vice versa. Interestingly, in baker's yeast, changes in telomere structures that perturb the distribution of heterochromatin factors within the nucleoplasm not only affect genes located in the subtelomeric regions but also some genes located far away from telomeres. It is thus expected, but not yet proven, that telomere shortening in mammalian cells primarily triggers large-scale transcriptomic changes throughout the genome.

Localizing a gene in a subtelomeric region might be an ancient evolutionary strategy that allows their reversible silencing and a fast adaptation to environmental changes. One can imagine that a single mutation or epimutation altering TPE would allow subtelomeric gene expression to proceed at full rein, increasing the chances for the cell to express a gene that would be important for adaptation.[6] Indeed, telomeres of budding yeast carrying a template mutation in the telomerase RNA gene replacing the yeast telomeric repeat sequence by the human-type sequence appear stable but have lost their capacity to silence.[132] Uncoupling the mechanisms of telomere protection from those required for TPE might provide a unique ability of rapid adaptation through large-scale transcriptomic changes without altering chromosome stability.

Acknowledgements

The work in Gilson lab is supported by the Ligue Nationale contre le Cancer (Equipe labellisée) and by the Association Française contre les Myopathies (AFM).

References

1. Jenuwein T, Allis CD. Translating the histone code. Science 2001; 293(5532):1074-1080.
2. Grewal SI, Jia S. Heterochromatin revisited. Nat Rev Genet 2007; 8(1):35-46.
3. Talbert PB, Henikoff S. Spreading of silent chromatin: inaction at a distance. Nat Rev Genet 2006; 7(10):793-803.
4. Bertuch AA, Lundblad V. The maintenance and masking of chromosome termini. Curr Opin Cell Biol 2006; 18(3):247-253.
5. Blasco MA. The epigenetic regulation of mammalian telomeres. Nat Rev Genet 2007; 8(4):299-309.
6. Teixeira MT, Gilson E. Telomere maintenance, function and evolution: the yeast paradigm. Chromosome Res 2005; 13(5):535-548.
7. Gehring WJ, Klemenz R, Weber U et al. Functional analysis of the white gene of Drosophila by P-factor-mediated transformation. EMBO J 1984; 3(9):2077-2085.
8. Hazelrigg T, Levis R, Rubin GM. Transformation of white locus DNA in drosophila: dosage compensation, zeste interaction and position effects. Cell 1984; 36(2):469-481.
9. Levis R, Hazelrigg T, Rubin GM. Effects of genomic position on the expression of transduced copies of the white gene of Drosophila. Science 1985; 229(4713):558-561.
10. Gottschling DE, Aparicio OM, Billington BL et al. Position effect at S. cerevisiae telomeres: reversible repression of Pol II transcription. Cell 1990; 63(4):751-762.
11. Mondoux MA, Zakian VA. Telomere position effect: Silencing near the end. In: De Lange T, Lundblad V, Blackburn EH, eds. Telomeres. 2nd ed. Cold Spring Harbor: Cold Spring Harbor Laboratory Press; 2006:261-316.

12. Aparicio OM, Billington BL, Gottschling DE. Modifiers of position effect are shared between telomeric and silent mating-type loci in S. cerevisiae. Cell 1991; 66(6):1279-1287.
13. Boulton SJ, Jackson SP. Components of the Ku-dependent nonhomologous end-joining pathway are involved in telomeric length maintenance and telomeric silencing. EMBO J 1998; 17(6):1819-1828.
14. Laroche T, Martin SG, Gotta M et al. Mutation of yeast Ku genes disrupts the subnuclear organization of telomeres. Curr Biol 1998; 8(11):653-656.
15. Kyrion G, Boakye KA, Lustig AJ. C-terminal truncation of RAP1 results in the deregulation of telomere size, stability and function in Saccharomyces cerevisiae. Mol Cell Biol 1992; 12(11):5159-5173.
16. Moretti P, Freeman K, Coodly L et al. Evidence that a complex of SIR proteins interacts with the silencer and telomere-binding protein RAP1. Genes Dev 1994; 8(19):2257-2269.
17. Tsukamoto Y, Kato J, Ikeda H. Silencing factors participate in DNA repair and recombination in Saccharomyces cerevisiae. Nature 1997; 388(6645):900-903.
18. Strahl-Bolsinger S, Hecht A, Luo K et al. SIR2 and SIR4 interactions differ in core and extended telomeric heterochromatin in yeast. Genes Dev 1997; 11(1):83-93.
19. Hecht A, Laroche T, Strahl-Bolsinger S et al. Histone H3 and H4 N-termini interact with SIR3 and SIR4 proteins: a molecular model for the formation of heterochromatin in yeast. Cell 1995; 80(4):583-592.
20. Renauld H, Aparicio OM, Zierath PD et al. Silent domains are assembled continuously from the telomere and are defined by promoter distance and strength and by SIR3 dosage. Genes Dev 1993; 7(7A):1133-1145.
21. Hoppe GJ, Tanny JC, Rudner AD et al. Steps in assembly of silent chromatin in yeast: Sir3-independent binding of a Sir2/Sir4 complex to silencers and role for Sir2-dependent deacetylation. Mol Cell Biol 2002; 22(12):4167-4180.
22. Mann RK, Grunstein M. Histone H3 N-terminal mutations allow hyperactivation of the yeast GAL1 gene in vivo. EMBO J 1992; 11(9):3297-3306.
23. Thompson JS, Ling X, Grunstein M. Histone H3 amino terminus is required for telomeric and silent mating locus repression in yeast. Nature 1994; 369(6477):245-247.
24. Wyatt HR, Liaw H, Green GR et al. Multiple roles for Saccharomyces cerevisiae histone H2A in telomere position effect, Spt phenotypes and double-strand-break repair. Genetics 2003; 164(1):47-64.
25. Carmen AA, Milne L, Grunstein M. Acetylation of the yeast histone H4 N terminus regulates its binding to heterochromatin protein SIR3. J Biol Chem 2002; 277(7):4778-4781.
26. Kimura A, Umehara T, Horikoshi M. Chromosomal gradient of histone acetylation established by Sas2p and Sir2p functions as a shield against gene silencing. Nat Genet 2002; 32(3):370-377.
27. Meneghini MD, Wu M, Madhani HD. Conserved histone variant H2A.Z protects euchromatin from the ectopic spread of silent heterochromatin. Cell 2003; 112(5):725-736.
28. Babiarz JE, Halley JE, Rine J. Telomeric heterochromatin boundaries require NuA4-dependent acetylation of histone variant H2A.Z in Saccharomyces cerevisiae. Genes Dev 2006; 20(6):700-710.
29. Zhang H, Richardson DO, Roberts DN et al. The Yaf9 component of the SWR1 and NuA4 complexes is required for proper gene expression, histone H4 acetylation and Htz1 replacement near telomeres. Mol Cell Biol 2004; 24(21):9424-9436.
30. Millar CB, Grunstein M. Genome-wide patterns of histone modifications in yeast. Nat Rev Mol Cell Biol 2006; 7(9):657-666.
31. Briggs SD, Xiao T, Sun ZW et al. Gene silencing: trans-histone regulatory pathway in chromatin. Nature 2002; 418(6897):498.
32. Liang G, Klose RJ, Gardner KE et al. Yeast Jhd2p is a histone H3 Lys4 trimethyl demethylase. Nat Struct Mol Biol 2007; 14(3):243-245.
33. Santos-Rosa H, Schneider R, Bannister AJ et al. Active genes are tri-methylated at K4 of histone H3. Nature 2002; 419(6905):407-411.
34. Gardner RG, Nelson ZW, Gottschling DE. Ubp10/Dot4p regulates the persistence of ubiquitinated histone H2B: distinct roles in telomeric silencing and general chromatin. Mol Cell Biol 2005; 25(14):6123-6139.
35. Cooper JP, Nimmo ER, Allshire RC et al. Regulation of telomere length and function by a Myb-domain protein in fission yeast. Nature 1997; 385(6618):744-747.
36. Kanoh J, Ishikawa F. spRap1 and spRif1, recruited to telomeres by Taz1, are essential for telomere function in fission yeast. Curr Biol 2001; 11(20):1624-1630.
37. Nimmo ER, Pidoux AL, Perry PE et al. Defective meiosis in telomere-silencing mutants of Schizosaccharomyces pombe. Nature 1998; 392(6678):825-828.
38. Park MJ, Jang YK, Choi ES et al. Fission yeast Rap1 homolog is a telomere-specific silencing factor and interacts with Taz1p. Mol Cells 2002; 13(2):327-333.
39. Wallrath LL, Elgin SC. Position effect variegation in Drosophila is associated with an altered chromatin structure. Genes Dev 1995; 9(10):1263-1277.

40. Ekwall K, Javerzat JP, Lorentz A et al. The chromodomain protein Swi6: a key component at fission yeast centromeres. Science 1995; 269(5229):1429-1431.
41. Nakayama J, Rice JC, Strahl BD et al. Role of histone H3 lysine 9 methylation in epigenetic control of heterochromatin assembly. Science 2001; 292(5514):110-113.
42. Volpe TA, Kidner C, Hall IM et al. Regulation of heterochromatic silencing and histone H3 lysine-9 methylation by RNAi. Science 2002; 297(5588):1833-1837.
43. Sugiyama T, Cam HP, Sugiyama R et al. SHREC, an effector complex for heterochromatic transcriptional silencing. Cell 2007; 128(3):491-504.
44. Gotta M, Laroche T, Formenton A et al. The clustering of telomeres and colocalization with Rap1, Sir3 and Sir4 proteins in wild-type Saccharomyces cerevisiae. J Cell Biol 1996; 134(6):1349-1363.
45. Maillet L, Gaden F, Brevet V et al. Ku-deficient yeast strains exhibit alternative states of silencing competence. EMBO Rep 2001; 2(3):203-210.
46. Taddei A, Hediger F, Neumann FR et al. Separation of silencing from perinuclear anchoring functions in yeast Ku80, Sir4 and Esc1 proteins. EMBO J 2004; 23(6):1301-1312.
47. Hediger F, Dubrana K, Gasser SM. Myosin-like proteins 1 and 2 are not required for silencing or telomere anchoring, but act in the Tel1 pathway of telomere length control. J Struct Biol 2002; 140(1-3):79-91.
48. Maillet L, Boscheron C, Gotta M et al. Evidence for silencing compartments within the yeast nucleus: a role for telomere proximity and Sir protein concentration in silencer-mediated repression. Genes Dev 1996; 10(14):1796-1811.
49. Marcand S, Buck SW, Moretti P et al. Silencing of genes at nontelomeric sites in yeast is controlled by sequestration of silencing factors at telomeres by Rap 1 protein. Genes Dev 1996; 10(11):1297-1309.
50. Andrulis ED, Neiman AM, Zappulla DC et al. Perinuclear localization of chromatin facilitates transcriptional silencing. Nature 1998; 394(6693):592-595.
51. Tham WH, Zakian VA. Telomeric tethers. Nature 2000; 403(6765):34-35.
52. Gartenberg MR, Neumann FR, Laroche T et al. Sir-mediated repression can occur independently of chromosomal and subnuclear contexts. Cell 2004; 119(7):955-967.
53. Lebrun E, Fourel G, Defossez PA et al. A methyltransferase targeting assay reveals silencer-telomere interactions in budding yeast. Mol Cell Biol 2003; 23(5):1498-1508.
54. Gerasimova TI, Byrd K, Corces VG. A chromatin insulator determines the nuclear localization of DNA. Mol Cell 2000; 6(5):1025-1035.
55. Magdinier F, Yusufzai TM, Felsenfeld G. Both CTCF-dependent and -independent insulators are found between the mouse T cell receptor alpha and Dad1 genes. J Biol Chem 2004; 279(24):25381-25389.
56. Yusufzai TM, Tagami H, Nakatani Y et al. CTCF tethers an insulator to subnuclear sites, suggesting shared insulator mechanisms across species. Mol Cell 2004; 13(2):291-298.
57. Ishii K, Arib G, Lin C et al. Chromatin boundaries in budding yeast: the nuclear pore connection. Cell 2002; 109(5):551-562.
58. Hediger F, Berthiau AS, van Houwe G et al. Subtelomeric factors antagonize telomere anchoring and Tel1-independent telomere length regulation. EMBO J 2006; 25(4):857-867.
59. Taddei A, Van Houwe G, Hediger F et al. Nuclear pore association confers optimal expression levels for an inducible yeast gene. Nature 2006; 441(7094):774-778.
60. Tham WH, Zakian VA. Transcriptional silencing at Saccharomyces telomeres: implications for other organisms. Oncogene 2002; 21(4):512-521.
61. Pryde FE, Louis EJ. Limitations of silencing at native yeast telomeres. EMBO J 1999; 18(9):2538-2550.
62. Zakian VA, Blanton HM. Distribution of telomere-associated sequences on natural chromosomes in Saccharomyces cerevisiae. Mol Cell Biol 1988; 8(5):2257-2260.
63. Louis EJ. The chromosome ends of Saccharomyces cerevisiae. Yeast 1995; 11(16):1553-1573.
64. Boscheron C, Maillet L, Marcand S et al. Cooperation at a distance between silencers and proto-silencers at the yeast HML locus. EMBO J 1996; 15(9):2184-2195.
65. Lebrun E, Revardel E, Boscheron C et al. Protosilencers in Saccharomyces cerevisiae subtelomeric regions. Genetics 2001; 158(1):167-176.
66. Fourel G, Revardel E, Koering CE et al. Cohabitation of insulators and silencing elements in yeast subtelomeric regions. EMBO J 1999; 18(9):2522-2537.
67. Chan CS, Tye BK. Organization of DNA sequences and replication origins at yeast telomeres. Cell 1983; 33(2):563-573.
68. Lieb JD, Liu X, Botstein D et al. Promoter-specific binding of Rap1 revealed by genome-wide maps of protein-DNA association. Nat Genet 2001; 28(4):327-334.
69. Foss M, McNally FJ, Laurenson P et al. Origin recognition complex (ORC) in transcriptional silencing and DNA replication in S. cerevisiae. Science 1993; 262(5141):1838-1844.

70. Triolo T, Sternglanz R. Role of interactions between the origin recognition complex and SIR1 in transcriptional silencing. Nature 1996; 381(6579):251-253.
71. Wyrick JJ, Holstege FC, Jennings EG et al. Chromosomal landscape of nucleosome-dependent gene expression and silencing in yeast. Nature 1999; 402(6760):418-421.
72. Fabre E, Muller H, Therizols P et al. Comparative genomics in hemiascomycete yeasts: evolution of sex, silencing and subtelomeres. Mol Biol Evol 2005; 22(4):856-873.
73. Gurevich R, Smolikov S, Maddar H et al. Mutant telomeres inhibit transcriptional silencing at native telomeres of the yeast Kluyveromyces lactis. Mol Genet Genomics 2003; 268(6):729-738.
74. Castano I, Pan SJ, Zupancic M et al. Telomere length control and transcriptional regulation of subtelomeric adhesins in Candida glabrata. Mol Microbiol 2005; 55(4):1246-1258.
75. Robyr D, Suka Y, Xenarios I et al. Microarray deacetylation maps determine genome-wide functions for yeast histone deacetylases. Cell 2002; 109(4):437-446.
76. Stone EM, Pillus L. Activation of an MAP kinase cascade leads to Sir3p hyperphosphorylation and strengthens transcriptional silencing. J Cell Biol 1996; 135(3):571-583.
77. Ai W, Bertram PG, Tsang CK et al. Regulation of subtelomeric silencing during stress response. Mol Cell 2002; 10(6):1295-1305.
78. De Las Penas A, Pan SJ, Castano I et al. Virulence-related surface glycoproteins in the yeast pathogen Candida glabrata are encoded in subtelomeric clusters and subject to RAP1- and SIR-dependent transcriptional silencing. Genes Dev 2003; 17(18):2245-2258.
79. Halme A, Bumgarner S, Styles C et al. Genetic and epigenetic regulation of the FLO gene family generates cell-surface variation in yeast. Cell 2004; 116(3):405-415.
80. Hansen KR, Burns G, Mata J et al. Global effects on gene expression in fission yeast by silencing and RNA interference machineries. Mol Cell Biol 2005; 25(2):590-601.
81. Raghuraman MK, Winzeler EA, Collingwood D et al. Replication dynamics of the yeast genome. Science 2001; 294(5540):115-121.
82. Kim SM, Dubey DD, Huberman JA. Early-replicating heterochromatin. Genes Dev 2003; 17(3):330-335.
83. Ferguson BM, Fangman WL. A position effect on the time of replication origin activation in yeast. Cell 1992; 68(2):333-339.
84. Stevenson JB, Gottschling DE. Telomeric chromatin modulates replication timing near chromosome ends. Genes Dev 1999; 13(2):146-151.
85. Makovets S, Herskowitz I, Blackburn EH. Anatomy and dynamics of DNA replication fork movement in yeast telomeric regions. Mol Cell Biol 2004; 24(9):4019-4031.
86. Poloumienko A, Dershowitz A, De J et al. Completion of replication map of Saccharomyces cerevisiae chromosome III. Mol Biol Cell 2001; 12(11):3317-3327.
87. Bianchi A, Shore D. Early replication of short telomeres in budding yeast. Cell 2007; 128(6):1051-1062.
88. Zappulla DC, Sternglanz R, Leatherwood J. Control of replication timing by a transcriptional silencer. Curr Biol 2002; 12(11):869-875.
89. Miklos GL, Nankivell RN. Telomeric satellite DNA functions in regulating recombination. Chromosoma 1976; 56(2):143-167.
90. Baudat F, Nicolas A. Clustering of meiotic double-strand breaks on yeast chromosome III. Proc Natl Acad Sci USA 1997; 94(10):5213-5218.
91. Kipling D, Wilson HE, Thomson EJ et al. Structural variation of the pseudoautosomal region between and within inbred mouse strains. Proc Natl Acad Sci USA 1996; 93(1):171-175.
92. Stavenhagen JB, Zakian VA. Yeast telomeres exert a position effect on recombination between internal tracts of yeast telomeric DNA. Genes Dev 1998; 12(19):3044-3058.
93. Therizols P, Fairhead C, Cabal GG et al. Telomere tethering at the nuclear periphery is essential for efficient DNA double strand break repair in subtelomeric region. J Cell Biol 2006; 172(2):189-199.
94. Mason JM, Biessmann H. The unusual telomeres of Drosophila. Trends Genet 1995; 11(2):58-62.
95. Mefford HC, Trask BJ. The complex structure and dynamic evolution of human subtelomeres. Nat Rev Genet 2002; 3(2):91-102.
96. Karpen GH, Spradling AC. Analysis of subtelomeric heterochromatin in the Drosophila minichromosome Dp1187 by single P element insertional mutagenesis. Genetics 1992; 132(3):737-753.
97. Pirrotta V. Chromatin complexes regulating gene expression in Drosophila. Curr Opin Genet Dev 1995; 5(4):466-472.
98. Boivin A, Gally C, Netter S et al. Telomeric associated sequences of Drosophila recruit polycomb-group proteins in vivo and can induce pairing-sensitive repression. Genetics 2003; 164(1):195-208.
99. Cryderman DE, Morris EJ, Biessmann H et al. Silencing at Drosophila telomeres: nuclear organization and chromatin structure play critical roles. EMBO J 1999; 18(13):3724-3735.

100. Donaldson KM, Lui A, Karpen GH. Modifiers of terminal deficiency-associated position effect variegation in Drosophila. Genetics 2002; 160(3):995-1009.
101. Fanti L, Giovinazzo G, Berloco M et al. The heterochromatin protein 1 prevents telomere fusions in Drosophila. Mol Cell 1998; 2(5):527-538.
102. Perrini B, Piacentini L, Fanti L et al. HP1 controls telomere capping, telomere elongation and telomere silencing by two different mechanisms in Drosophila. Mol Cell 2004; 15(3):467-476.
103. Cowell IG, Aucott R, Mahadevaiah SK et al. Heterochromatin, HP1 and methylation at lysine 9 of histone H3 in animals. Chromosoma 2002; 111(1):22-36.
104. Savitsky M, Kravchuk O, Melnikova L et al. Heterochromatin protein 1 is involved in control of telomere elongation in Drosophila melanogaster. Mol Cell Biol 2002; 22(9):3204-3218.
105. Melnikova L, Biessmann H, Georgiev P. The Ku protein complex is involved in length regulation of Drosophila telomeres. Genetics 2005; 170(1):221-235.
106. Oikemus SR, McGinnis N, Queiroz-Machado J et al. Drosophila atm/telomere fusion is required for telomeric localization of HP1 and telomere position effect. Genes Dev 2004; 18(15):1850-1861.
107. Shanower GA, Muller M, Blanton JL et al. Characterization of the grappa gene, the Drosophila histone H3 lysine 79 methyltransferase. Genetics 2005; 169(1):173-184.
108. Savitsky M, Kwon D, Georgiev P et al. Telomere elongation is under the control of the RNAi-based mechanism in the Drosophila germline. Genes Dev 2006; 20(3):345-354.
109. Ofir R, Wong AC, McDermid HE et al. Position effect of human telomeric repeats on replication timing. Proc Natl Acad Sci USA 1999; 96(20):11434-11439.
110. Bayne RA, Broccoli D, Taggart MH et al. Sandwiching of a gene within 12 kb of a functional telomere and alpha satellite does not result in silencing. Hum Mol Genet 1994; 3(4):539-546.
111. Sprung CN, Sabatier L, Murnane JP. Effect of telomere length on telomeric gene expression. Nucleic Acids Res 1996; 24(21):4336-4340.
112. Baur JA, Zou Y, Shay JW et al. Telomere position effect in human cells. Science 2001; 292(5524):2075-2077.
113. Koering CE, Pollice A, Zibella MP et al. Human telomeric position effect is determined by chromosomal context and telomeric chromatin integrity. EMBO Rep 2002; 3(11):1055-1061.
114. Pedram M, Sprung CN, Gao Q et al. Telomere position effect and silencing of transgenes near telomeres in the mouse. Mol Cell Biol 2006; 26(5):1865-1878.
115. Garcia-Cao M, O'Sullivan R, Peters AH et al. Epigenetic regulation of telomere length in mammalian cells by the Suv39h1 and Suv39h2 histone methyltransferases. Nat Genet 2004; 36(1):94-99.
116. Sharma GG, Hwang KK, Pandita RK et al. Human heterochromatin protein 1 isoforms HP1(Hsalpha) and HP1(Hsbeta) interfere with hTERT-telomere interactions and correlate with changes in cell growth and response to ionizing radiation. Mol Cell Biol 2003; 23(22):8363-8376.
117. Netzer C, Rieger L, Brero A et al. SALL1, the gene mutated in Townes-Brocks syndrome, encodes a transcriptional repressor which interacts with TRF1/PIN2 and localizes to pericentromeric heterochromatin. Hum Mol Genet 2001; 10(26):3017-3024.
118. Kaminker P, Plachot C, Kim SH et al. Higher-order nuclear organization in growth arrest of human mammary epithelial cells: a novel role for telomere-associated protein TIN2. J Cell Sci 2005; 118 (Pt 6):1321-1330.
119. Gao Q, Reynolds GE, Innes L et al. Telomeric Transgenes are Silenced in Adult Mouse Tissues and Embryo Fibroblasts, but are Expressed in Embryonic Stem Cells. Stem Cells 2007 [doi:10.1634/stemcells.2007-0478].
120. Luderus ME, van Steensel B, Chong L et al. Structure, subnuclear distribution and nuclear matrix association of the mammalian telomeric complex. J Cell Biol 1996; 135(4):867-881.
121. Gilson E, Laroche T, Gasser SM. Telomeres and the functional architecture of the nucleus. Trends Cell Biol 1993; 3(4):128-134.
122. Masny PS, Bengtsson U, Chung SA et al. Localization of 4q35.2 to the nuclear periphery: is FSHD a nuclear envelope disease? Hum Mol Genet 2004; 13(17):1857-1871.
123. Tam R, Smith KP, Lawrence JB. The 4q subtelomere harboring the FSHD locus is specifically anchored with peripheral heterochromatin unlike most human telomeres. J Cell Biol 2004; 167(2):269-279.
124. de Lange T. Human telomeres are attached to the nuclear matrix. EMBO J 1992; 11(2):717-724.
125. Barry JD, Ginger ML, Burton P et al. Why are parasite contingency genes often associated with telomeres? Int J Parasitol 2003; 33(1):29-45.
126. Borst P, Ulbert S. Control of VSG gene expression sites. Mol Biochem Parasitol 2001; 114(1):17-27.
127. Stringer JR, Keely SP. Genetics of surface antigen expression in Pneumocystis carinii. Infect Immun 2001; 69(2):627-639.
128. Duraisingh MT, Voss TS, Marty AJ et al. Heterochromatin silencing and locus repositioning linked to regulation of virulence genes in Plasmodium falciparum. Cell 2005; 121(1):13-24.

129. Freitas-Junior LH, Hernandez-Rivas R, Ralph SA et al. Telomeric heterochromatin propagation and histone acetylation control mutually exclusive expression of antigenic variation genes in malaria parasites. Cell 2005; 121(1):25-36.
130. Ralph SA, Scherf A. The epigenetic control of antigenic variation in Plasmodium falciparum. Curr Opin Microbiol 2005; 8(4):434-440.
131. Voss TS, Healer J, Marty AJ et al. A var gene promoter controls allelic exclusion of virulence genes in Plasmodium falciparum malaria. Nature 2006; 439(7079):1004-1008.
132. Brevet V, Berthiau AS, Civitelli L et al. The number of vertebrate repeats can be regulated at yeast telomeres by Rap1-independent mechanisms. EMBO J 2003; 22(7):1697-1706.

CHAPTER 10

Cancer as a Microevolutionary Process Affecting Telomere Structure and Dynamics: The Contribution of Telomeres to Cancer

J. Arturo Londoño-Vallejo*

Abstract

Telomeres play fundamental roles in genome stability, nuclear architecture and chromosome pairing during meiosis. They shorten at every cell division and may be re-elongated or not depending on the presence of the dedicated enzyme, telomerase. Since in most human somatic cells telomerase is not expressed, shortening of telomeres during development and aging is the rule. Short telomeres being, under physiological conditions, incompatible with extended cell proliferation, telomere length defines the proliferation potential of a cell and operates as a mechanism to prevent uncontrolled cell growth. Conversely, in cells in which proliferation checkpoints have been abolished, shortening of telomeres causes chromosomes to fuse and to initiate cycles of breakage-fusion-bridge thus becoming a strong driving force for genome instability. In vitro, transformed cells with highly unstable genomes because of severe telomere shortening accumulate deleterious genetic changes and die (crisis). At the same time, random genetic or epigenetic changes may allow cells to acquire a telomere maintenance mechanism (as well as other tumor phenotypes) and to become immortal. Although telomere shortening and other types of telomere dysfunction probably contribute to the genome instability detected in early tumors in vivo, the direct contributions of dysfunctional telomeres to the acquisition of tumor phenotypes in humans remain largely unspecified.

Introduction

Telomeres, the repeated nucleoprotein structures (5'-$(T_2AG_3)_n$-3' in all vertebrates) found at the end of linear chromosomes, are absolutely essential to preserve chromosome structure and function.[1] At the same time, they constitute one of the most dynamic structures in all genomes, both at the scale of species evolution as well as at the individual level.[2] During species evolution, the fusion of primitive chromosomes, or the stabilization of new ends resulting from fission events, as a consequence of the loss or the gain of telomeric sequences, respectively, have contributed to the shaping of karyotypes.[3] Likewise, telomeres play an important role in human cells that go into uncontrolled proliferation, mediating chromosome fusions and inducing cycles of genome instability.[4] In the last case, the resulting genome rearrangements may participate in the acquisition of new phenotypes and probably contribute to the emergence or progression of genetic diseases,

*J. Arturo Londoño-Vallejo—UMR7147, Institut Curie-UPMC-CNRS, 26, rue d'Ulm, 75248 Paris, France. Email: arturo.londono@curie.fr

Origin and Evolution of Telomeres, edited by Jozef Nosek and Ľubomír Tomáška.

such as cancer. Although a great deal has been learned about telomere biology during the last decades, numerous questions remain unanswered with regard to its precise role in the evolution of tumor cells.

Telomere Length Dynamics and Cell Proliferation

The structure of telomeres in all organisms is pretty much conserved: a double stranded repeated region, with a G-rich strand running 5' > 3', terminated in a 3' G-rich overhang -an essential feature of telomere function- of variable length.[1] Telomere DNA sequences, including the overhang, serve also as a platform for the assembly of specific macromolecular nucleoprotein complexes, which cap the chromosome extremity.[5] In vertebrates, it has been proposed that this protein complex, dubbed shelterin,[6] promotes the invasion of the double strand repeated sequence by the overhang, forming a T-loop that masks the chromosome end[7] and protects it from the double strand break (DSB) sensing and repairing devices operating in normal cells.[8] "Uncapping" of one telomere, because of damage (loss) of telomere sequences or because of destabilization of the protein complex, triggers a DNA damage response and an attempt by the cell to repair the unprotected extremity.[9-11]

Sources of telomere damage that may lead to sudden telomere shortening and uncapping in human cells are multiple (exogenous, such as UV irradiation, or endogenous, such as reactive oxygen species). Yet, the universal source of telomere shortening is cell proliferation since replication of telomeres by conventional mechanisms is inevitably incomplete (the end-replication problem). Telomerase, the unique enzyme in the cell able to add telomeric repeats de novo to the 3' end, may counteract this loss.[12] However, since the expression of the enzyme is highly regulated and most somatic cells do not possess any telomerase activity, proliferating cells undergo sustained telomere shortening up to a point where telomeres within the cell become uncapped.[13] Under normal conditions, shortened telomeres induce, together with the damage response, a cellular arrest, which may become permanent (senescence) or lead to apoptosis (depending on the cell), if telomere uncapping persists.[14]

How short a telomere should be in order to become uncapped and be recognized as a DSB remains undetermined. Initial studies using Southern blotting techniques indicated that cells enter mitotic senescence when the mean length of telomere restriction fragments (TRF) reaches around 4 kb.[15] In fact, senescent cells carry very short telomeres as recently revealed by a study using a PCR-based technique.[16] This is because telomere lengths are very heterogeneous within cells, with chromosome arms bearing either short or long telomeres.[17] Given this heterogeneity and the fact that shortening rate is similar for long and short telomeres,[16,18] it is expected that only very few telomeres will become uncapped at the same time so that the DNA damage signal most probably originates from a limited number of extremities.[19,20] Therefore the number of divisions a cell is able to make will depend on the initial length of the shortest telomeres rather than on mean telomere length.[21] Thus, the so-called "mitotic clock" constitutes a strong mechanism against unlimited proliferation.[22]

Telomere Length Dynamics and Aging

In humans, mean telomere lengths, as measured in different tissues by Southern blotting or in situ hybridization techniques, shorten with age.[23] Numerous associations between aging phenotypes and short telomeres have been described and therefore it is tempting to draw a direct link between mitotic senescence and organismal aging, but a formal link remains to be established.[24] Even so, patients with mutations in the gene coding for the telomerase RNA exhibit a number of manifestations of premature aging, including a predisposition to develop cancer.[25] In addition, the higher incidence of cancer in older populations has prompted the hypothesis that shortening of telomeres promotes tumor development and several studies have found that patients with shorter telomeres in peripheral blood cells have a higher risk of developing carcinomas.[26] Alternatively, cancer prone or carcinogen treated mice tend to form fewer tumors if they carry very short telomeres, rather suggesting that shortening of telomeres by itself has a protective role against uncontrolled proliferation.[27,28] Nevertheless, in mice carrying mutations in the gene coding for p53, a major tumor suppressor,[29] short telomeres not only increase the incidence of tumors but shift their spectrum towards epithelial carcinomas, of the type associated with aging in humans.[30,31] Consequently, short telomeres may have both protective and deleterious effects depending on the genetic context.

A fascinating aspect of the relationship between aging and cancer is the proposed contribution of senescent stromal fibroblast cells to the progression of nearby tumors from epithelial origin.[32] Stromal cells that enter senescence because of telomere shortening are able to modify the tissue architecture and to secrete factors that stimulate proliferation of surrounding cells.[33] Thus, although telomere induced senescence has a positive role early in life, preventing the development of tumors, the accumulation of telomere dysfunctional cells with aging may have a harmful impact promoting age-related diseases including cancer.[34]

Genomic Instability Pathways Initiated by Telomeres

In vitro models of oncogenesis have shown that, if the pathways responsible for growth arrest in response to short telomeres are disabled (for instance, through the introduction of viral proteins that inactivate p53 and Rb proteins) the transformed cell continues to divide and telomeres shorten further (Fig. 1).[35,36] Unprotected chromosome ends then become the substrates of repair activities that result in chromosome fusions, most likely mediated by non homologous end-joining mechanisms,[37] perhaps without further processing.[38] Telomeric fusions create dicentric chromosomes whose segregation, during the next cell division, is problematic since their centromeres may be brought to opposite poles of the cell during anaphase. These anaphase bridges have to be torn

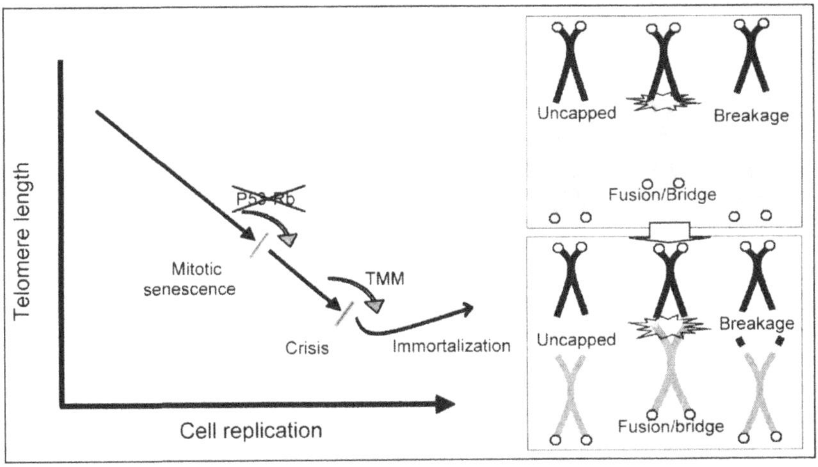

Figure 1. Left) Telomere barriers to tumor development. In the absence of telomerase (i.e., most somatic cells), telomere length decreases with cell replication. The presence of short telomeres triggers a signaling pathway mediated by the tumor suppressors p53 and Rb, leading to growth arrest (mitotic senescence). If these pathways are inactivated, cells continue to replicate and telomeres shorten further, destabilizing chromosome ends. Generalized genome instability will invariably lead to death (crisis), unless a telomere maintenance mechanism (TMM) is acquired by the cell, which then becomes immortal. Right) Telomere contributions to tumor development. Telomere uncapping due to telomere progressive shortening or sudden loss elicits a DNA damage response and repair reaction that fuses together unprotected extremities (top). Fused chromosomes will break apart if they segregate to opposite sides, initiating cycles of breakage-fusion-bridge (BFB) in daughter cells (bottom). Generalization of this phenomenon to many chromosome extremities leads to extensive chromosome fragmentation and rearrangements, as well as to mitotic disturbances presumably leading to loss or gain of whole chromosomes.

apart to allow the mitosis to proceed, each daughter cell then inheriting a chromosome with a DSB at one end, which will be again fused to another uncapped extremity, reinitiating a cycle of breakage/fusion/bridge (BFB) (Fig. 1).[39]

During the first stages of proliferation-driven telomere instability (precrisis), a few telomeres, the shortest ones, may contribute to BFBs. If only one dysfunctional telomere is present, it may become fused to itself after replication (sister chromatid fusion) and initiate instability.[40,41] However, since telomeres become shorter and shorter with cell replication, the uncapping process rapidly affects many chromosome arms, which simultaneously become exposed to repair activities. At the beginning of the precrisis period, cells harboring similar telomere length distributions carry fusions involving the same chromosome arms and therefore are also expected to be subjected to similar karyotype evolutions.[41-44] However, both the multiplication of extremities available for fusion reactions and the stochastic nature of BFB cycles should allow for genetic divergence in the lineage.

Cells undergoing a process of telomere-driven genome instability will rapidly accumulate genomic changes, mostly gains and losses, through nonreciprocal translocations that cause deletions and amplifications. In advanced precrisis, BFB cycles may lead to extensive genome fragmentation with repair reactions that fuse together genome fragments coming from different chromosomes, adding a dimension of complex structural abnormalities to the ploidy changes. Duplication of whole chromosomes or even whole genomes (tetraploidization) may also accompany the process,[42,45] at the end of which, the number of unstable chromosomes becomes too high and cells enter mitotic catastrophe and die (crisis).[46] Therefore, crisis represents another strong mechanism acting against uncontrolled cell proliferation.[14]

To escape from assured death, cells must acquire a mechanism of telomere maintenance, which is most often achieved through the reactivation of telomerase. Nonetheless, spontaneous immortalization of transformed cells is a rare event (10^{-7}),[47,48] so that rescued post-crisis cultures tend to be of clonal origin, as often evidenced by a certain degree of karyotypic homogeneity within, but not between, independently obtained cell lines. Although re-expression of telomerase tends to stabilize chromosome ends and freeze karyotypic progression,[35] a complete stabilization may require time and a minimum level of expression of the enzyme, both conditions allowing for karyotype divergence in siblings. Also, it is possible that by the time telomerase is re-expressed, at least part of the unprotected chromosome ends in the cell correspond to interstitial DSBs and may not be recognized as natural substrates by telomerase. Therefore, the enzyme, although proficient in adding telomere repeats at telomeric extremities, may be much less competent in healing interstitial breaks and preventing more BFBs.

In tumor cells expressing telomerase a similar phenomenon has been observed: loss of one single telomere may provoke a sister chromatid fusion followed by a BFB event, which recurs as long as there is an uncapped extremity.[49] In this case also, telomerase activity is unable to prevent the instability suggesting that the uncapping event is the consequence of a DSB that eliminates all or most telomere repeats, making it undistinguishable from an interstitial DSB on which telomerase is mostly inefficient. Such episodes of telomeric DSB drive further genomic instability and may accelerate the mutation process.

In the case of acquisition of an alternative mechanism of telomere maintenance, based on recombination (ALT),[50] some level of chromosome instability often persists, perhaps intrinsic to the recombination mechanisms at play in this case and to the presence of chromosome ends with extremely short telomeres.[51] The expression of telomerase in cells that have been immortalized by ALT mechanisms does not interfere with recombination events at telomeres nor prevents further karyotypic evolution,[52,53] suggesting that, here also, DSBs (subtelomeric or strictly interstitial) are source of genome rearrangements. Whether these DSBs may serve as substrates for ALT telomere recombination-dependent replication is not known.

Modifications of the telomere nucleoprotein complex may also influence karyotype evolution. For instance, inactivation of TRF2 in cells leads to telomere uncapping and rampant chromosome fusions in the presence of telomere repeats.[54] However, permanent inactivation of TRF2 induce severe growth defect probably due to the impossibility for chromosomes to segregate[38] and therefore BFBs are presumably not initiated, preventing further genome rearrangements. Interestingly, detection of interstitial telomeric sequences by in situ hybridization is common in both in vitro immortalized cells using ALT mechanisms and in tumor-derived telomerase negative cell lines.[52] The mechanism of these telomeric fusions is not known but they could mark a transient deficiency in TRF2 leading to limited telomere driven genome instability.

Being Immortal Is Not Enough

The forced expression of telomerase in certain types of human cells has been shown to induce immortalization in vitro.[55] These cells are otherwise phenotypically normal and unable to form tumors. On the other hand, reactivation of a telomere maintenance mechanism that allows indefinite replication potential constitutes a hallmark of tumor cells. How telomere maintenance mechanisms are reactivated during crisis (or, for that matter, at any moment during the tumorigenesis process in vivo) is an open question. Amplification of the hTERT locus (itself at a telomeric position on human chromosome 5p) is a frequent finding both in vitro[42] and in vivo,[56,57] and duplication/translocation of the locus have been related to the reactivation of the enzyme.[42,58] In addition, the promoter of the hTERT gene is a target for numerous oncogenes and tumor suppressors[59] whose genes may be, in the course of the telomere-driven instability, affected in their number of copies and perhaps their expression.

Another important question refers to the moment of reactivation of telomerase, since it may significantly affect the way transformed cells or tumors progress. If telomerase reactivation occurs early during precrisis, cells, although immortal, may not have accumulated sufficient genetic changes to become fully tumorigenic. Conversely, a late telomerase reactivation will have allowed cells to accrue chromosome rearrangements, some of which may contribute to tumor phenotypes. In fact, it is likely that, contrary to most point mutations, gains and losses of large genome fragments or whole chromosomes are not neutral and although some of these changes are compatible with immortalization, they may prevent tumor formation under a highly selective environment. It is well recognized that spontaneously immortalized post-crisis cell lines obtained in vitro after transformation with viral oncogenes such as ER-SV40 seldom have a tumorigenic potential and most often need the introduction of oncogenic forms of RAS to become tumor proficient.[60] This is most likely due to the fact that, in vitro, the only absolutely required genetic change in transformed cells to keep proliferation going is the reactivation of a telomere maintenance mechanism, whereas the generation of other hallmarks found in tumor cells in vivo requires many more genetic changes. Nonetheless, the fact that in vitro immortalized postcrisis cells occasionally have tumorigenic potential shows that not only telomere-driven instability is compatible with, but may be sufficient to generate a tumor phenotype.

Telomere-Driven Genome Instability in Vivo

The karyotypes of in vitro immortalized, postcrisis cells are reminiscent of those found in tumors in vivo. Recent work on several types of human tumors has clearly established that shortening of telomeres occurs in tumors at early stages.[61-63] Moreover, the occurrence of short telomeres may be contemporaneous to a burst of chromosome instability.[64-66] In agreement with this, it has been showed that early tumor lesions accumulate chromosome breaks towards the end of chromosomes whereas later, more advanced lesions do not show this bias.[67] The available data has been taken as evidence that, in vivo, dysfunctional short telomeres induce chromosome fusions followed by BFB cycles in initial stages of tumor development. Further telomere shortening and the emergence of double strand breaks likely contribute to amplify the BFB cycles, which then can affect any chromosome and any region.[68] Similar karyotype evolutions are seen in tumors from mice with short telomeres.[69] In these models, tumor cells go through a period of chromosome instability and accumulate karyotypical aberrations very similar to the ones seen in human tumors. Moreover, some of the chromosome rearrangements found in full grown tumors appear to affect regions coding for well known oncogenes, suggesting a direct contribution of such rearrangements to cancer development.[70] However, whether or not human cells go through a period of crisis in vivo remains hypothetical.[65]

Genetic analyses of cancer specimens obtained from patients often show intratumor heterogeneity, which can be also observed at the karyotypic level, suggesting on-going chromosome instability. Most of these tumors are telomerase positive and therefore telomeres might not be implicated in this type of instability. However, in some tumors in which telomerase activity is absent, karyotypes tend to be highly unstable, suggesting persistent BFB cycles very much like ALT cells in vitro.[71]

As noted before, defects in telomere capping may result from mutations affecting telomeric proteins. Since homozygous deletions of TRF2 are embryonic lethal or, when induced, provoke permanent cell arrest, it has not been possible to determine the long-term impact of the genome instability caused by the absence of TRF2. Otherwise, mice that constitutively overexpress TRF2 in skin epithelium have a high incidence of cancer,[72] which increases even further when mice already carry very short telomeres.[73] At the molecular level, overexpression of TRF2 alone leads to shortening of telomeres and genome instability. On the other hand, the presence of telomeric sequences at some of the chromosome fusions and other signs of telomere dysfunction[72,73] suggest that overexpression of the wild type protein exerts a sort of dominant-negative effect, perturbing the normal function of the endogenous protein. Recently, mouse models of inactivation affecting another telomeric protein, Pot1, have become available and suggest that this protein may have telomere-independent roles in genome instability.[74] In humans, studies in tumors have shown that levels of telomeric proteins may be altered and correlated to certain tumor phenotypes.[75-80] How these proteins contribute to different aspects of tumor biology and whether or not this contribution is related to telomere maintenance are pending questions.

Telomere Instability as a Mutator Phenotype: One Train May Hide Another

Cancer cells carry mutations not only in genes directly implicated in this process but also in many other loci in the genome.[81] Since the number of accumulated mutations cannot be accounted for by the low mutations rates observed in normal somatic cells, it has been proposed that tumor cells acquire, early during transformation, a mutator phenotype (Fig. 2).[82] At the same time, an increase in the mutation rate might not always be beneficial, as most nonneutral mutations are thought to be deleterious.[83] However, under in vivo selection, rare advantageous mutations may rapidly become fixed in the population, together with many others as passengers.[84]

Two mutator phenotypes have been well identified in tumor cells. Mutations affecting DNA repair factors such as MLH1 or MSH2 lead to a mutator phenotype characterized by instability of microsatellites (MIN).[85] Most of these sequences are found in non coding regions and therefore may be neutral, but some are found within or nearby genes, consequently modifying the expression landscape in these cells.[86] MIN has been directly connected to the development of certain types of human tumors, mainly proximal colon cancers.[87] On the other hand, chromosome instability phenotypes (CIN), presumably resulting from mutations in particular genes required for correct chromosome segregation,[88] have been connected to other types of tumors and particularly cancers in the distal colon.[87] However, such mutations tend to be rare in human tumors, as a recent study showed.[89] Alternatively, telomere dysfunction due to critical shortening may very well be the major source of chromosome instability in these tumors.[90] In fact, telomere-driven crisis formally corresponds to a mutator phenotype since it provides a strong driving force for the accumulation of genetic changes in a few generations. Furthermore, more than any other mutator phenotype with mutation rates high enough to affect cellular fitness, telomere-driven crisis invariably leads to cell death, providing a strong bottleneck that favors the selection of clones having, on one side, reactivated telomerase and, on the other, accumulated less deleterious changes (Fig. 2). Whereas in vitro the only required phenotypic change to achieve immortalization of pretumoral cells is the reactivation of telomerase, the harsh environment in vivo provides further selection pressure to select for other changes allowing tumor progression. These changes may be brought about through

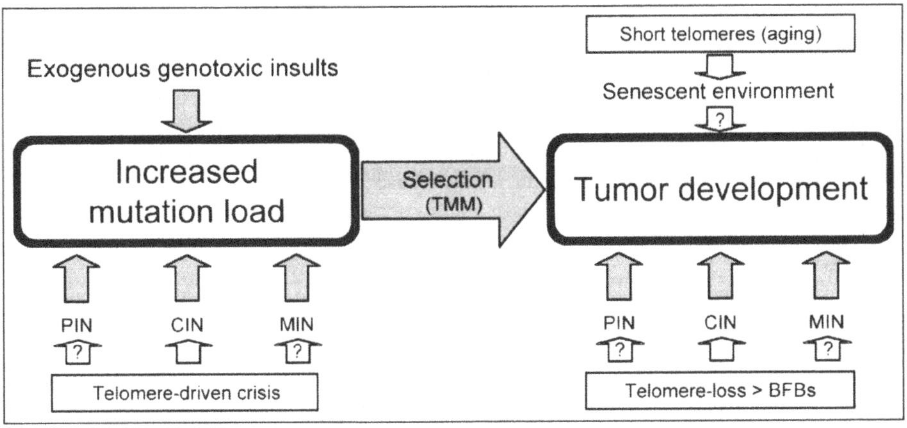

Figure 2. Telomere instability, mutator phenotypes and cancer. Dividing cells typically accumulate mutations at a rate not high enough to affect normal functions. Exposure to genotoxic agents often contributes to the mutation load, but cells may also acquire mutator phenotypes that lead either to accumulation of point mutations (PIN), to microsatellite instability (MIN) or to chromosome instability (CIN). Continued shortening of telomeres in cells that escaped from senescent arrest leads to a severe CIN phenotype. Cells carrying advantageous mutations, including those allowing the activation of a telomere maintenance mechanism (TMM), will be selected and be able to form tumors, a process perhaps facilitated by a modified environment due to aged (senescent) stromal cells. In spite of the presence of a TMM in tumor cells, the sudden loss of telomere sequences or the alteration in the shelterin structure may lead to recurrent cycles of BFB. Hypothetically, widespread telomere-driven genome instability might interfere with cell responses against other DNA damages, thereby favoring other mutator phenotypes.

telomere-driven genome instability but may also occur through processes independent, at least in their intimate mechanism, of telomeres.

A third mutator process (PIN) has been proposed to exist in cancer cells and to increase the rates of point mutations (base substitutions, insertions and deletions) affecting, amongst many other genes, key targets during early transformation.[91] Recent observations have given support to this hypothesis although the molecular mechanisms or the factors involved remain entirely unexplored.[92] It has been argued that this phenotype may be related, at least partially, to malfunctions in proofreading activities of polymerases[93] and at least one mouse model supports this hypothesis.[94] However, other dysfunctions in repair activities such as nucleotide and base excision repair (NER and BER, respectively) also may lead to point mutator phenotypes.[95] Interestingly, at least some of these DNA damage responses share the same signaling pathways or use identical factors at the repair level (BRCA1 may be such an example).[96] This means that overwhelming DNA damage of one type may encroach in the response to other type of damage. It is then reasonable to hypothesize that during telomere-driven crisis, cells may be using costly resources to repair the numerous DSBs, thus facilitating the surfacing of dysfunctions in other fronts and increasing the chances for a transient mutator phenotype to appear (Fig. 2). A coexistence of mutator phenotypes with telomere-driven genome instability in human cells has not been reported, but analogous situations have been already evidenced in model organisms.[97]

Telomere Instability and Epigenetic Changes

Lastly, amongst the changes brought about by telomere instability may be modifications at the level of the chromatin. Senescence due to telomere shortening is accompanied by changes in gene expression profiles.[98] Although genes in subtelomeric regions do not seem preferentially affected, it is possible either that they have been underrepresented in the studies carried out thus far or that telomeres have not shortened enough in these cells to induce broad expression changes.[99,100] In addition, little is known about the effect of critical telomere shortening on whole genome or subtelomeric expression profiles in humans or during telomere driven-genome instability. It is plausible that the loss of telomere repeats leads to changes in the acetylation/methylation patterns of histones as well as the methylation status of DNA.[101,102] Moreover, fusions of telomeric extremities to extremities coming from interstitial double strand breaks may modify in a deep way the chromatin structure of the latter. Although very little research has been conducted on this aspect, the impact of telomere instability on the epigenetic status of subtelomeric and other regions of the genome is likely to be significant.

Conclusion

The somatic evolution of cancer cells reflects the action, at the shortest timescale, of natural selection mechanisms.[103] It has been estimated that four mutation steps may be all that is necessary for the development of cancer,[104] although this view most probably underestimates the complexity introduced by mutator phenotypes, which increase mutation rates by several orders of magnitude.[105,106] Models of somatic evolution propose that genome instability phenotypes are often the causative mutation in early onset tumors while acquisition of genome instability late in the pathway confers little selective advantage.[107] Telomere-driven genome instability occurs early during transformation processes and, in all probability, is a frequent event in vivo, thus representing the most pervasive mutator phenotype of all. Defining both the precise contribution of telomere instability to the development of human cancer and its connection to other mutator phenotypes constitute major objectives in telomere research in the next years.

Acknowledgements

I thank Silvia Bacchetti for the numerous discussions on this subject throughout the years and for reviewing this manuscript. Work in the author's laboratory is supported by grants from the Association pour la Recherche contre le Cancer (ARC), from the Institut National du Cancer (INCa) and from the Ligue contre le Cancer.

References

1. Blackburn EH. Telomere states and cell fates. Nature 2000; 408(6808):53-56.
2. Eichler EE, Sankoff D. Structural dynamics of eukaryotic chromosome evolution. Science 2003; 301(5634):793-797.
3. Murphy WJ, Larkin DM, Everts-van der Wind A et al. Dynamics of mammalian chromosome evolution inferred from multispecies comparative maps. Science 2005; 309(5734):613-617.
4. Ishikawa F. Telomere crisis, the driving force in cancer cell evolution. Biochem Biophys Res Commun 1997; 230(1):1-6.
5. Liu D, O'Connor MS, Qin J et al. Telosome, a mammalian telomere-associated complex formed by multiple telomeric proteins. J Biol Chem 2004; 279(49):51338-51342.
6. de Lange T. Shelterin: the protein complex that shapes and safeguards human telomeres. Genes Dev 2005; 19(18):2100-2110.
7. Griffith JD, Comeau L, Rosenfield S et al. Mammalian telomeres end in a large duplex loop. Cell 1999; 97(4):503-514.
8. de Lange T. Protection of mammalian telomeres. Oncogene 2002; 21(4):532-540.
9. Takai H, Smogorzewska A, de Lange T. DNA damage foci at dysfunctional telomeres. Curr Biol 2003; 13(17):1549-1556.
10. d'Adda di Fagagna F, Reaper PM, Clay-Farrace L et al. A DNA damage checkpoint response in telomere-initiated senescence. Nature 2003; 426(6963):194-198.
11. Lou Z, Chen J. Cellular senescence and DNA repair. Exp Cell Res 2006; 312(14):2641-2646.
12. Cech TR, Lingner J. Telomerase and the chromosome end replication problem. Ciba Found Symp 1997; 211:20-28.
13. Reaper PM, di Fagagna F, Jackson SP. Activation of the DNA damage response by telomere attrition: a passage to cellular senescence. Cell Cycle 2004; 3(5):543-546.
14. Shay JW, Wright WE. Senescence and immortalization: role of telomeres and telomerase. Carcinogenesis 2005; 26(5):867-874.
15. Levy MZ, Allsopp RC, Futcher AB et al. Telomere end-replication problem and cell aging. J Mol Biol 1992; 225(4):951-960.

16. Baird DM, Rowson J, Wynford-Thomas D et al. Extensive allelic variation and ultrashort telomeres in senescent human cells. Nat Genet 2003; 33(2):203-207.
17. Lansdorp PM, Verwoerd NP, van de Rijke FM et al. Heterogeneity in telomere length of human chromosomes. Hum Mol Genet 1996; 5(5):685-691.
18. Londoño -Vallejo JA, DerSarkissian H, Cazes L et al. Differences in telomere length between homologous chromosomes in humans. Nucleic Acids Res 2001; 29(15):3164-3171.
19. Hao LY, Strong MA, Greider CW. Phosphorylation of H2AX at short telomeres in T-cells and fibroblasts. J Biol Chem 2004; 279(43):45148-45154.
20. Zou Y, Sfeir A, Gryaznov SM et al. Does a sentinel or a subset of short telomeres determine replicative senescence? Mol Biol Cell 2004; 15(8):3709-3718.
21. Hemann MT, Strong MA, Hao LY et al. The shortest telomere, not average telomere length, is critical for cell viability and chromosome stability. Cell 2001; 107(1):67-77.
22. Wright WE, Shay JW. Cellular senescence as a tumor-protection mechanism: the essential role of counting. Curr Opin Genet Dev 2001; 11(1):98-103.
23. von Zglinicki T, Martin-Ruiz CM. Telomeres as biomarkers for ageing and age-related diseases. Curr Mol Med 2005; 5(2):197-203.
24. Boukamp P. Ageing mechanisms: the role of telomere loss. Clin Exp Dermatol 2001; 26(7):562-565.
25. Vulliamy T, Dokal I. Dyskeratosis congenita. Semin Hematol 2006; 43(3):157-166.
26. DePinho RA. The age of cancer. Nature. 2000; 408(6809):248-254.
27. Gonzalez-Suarez E, Samper E, Flores JM et al. Telomerase-deficient mice with short telomeres are resistant to skin tumorigenesis. Nat Genet 2000; 26(1):114-117.
28. Greenberg RA, Chin L, Femino A et al. Short dysfunctional telomeres impair tumorigenesis in the INK4a(delta2/3) cancer-prone mouse. Cell 1999; 97:515-525.
29. Attardi LD. The role of p53-mediated apoptosis as a crucial anti-tumor response to genomic instability: lessons from mouse models. Mutat Res 2005; 569(1-2):145-157.
30. Chin L, Artandi SE, Shen Q et al. p53 deficiency rescues the adverse effects of telomere loss and cooperates with telomere dysfunction to accelerate carcinogenesis. Cell 1999; 97(4):527-538.
31. Artandi SE, Chang S, Lee SL et al. Telomere dysfunction promotes nonreciprocal translocations and epithelial cancers in mice. Nature 2000; 406(6796):641-645.
32. Krtolica A, Campisi J. Cancer and aging: a model for the cancer promoting effects of the aging stroma. Int J Biochem Cell Biol 2002; 34(11):1401-1414.
33. Krtolica A, Parrinello S, Lockett S et al. Senescent fibroblasts promote epithelial cell growth and tumorigenesis: a link between cancer and aging. Proc Natl Acad Sci USA 2001; 98(21):12072-12077.
34. Campisi J. Senescent cells, tumor suppression and organismal aging: good citizens, bad neighbors. Cell 2005; 120(4):513-522.
35. Counter CM, Avilion AA, LeFeuvre CE et al. Telomere shortening associated with chromosome instability is arrested in immortal cells which express telomerase activity. EMBO J 1992; 11(5):1921-1929.
36. Shay JW, Wright WE, Brasiskyte D et al. E6 of human papillomavirus type 16 can overcome the M1 stage of immortalization in human mammary epithelial cells but not in human fibroblasts. Oncogene 1993; 8(6):1407-1413.
37. Smogorzewska A, Karlseder J, Holtgreve-Grez H et al. DNA ligase IV-dependent NHEJ of deprotected mammalian telomeres in G1 and G2. Curr Biol 2002; 12(19):1635-1644.
38. Celli GB, de Lange T. DNA processing is not required for ATM-mediated telomere damage response after TRF2 deletion. Nat Cell Biol 2005; 7(7):712-718.
39. Lundblad V. Genome instability: McClintock revisited. Curr Biol 2001; 11(23):R957-960.
40. Lo AW, Sabatier L, Fouladi B et al. DNA amplification by breakage/fusion/bridge cycles initiated by spontaneous telomere loss in a human cancer cell line. Neoplasia 2002; 4(6):531-538.
41. Soler D, Genesca A, Arnedo G et al. Telomere dysfunction drives chromosomal instability in human mammary epithelial cells. Genes Chromosomes Cancer 2005; 44(4):339-350.
42. Der-Sarkissian H, Bacchetti S, Cazes L et al. The shortest telomeres drive karyotype evolution in transformed cells. Oncogene 2004; 23:1221-1228.
43. Deng W, Tsao SW, Guan XY et al. Distinct profiles of critically short telomeres are a key determinant of different chromosome aberrations in immortalized human cells: whole-genome evidence from multiple cell lines. Oncogene 2004; 23(56):9090-9101.
44. Deng W, Tsao SW, Guan XY et al. Role of short telomeres in inducing preferential chromosomal aberrations in human ovarian surface epithelial cells: A combined telomere quantitative fluorescence in situ hybridization and whole-chromosome painting study. Genes Chromosomes Cancer 2003; 37(1):92-97.
45. Stewart N, Bacchetti S. Expression of SV40 large T antigen, but not small t antigen, is required for the induction of chromosomal aberrations in transformed human cells. Virology 1991; 180(1):49-57.
46. Macera-Bloch L, Houghton J, Lenahan M et al. Termination of lifespan of SV40-transformed human fibroblasts in crisis is due to apoptosis. J Cell Physiol 2002; 190(3):332-344.

47. Shay JW, Wright WE. Quantitation of the frequency of immortalization of normal human diploid fibroblasts by SV40 large T-antigen. Exp Cell Res 1989; 184(1):109-118.
48. Imai S, Saito F, Ikeuchi T et al. Escape from in vitro aging in SV40 large T antigen-transformed human diploid cells: a key event responsible for immortalization occurs during crisis. Mech Ageing Dev 1993; 69(1-2):149-158.
49. Sabatier L, Ricoul M, Pottier G et al. The loss of a single telomere can result in instability of multiple chromosomes in a human tumor cell line. Mol Cancer Res 2005; 3(3):139-150.
50. Murnane JP, Sabatier L, Marder BA et al. Telomere dynamics in an immortal human cell line. EMBO J 1994; 13(20):4953-4962.
51. Henson JD, Neumann AA, Yeager TR et al. Alternative lengthening of telomeres in mammalian cells. Oncogene 2002; 21(4):598-610.
52. Cerone MA, Londoño-Vallejo JA, Bacchetti S. Telomere maintenance by telomerase and by recombination can coexist in human cells. Hum Mol Genet 2001; 10(18):1945-1952.
53. Perrem K, Colgin LM, Neumann AA et al. Coexistence of alternative lengthening of telomeres and telomerase in hTERT-transfected GM847 cells. Mol Cell Biol 2001; 21(12):3862-3875.
54. van Steensel B, Smogorzewska A, de Lange T. TRF2 protects human telomeres from end-to-end fusions. Cell 1998; 92(3):401-413.
55. Bodnar AG, Ouellette M, Frolkis M et al. Extension of life-span by introduction of telomerase into normal human cells. Science 1998; 279:349-352.
56. Takuma Y, Nouso K, Kobayashi Y et al. Telomerase reverse transcriptase gene amplification in hepatocellular carcinoma. J Gastroenterol Hepatol 2004; 19(11):1300-1304.
57. Mosse YP, Greshock J, Margolin A et al. High-resolution detection and mapping of genomic DNA alterations in neuroblastoma. Genes Chromosomes Cancer 2005; 43(4):390-403.
58. Nowak T, Januszkiewicz D, Zawada M et al. Amplification of hTERT and hTERC genes in leukemic cells with high expression and activity of telomerase. Oncol Rep 2006; 16(2):301-305.
59. Janknecht R. On the road to immortality: hTERT upregulation in cancer cells. FEBS Lett 2004; 564(1-2):9-13.
60. Hahn WC, Counter CM, Lundberg AS et al. Creation of human tumour cells with defined genetic elements. Nature 1999; 400:464-468.
61. Meeker AK, Argani P. Telomere shortening occurs early during breast tumorigenesis: a cause of chromosome destabilization underlying malignant transformation? J Mammary Gland Biol Neoplasia 2004; 9(3):285-296.
62. Meeker AK, Hicks JL, Iacobuzio-Donahue CA et al. Telomere length abnormalities occur early in the initiation of epithelial carcinogenesis. Clin Cancer Res 2004; 10(10):3317-3326.
63. Meeker AK, Hicks JL, Gabrielson E et al. Telomere shortening occurs in subsets of normal breast epithelium as well as in situ and invasive carcinoma. Am J Pathol 2004; 164(3):925-935.
64. Romanov SR, Kozakiewicz BK, Holst CR et al. Normal human mammary epithelial cells spontaneously escape senescence and acquire genomic changes. Nature 2001; 409(6820):633-637.
65. DePinho RA, Polyak K. Cancer chromosomes in crisis. Nat Genet 2004; 36(9):932-934.
66. Chin K, de Solorzano CO, Knowles D et al. In situ analyses of genome instability in breast cancer. Nat Genet 2004; 36(9):984-988.
67. Gisselsson D, Pettersson L, Hoglund M et al. Chromosomal breakage-fusion-bridge events cause genetic intratumor heterogeneity. Proc Natl Acad Sci USA 2000; 97(10):5357-5362.
68. Gisselsson D. Chromosome instability in cancer: how, when and why? Adv Cancer Res 2003; 87:1-29.
69. Chang S, Khoo CM, Naylor ML et al. Telomere-based crisis: functional differences between telomerase activation and ALT in tumor progression. Genes Dev 2003; 17(1):88-100.
70. O'Hagan RC, Chang S, Maser RS et al. Telomere dysfunction provokes regional amplification and deletion in cancer genomes. Cancer Cell 2002; 2(2):149-155.
71. Montgomery E, Argani P, Hicks JL et al. Telomere lengths of translocation-associated and nontranslocation-associated sarcomas differ dramatically. Am J Pathol 2004; 164(5):1523-1529.
72. Munoz P, Blanco R, Flores JM et al. XPF nuclease-dependent telomere loss and increased DNA damage in mice overexpressing TRF2 result in premature aging and cancer. Nat Genet 2005; 37(10):1063-1071.
73. Blanco R, Munoz P, Flores JM et al. Telomerase abrogation dramatically accelerates TRF2-induced epithelial carcinogenesis. Genes Dev 2007; 21(2):206-220.
74. Wu L, Multani AS, He H et al. Pot1 deficiency initiates DNA damage checkpoint activation and aberrant homologous recombination at telomeres. Cell 2006; 126(1):49-62.
75. Bellon M, Datta A, Brown M et al. Increased expression of telomere length regulating factors TRF1, TRF2 and TIN2 in patients with adult T-cell leukemia. Int J Cancer 2006; 119(9):2090-2097.
76. Ning H, Li T, Zhao L et al. TRF2 promotes multidrug resistance in gastric cancer cells. Cancer Biol Ther 2006; 5(8):950-956.

77. Oh BK, Kim YJ, Park C et al. Up-regulation of telomere-binding proteins, TRF1, TRF2 and TIN2 is related to telomere shortening during human multistep hepatocarcinogenesis. Am J Pathol 2005; 166(1):73-80.
78. Yamada M, Tsuji N, Nakamura M et al. Down-regulation of TRF1, TRF2 and TIN2 genes is important to maintain telomeric DNA for gastric cancers. Anticancer Res 2002; 22(6A):3303-3307.
79. Yamada K, Yagihashi A, Yamada M et al. Decreased gene expression for telomeric-repeat binding factors and TIN2 in malignant hematopoietic cells. Anticancer Res 2002; 22(2B):1315-1320.
80. Lin X, Gu J, Lu C et al. Expression of telomere-associated genes as prognostic markers for overall survival in patients with non-small cell lung cancer. Clin Cancer Res 2006; 12(19):5720-5725.
81. Loeb LA, Christians FC. Multiple mutations in human cancers. Mutat Res 1996; 350(1):279-286.
82. Loeb LA. Cancer cells exhibit a mutator phenotype. Adv Cancer Res 1998; 72:25-56.
83. Sniegowski PD, Gerrish PJ, Johnson T et al. The evolution of mutation rates: separating causes from consequences. Bioessays 2000; 22(12):1057-1066.
84. Merlo LM, Pepper JW, Reid BJ et al. Cancer as an evolutionary and ecological process. Nat Rev Cancer 2006; 6(12):924-935.
85. Woerner SM, Kloor M, von Knebel Doeberitz M et al. Microsatellite instability in the development of DNA mismatch repair deficient tumors. Cancer Biomark 2006; 2(1-2):69-86.
86. di Pietro M, Sabates Bellver J, Menigatti M et al. Defective DNA mismatch repair determines a characteristic transcriptional profile in proximal colon cancers. Gastroenterology 2005; 129(3):1047-1059.
87. Jeter JM, Kohlmann W, Gruber SB. Genetics of colorectal cancer. Oncology (Williston Park) 2006; 20(3):269-276; discussion 285-266, 288-269.
88. Komarova NL, Lengauer C, Vogelstein B et al. Dynamics of genetic instability in sporadic and familial colorectal cancer. Cancer Biol Ther 2002; 1(6):685-692.
89. Sjoblom T, Jones S, Wood LD et al. The consensus coding sequences of human breast and colorectal cancers. Science 2006; 314(5797):268-274.
90. Gisselsson D. Mitotic instability in cancer: is there method in the madness? Cell Cycle 2005; 4(8):1007-1010.
91. Venkatesan RN, Bielas JH, Loeb LA. Generation of mutator mutants during carcinogenesis. DNA Repair (Amst) 2006; 5(3):294-302.
92. Bielas JH, Loeb KR, Rubin BP et al. Human cancers express a mutator phenotype. Proc Natl Acad Sci USA 2006; 103(48):18238-18242.
93. Venkatesan RN, Hsu JJ, Lawrence NA et al. Mutator phenotypes caused by substitution at a conserved motif A residue in eukaryotic DNA polymerase delta. J Biol Chem 2006; 281(7):4486-4494.
94. Goldsby RE, Hays LE, Chen X et al. High incidence of epithelial cancers in mice deficient for DNA polymerase delta proofreading. Proc Natl Acad Sci USA 2002; 99(24):15560-15565.
95. Charames GS, Bapat B. Genomic instability and cancer. Curr Mol Med 2003; 3(7):589-596.
96. Deng CX, Wang RH. Roles of BRCA1 in DNA damage repair: a link between development and cancer. Hum Mol Genet 2003; 12(Spec No 1):R113-R123.
97. Hackett JA, Feldser DM, Greider CW. Telomere dysfunction increases mutation rate and genomic instability. Cell 2001; 106(3):275-286.
98. Zhang H, Pan KH, Cohen SN. Senescence-specific gene expression fingerprints reveal cell-type-dependent physical clustering of up-regulated chromosomal loci. Proc Natl Acad Sci USA 2003; 100(6):3251-3256.
99. Zhang H, Herbert BS, Pan KH et al. Disparate effects of telomere attrition on gene expression during replicative senescence of human mammary epithelial cells cultured under different conditions. Oncogene 2004; 23(37):6193-6198.
100. Ning Y, Xu JF, Li Y et al. Telomere length and the expression of natural telomeric genes in human fibroblasts. Hum Mol Genet 2003; 12(11):1329-1336.
101. Benetti R, Garcia-Cao M, Blasco MA. Telomere length regulates the epigenetic status of mammalian telomeres and subtelomeres. Nat Genet 2007; 39(2):243-250.
102. Gonzalo S, Jaco I, Fraga MF et al. DNA methyltransferases control telomere length and telomere recombination in mammalian cells. Nat Cell Biol 2006; 8(4):416-424.
103. Nunney L. Lineage selection and the evolution of multistage carcinogenesis. Proc Biol Sci 1999; 266(1418):493-498.
104. Luebeck EG, Moolgavkar SH. Multistage carcinogenesis and the incidence of colorectal cancer. Proc Natl Acad Sci USA 2002; 99(23):15095-15100.
105. Little MP, Li G. Stochastic modelling of colon cancer: is there a role for genomic instability? Carcinogenesis 2007; 28(2):479-487.
106. Nowak MA, Michor F, Komarova NL et al. Evolutionary dynamics of tumor suppressor gene inactivation. Proc Natl Acad Sci USA 2004; 101(29):10635-10638.
107. Spencer SL, Gerety RA, Pienta KJ et al. Modeling somatic evolution in tumorigenesis. PLoS Comput Biol 2006; 2(8):e108.

Chapter 11

Prokaryotic Telomeres:
Replication Mechanisms and Evolution

Sherwood R. Casjens* and Wai Mun Huang

Abstract

Two types of bacterial telomeres of linear genomes are known. One type involves the covalent attachment of a terminal protein to each of the 5′-ends and the protective terminal protein is part of the priming complex in new rounds of DNA replication. The second type is a protein free DNA end in which one strand of the DNA duplex turns around and becomes its own complement. In the latter case, the telomere ends are copied into inverted repeats as replication proceeds around the hairpin ends and the resultant two halves of the inverted repeat are then resolved by a dimeric protein called protelomerase to form two new hairpin ends. Both of these telomere systems are found in bacteria and bacteriophages.

Introduction

Most prokaryote chromosomes, of both bacteria and bacteriophages, replicate as circular molecules, thereby avoiding altogether the problems and issues relating to end shortening during replication and the ends of these DNA molecules are not exposed. However, there are linear DNAs that are stably maintained in bacterial cells and which are replicated without a conventional theta-form intermediate.[1-4] These include linear bacterial chromosomes, plasmids and bacteriophage genomes. The ends of these molecules are called telomeres, but they are unrelated in sequence and structure to those of the eukaryote chromosomes. The prokaryotes have devised two general strategies for replicating linear genomes, (i) replication by "protein priming" in which a protein acts as the primer to which the complement of the terminal nucleotide of the template strand is added and (ii) the utilization of covalently-closed hairpin telomeres in which one strand simply turns around and becomes the other strand. The former type is found in some phages that infect diverse bacteria and in the chromosomes and plasmids of some actinomycete bacteria. The latter type is found in all of the chromosomes and many of the plasmids of the *Borrelia* spirochetes and in some chromosomes and plasmids of proteobacteria. We also note that organelle DNAs in some eukaryotes, which may have a prokaryotic origin, are also linear and can utilize both these replication strategies, but we do not review these here (see refs. 1 and 5 and references therein).

Prokaryotic Telomeres with Covalently Bound Terminal Proteins

Protein-Primed Replication by Bacteriophages

This mode of replication has been studied extensively by Salas and colleagues in *Bacillus subtilis* phage ϕ29.[6,7] This phage utilizes a protein-primed strategy for replication of its linear chromosome during its lytic growth. Briefly, the phage-encoded DNA polymerase causes the formation of a

*Corresponding Author: Sherwood R. Casjens—Pathology Department, University of Utah Medical School Emma Eccles Jones Medical Research Building, Room 5200 Salt Lake City, UT 84112 USA. Email: sherwood.casjens@path.utah.edu

Origin and Evolution of Telomeres, edited by Jozef Nosek and Ľubomír Tomáška.

covalent linkage between the first nucleotide (dAMP) and serine 232 of the terminal protein (TP). Interestingly, the incorporation of this dAMP is templated by nucleotide number two from the 3′-end of the virus DNA's template strand and then the complex slides back (3′-direction on the template) one nucleotide to "recover" the information of the terminal nucleotide (also a T) and the DNA polymerase then proceeds to copy the template strand. The elongating polymerase dissociates from the terminal protein after incorporation of about 6 nucleotides. Such protein-primed initiation occurs at both ends of the molecule and when the two elongating polymerases meet, the two template strands are separated and replication continues until both parental strands are completely copied by only leading strand synthesis, creating two daughter dsDNA molecules, each of which has one parental strand with its 5′-covalently attached parental TP and one newly synthesized daughter strand with its 5′-daughter TP (Fig. 1). The "telomeres" of phage ϕ29 are blunt-ended with 3′-TTTCAT present at the 3′-end of both strands. Although this sequence appears to form part of the recognition site for polymerase initiation, the parental TP is also an important part in the polymerase recognition complex.[6]

Terminal Protein-Linked Telomeres in Bacteria

Protein-primed replication is also successfully used by various plasmids and chromosomes in the actinomycete bacteria. These have been studied in less detail than the ϕ29 type phages, but they appear to use a mechanism that is reminiscent of the phage systems to replicate the ends of their linear DNA molecules. However, rather than copying the entire DNA molecule by bidirectional single-strand replacement from terminal origins, these bacterial replicons also have an internal bidirectional replication origin that is responsible for replication of the bulk of the DNA and putatively protein-primed replication is only used to replicate the termini[8,9] (Fig. 1). The details of this mode of telomere replication are still being worked out and they may be somewhat different on different replicons[10-12] and it has not yet been conclusively documented that protein-priming actually occurs. Indeed alternate models in which the TP is inserted into a continuous DNA strand have been suggested.[13] Complex palindromic sequences in the identical telomeric sequences at both ends are important[13,14] and several additional proteins including DNA polymerase I and DNA topoisomerase I are involved.[15]

Protein-Linked Telomeres and Evolution

Other phages that utilize the protein-primed replication strategy are *B. subtilis* phages B103 and GA-1 (close relatives of ϕ29) which have 3′-TTTCAT and 3′-TTTATCT short terminal sequences present at the ends of both strands, respectively[16,17] and the related *Streptococcus pneumoniae* phage Cp-1 which has a longer terminal 236 bp sequence present at both ends.[18] *Escherichia coli* phage PRD1 has a very different overall lifestyle from the above phages, but utilizes a similar replication mechanism and has 110 bp long inverted terminal repeats[19] and the eukaryotic adenoviruses also utilize protein-primed replication.[20] Thus, this replication strategy is used by two different very distantly related viruses that infect Gram positive Firmicutes and Gram negative Proteobacteria, as well as in the eukaryote adenoviruses. In addition, the linear cellular replicons in the actinomycetes probably utilize a similar TP based mechanism. Was this strategy invented only once, or are these examples of convergent evolution?

These linear replicons all have terminal inverted repeats, but they have variable lengths and sequences ranging from six bp (phage ϕ29) to many thousands of bps (*Streptomyces coelicolor* A3(2) chromosome).[21] It is interesting to note that in the few cases examined in detail, they all use a sliding back mechanism (above) for error-checking the first nucleotide that is linked to the terminal protein, but they do not all use the same nucleotide position as template for the incorporation of the first templated nucleotide before the complex slides back (e.g., bp number 2 in ϕ29 and number 4 in PRD1).[19] The TPs of the phages form a group of related proteins,[6] but not all the bacterial TPs are obviously related to one another[10] and the adenovirus TP proteins are not straightforwardly related to the others.[22] It has therefore been suggested that these similar systems are the result of several convergent evolutionary paths; however, on the other hand, ϕ29 and adenovirus DNA polymerases have conserved domains involved in binding to their respective

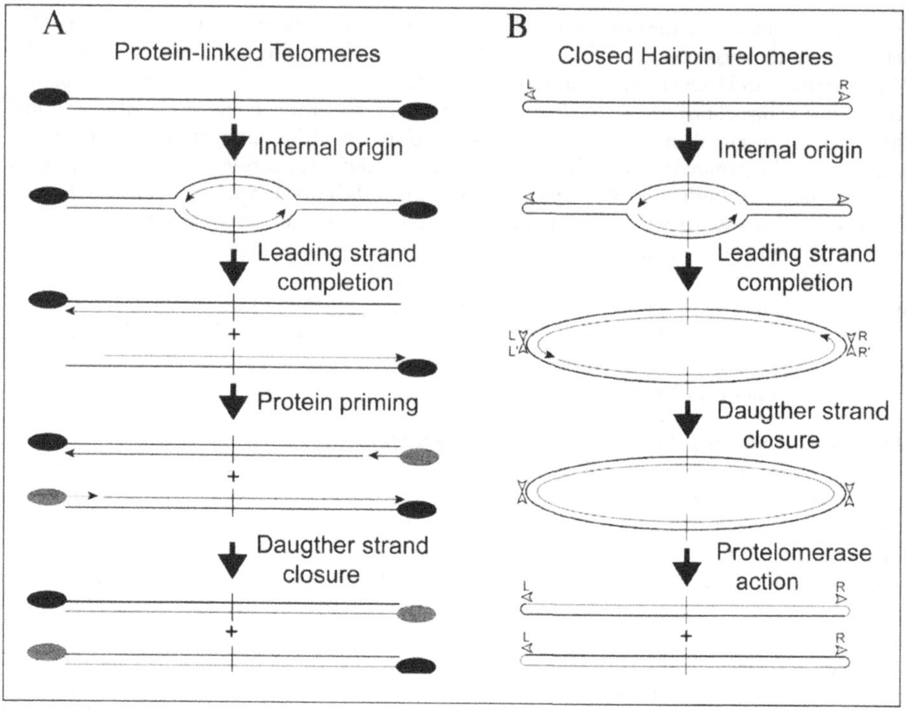

Figure 1. Prokaryote telomeres and their replication. A) Protein-linked telomeres. The black ovals represent the parental terminal proteins and gray ovals daughter terminal proteins. B) Closed hairpin telomeres. Open arrowheads denote the telomere sequence and protelomerase recognition site. In the figure black lines are parental DNA strands and gray lines are daughter strands with small black arrowheads denoting their 3'-ends and direction of synthesis. Thin vertical lines mark the location of the internal origin. Note that bacteriophages such as φ29 do not have an internal origin and are completely replicated from the protein-priming site.

TPs.[23,7] More structural information on the proteins involved is needed to determine whether the different protein-linked telomere systems are all descendent from a very ancient common ancestor or whether they result from convergent evolution.

Prokaryote Telomeres with Covalently-Closed Terminal Hairpins

Covalently-Closed Hairpin Replication by Bacteriophages

Unlike the protein-primed replication discussed above, the bacterial closed hairpin DNA ends do not serve as replication origins and replication is driven solely by internal origins. However, semiconservative replication of these linear DNA molecules with closed hairpin ends cannot be completed because the two parental strands are covalently linked to one another through the loop of nucleotides in their hairpin ends. A mechanism must therefore exist to allow separation of the two strands. The current working model for this type of replication is diagrammed in Figure 1, where an internal origin programs bidirectional replication. The replication complex can readily traverse the hairpin ends to generate a pair of inverted repeating sequences; these are marked with a pair of arrowheads and RR' and LL' in Figure 1. The completed replication will generate a head-to-head dimer circle. The enzyme that performs the unique function of separating the two daughter chromosomes has been called protelomerase (for prokaryotic telomerase[24]) in most systems or ResT (for resolution of telomeres[25]) in the *Borrelia* system.

Three temperate bacteriophages are known in which the prophage exists as a non-integrated, linear plasmid with closed hairpin ends. These are N15 (in *E. coli*),[26] YP54 (in *Yersinia enterocolitica*)[27] and ϕKO2 (in *Klebsiella oxytoca*).[28] The genes in these phages' genomes are arranged in the same order as the phage lambda gene functions and they have varying sequence homologies to one another that include a replicase gene (*repA*) and the protelomerase gene. The linear N15 and PY54 plasmid replication origins have been identified to lie inside the *repA* genes.[29-31]

Thus, these three phages appear to replicate their prophage plasmid DNAs in the same manner. Replication forks are thought to move outward in both directions from the asymmetrically positioned origin (which is most likely recognized by the RepA protein). When a fork reaches a hairpin end, replication can readily copy the single-stranded parental template and traverse the hairpin ends. After passing the hairpin, the newly synthesized leading and lagging strands are expected to converge and join together, forming a continuous completed daughter strand (Fig. 1). Once the hairpin telomere is converted to double strand DNA (Fig. 1; LL' and RR'), it becomes a substrate for the protelomerase, which converts the duplicated telomeres into two hairpins (see below for mechanism of the conversion). In phage N15 and ϕKO2 the protelomerase is required for maintenance of the linear form of the prophage plasmid[32] (W. Huang, unpublished results). Protelomerase should be able to perform its reaction as soon as its substrate is formed, so most likely such resolution at the two ends of the parental molecule need not be synchronized and replication intermediates such as Y-shaped molecules where only one end is resolved, rather than full dimer circles, have been observed.[29] DNA replication during the lytic growth of these phages presumably must have a different mechanism, since the two prophage ends are covalently joined in virion DNA.[33]

Protelomerase has been purified from all three phage systems and in each case it faithfully carries out the above reaction with the expected sequence specificity and without cofactors.[34,35] The reaction catalyzed by the phage ϕKO2 protelomerase is illustrated in Figure 2. It uses a concerted breakage-rejoining mechanism that is similar to that of the tyrosine-recombinases s (e.g., topoisomerase IB) to create the hairpin telomere.[35] Two molecules of protelomerase begin the reaction by making a pair of nicks (one in each strand) 6 bp apart and 3 bp on either side of the center of dyad symmetry of the target site. As a reaction intermediate, tyrosine 425 of each enzyme subunit is covalently attached at each nick forming a 3'-phosphoryl-protein-DNA complex and a free 5'-OH end.[35] The 6 bps between the nicks are then separated and each of the resulting 6 nucleotide single-strands loops back and is rejoined to the 3'-phosphoryl group bound to the other protelomerase subunit as an intra-DNA strand reaction, thus forming the covalently closed hairpins. This reaction is catalyzed accurately and efficiently by purified systems in vitro, but it is not known how its actions are regulated in vivo, or whether there are other proteins that may facilitate hairpin telomere formation in the cell. Recently, the x-ray structure of the cleavage complex of the ϕKO2 protelomerase has been solved (H. Aihara, W. Huang and T. Ellenberger, unpublished results). It shows a pair of interlocked protelomerase subunits interacting with the nicked and distorted duplex target DNA. The pairings in the central 6 bp containing the nucleotides which will eventually loop back to form the hairpins are markedly deformed. This structure strongly supports the model described above.

Covalently-Closed Hairpin Replication in Bacteria

All members of the *Borrelia* spirochete genus studied to date have hairpin-tipped linear chromosomes and plasmids.[36] These bacteria cause relapsing fevers and Lyme disease in humans. The *B. burgdorferi* strain B31 protelomerase (called ResT) shares approximately 22% amino acid sequence identity with the phage enzymes and is smaller by about 150 residues. Its action has been investigated both in vivo and in vitro in some detail and the overall cutting-rejoining mechanism which involves a central 6 bp for the strand exchange, appears to be similar to that of the phage enzymes discussed above.[25,37-39] A hairpin binding module in addition to the catalytic module has been proposed for the *Borrelia* enzyme to generate a preformed hairpin within its target site to

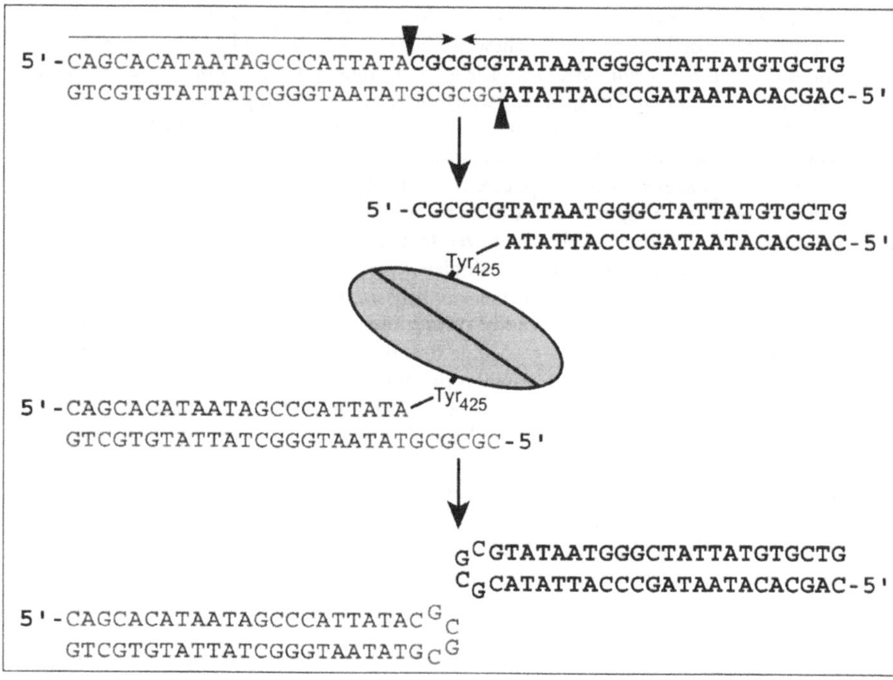

Figure 2. The protelomerase catalyzed reaction. The phage ϕKO2 protelomerase target site shown on top has arrows indicating the dyad symmetry of the target site and black triangles mark the positions of cleavage by protelomerase. In the middle, a protelomerase dimer (gray oval) is covalently bound to the 3-ends after it has cleaved the target. Below, the two closed hairpin telomeres formed after religation and protelomerase release.

initiate the reaction.[40] However, such a model is not consistent with the structure of the ϕKO2 protein-DNA complex.

Unlike the phage situations, where only the two different ends of the genome are processed, *Borrelia* bacteria harbor as many as twelve linear plasmids in addition to their ~900 kbp linear chromosome.[41] These multiple telomeres do not have identical nucleotide sequences, but in the cases where the terminal sequence is known, all of the telomeres have some similarity in sequence in their terminal 21 bp.[42] To date the sequences of 24 telomeres represent 20 different telomeric sequences are known (Casjens S, Huang W and Fraser C, unpublished results). There is only one recognizable protelomerase gene in the *Borrelia* genome[41] and it seems certain that all of the telomeres are created by the same enzyme, ResT. Preliminary analysis of these telomeric sequences suggests that a conserved 5'-TAGTA (except in one case where it is 5'-TATTA) is always present 14 bp from the end of the linear DNA molecule[42] (Casjens S, Huang W and Fraser C, unpublished results). Most of the remaining bp in the terminal region appear to have little sequence restriction beyond the very high A+T content of the region, so it seems reasonable to postulate that the 5'-TAGTA forms at least part of the recognition sequence for this enzyme.

The α-Proteobacteria *Agrobacterium tumefaciens*, the causative agent of crown gall disease of plants, is the only other bacteria that is currently known to have a chromosome with hairpin telomeres (it also has a circular chromosome).[43,44] Its circular chromosome encodes a protein (open reading frame Atu_2523) with 23% identity to the *Borrelia* protelomerase; however, its reported genome sequence did not include the telomeric sequences since hairpin-ended DNA fragments cannot be cloned.[44,45] We recently cloned, expressed, purified and characterized the *Agrobacterium* protelomerase and sequenced the telomeres of its linear chromosome (Huang W, Aron J and

Casjens S, unpublished results). Like the other protelomerases, this enzyme accurately regenerates *Agrobacterium* hairpin telomeres from the putative replication intermediate target sequence (novel joints on the head-to-head dimer circle; Fig. 1) and utilizes a mechanism that is similar to the phage and *Borrelia* enzymes in which the central 6 bp are involved in the strand-exchange for hairpin formation.

To date, bacterial linear hairpin chromosomal telomeres have been found either as the only chromosome configuration as in the case of *Borrelia* species, or co-existing with another circular chromosome as in the two chromosomes of *Agrobacterium tumefaciens* (biovar 1). In both organisms, the composition of the replication machinery required for bidirectional DNA initiation and elongation are principally conserved. These include the highly conserved DnaA and DnaE and their auxiliary proteins. However, a difference appears to be in the way in which the postreplicated chromosomes are separated. Organisms with circular chromosomes such as *E. coli* or *B. subtilis*, employ the *dif*/XerCD system, which separates interlocked daughter circles, for their resolution.[46] On the other hand *Borrelias* do not have a known *dif* site, nor do they encode XerCD proteins and instead they utilize the protelomerase/resolvase system, described here, to separate the covalently joined daughter chromosomes. *Agrobacterium tumefaciens C58* carries both the *xerCD* genes and a protelomerase gene for the circular and linear modes of resolution, respectively. It is therefore of interest to note that recently, the *E. coli* circular chromosome was made linear without causing an appreciable growth defect when the linear hairpin end generating system from the bacteriophage N15 was introduced.[47] This interesting observation certainly lends support to the notion that protelomerase-linear hairpin generating system is the functional equivalent of the *dif*/xerCD resolution system of circular chromosomes and that there aren't any important players in the protelomerase system that were not previously recognized.

Hairpin Telomeres and Evolution

Hairpin telomeres are present in some mitchondrial genomes,[48] some virus chromosomes[49] and even nuclear chromosomes of some yeast mutants,[50] but with the exception of some aglal viruses (see below) these do not appear to be closely evolutionarily related to the bacterial hairpin telomeres and their resolution system. The hairpin telomeres found in bacteria are generated by a very simple system involving a single protein and its specific cognate target sequence. These targets are short inverted repeats of less than 60 bp (two half sites of 30 bp or less). This elegant and efficient linear replicon system allows the bacteria or the bacteriophages to avoid the peril of exposed free DNA ends and end-shortening issues due to the primer requirement and exclusive 5′ to 3′ direction of DNA polymerizing activities. The relatively low amino acid sequence identity between the different protelomerases and their apparently unrelated target sequences certainly indicates that hairpin telomeres and the enzyme system that generates them have existed for a very long time.

Yet the presence of such systems is rare compared with the vast majority of circular chromosomes and plasmids in prokaryotes in nature. This rarity suggests that either there must be an inherent disadvantage (high cost) to maintaining a hairpin-ended linear chromosomes and plasmids, so only a few remain from an era when they were more common, or they have arrived late in evolution and been moved around by horizontal transfer to achieve their current distribution. To date there are less than ten convincing cases among the hundreds of completely sequenced bacteria and bacteriophage genomes. In as much as inverted repeats are common in genomes, the recognition of unresolved target sites for protelomerases in sequenced genomes is by no means straightforward without concomitant biochemical analyses. On the other hand, the protelomerase-proteins are easily recognized with conserved signature amino acids as an expanded set of motifs described for the well characterized tyrosine-recombinase superfamily.[35,37] Based on the presence of the unique protein motifs, it has recently been found that protelomerase-like genes are also present in large DNA phycodnaviridae viruses that infect eukaryotic fresh water and marine alga.[51,52] These groups of unicellular alga are generally believed to be near the root of the eukaryotic phylogenic branches and they often carry genes characteristic of both eukaryotic and prokaryotic in nature. Although the presence of hairpin end generating systems in these eukaryotic phycodnaviridae viruses remains to be experimentally verified, it raises the possibility that the hairpin generating systems are "ancient". Thus, since horizontal transfer among eukaryotes is rare compared to bacteria, such systems may have existed in ancient systems before the split of prokaryotes and eukaryotes. If this were true, why was it lost in most of the present day organisms? Hairpin ends, when inappropriately formed (i.e., at stages of the life cycles other then during chromosome or plasmid segregation where two half sites come together) will generate breaks in the genome that are difficult to repair, as known repair enzymes all require a free 3′ or 5′ nucleotide entrance for activity. Hence it is reasonable to assume that the activity of protelomerase must be tightly regulated, but at present very little is known about its regulation. Perhaps it is the high "cost" of its regulation is one of the reasons for its rarity in present organisms, whether that rarity is due to loss from ancient systems or failure to spread widely after a more recent origin.

Concluding Remarks

There are two different types of prokaryote telomeres, the terminal protein-linked and covalently-closed hairpin telomeres, which are replicated by very different mechanisms. The terminal protein that is covalently bound to the tips of the former type acts as the primer for complete copying of the 3′-end of the template DNA strands (shown directly only in the virus systems). The amino acid sequences of these terminal proteins and the nucleotide sequences of the telomeric regions are highly variable and it has been suggested that these telomeres (and their replication machinery) may have arisen several times independently during evolution in prokaryotes and eukaryotes. In support this idea, the fact that there are multiple interacting molecular components involved (the telomere itself, the terminal protein, DNA polymerase, DNA topoisomerase and other proteins that must specifically interact with terminal protein) which are not genetically linked, at least in cases that have been studied, makes horizontal transfer of this apparatus seem less likely.

On the other hand, the phage and bacterial protelomerase enzymes that create hairpin telomeres are all homologous and so are thought to have arisen only once. To date, in bacteria this apparatus has only been found in *Borrelia* and *Agrobacterium*. These two genera are very distant. Since it appears not to have been "invented" independently in the phages or these two bacterial branches, either it was ancestral and the larger majority of bacteria have lost it, or it was horizontally transferred between the spirochetes and the proteobacteria in the more recent past. Parsimony suggests that the latter may be the likely evolutionary history and the fact that the hairpin telomere resolution machinery is "simple", containing only the target site and one protein, makes horizontal transfer seem rather reasonable. In addition, phage are by their nature mediators of horizontal transfer[53] and the presence of hairpin telomere machinery in phage suggests that perhaps such phages might have been responsible for their horizontal transfer.

A common feature of both types of telomeres discussed here is that their locations are programmed by the local nucleotide sequences, so that if that sequence is transferred to another genomic location it will program a telomere at that location. It appears that telomere rearrangement has happened with both types. In *Borrelia* there is evidence of numerous past rearrangements where one hairpin telomeric region appears to have replaced another[41,42] and in *Streptomyces* there appear to be internal "pseudo-telomeres" that may represent partly damaged telomeres to which new protein-linked telomeric regions have been added.[54,55] It thus appears that the presence of telomeres can add to the long term genetic instability of nearby sequences.

The question remains as to why some prokaryote DNA molecules are replicated by one or the other of these linear strategies, while the majority of DNA molecules in bacteria replicate as circles. Is there some inherent advantage or disadvantage to the linear strategies in some situations, are linear molecules relics of an earlier "linear age", or are these mechanisms randomly extant simply because they "work"? It seems most reasonable to postulate that telomeres do give advantage in some evolutionary niches and we hope that further research in this area will eventually be able to solve this mystery.

References

1. Hinnebusch J, Tilly K. Linear plasmids and chromosomes in bacteria. Mol Microbiol 1993; 10:917-922.
2. Huang WM, Ruan Q, Casjens S. Hairpin telomeres of inear bacterial chromosomes and plasmids: how to make them. In: Cabello F, Hulinska D, Godfrey H, eds. Molecular Biology of Spirochetes. Amsterdam: IOS Press; 2006:299-308.
3. Meinhardt F, Schaffrath R, Larsen M. Microbial linear plasmids. Appl Microbiol Biotechnol 1997; 47:329-336.
4. Casjens S. Evolution of the linear DNA replicons of the Borrelia spirochetes. Curr Opin Microbiol 1999; 2:529-534.
5. Nosek J, Tomaska L, Fukuhara H et al. Linear mitochondrial genomes: 30 years down the line. Trends Genet 1998; 14:184-188.
6. Meijer WJ, Horcajadas JA, Salas M. ϕ29 family of phages. Microbiol Mol Biol Rev 2001; 65:261-287
7. Kamtekar S, Berman AJ, Wang J et al. The ϕ29 DNA polymerase:protein-primer structure suggests a model for the initiation to elongation transition. EMBO J 2006; 25:1335-1343.
8. Chang PC, Cohen SN. Bidirectional replication from an internal origin in a linear streptomyces plasmid. Science 1994; 265:952-954.
9. Musialowski MS, Flett F, Scott GB et al. Functional evidence that the principal DNA replication origin of the Streptomyces coelicolor chromosome is close to the dnaA-gyrB region. J Bacteriol 1994; 176:5123-5125.
10. Huang CH, Tsai HH, Tsay YG et al. The telomere system of the Streptomyces linear plasmid SCP1 represents a novel class. Mol Microbiol 2007; 63:1710-1718.
11. Stoll A, Redenbach M, Cullum J. Identification of essential genes for linear replication of an SCP1 composite plasmid. FEMS Microbiol Lett 2007; 270:146-154.
12. Zhang R, Yang Y, Fang P et al. Diversity of telomere palindromic sequences and replication genes among Streptomyces linear plasmids. Appl Environ Microbiol 2006; 72:5728-5733.
13. Qin Z, Cohen SN. Replication at the telomeres of the Streptomyces linear plasmid pSLA2. Mol Microbiol 1998; 28:893-903.
14. Huang CH, Lin YS, Yang YL et al. The telomeres of Streptomyces chromosomes contain conserved palindromic sequences with potential to form complex secondary structures. Mol Microbiol 1998; 28:905-916.
15. Bao K, Cohen SN. Reverse transcriptase activity innate to DNA polymerase I and DNA topoisomerase I proteins of Streptomyces telomere complex. Proc Natl Acad Sci USA 2004; 101:14361-14366.
16. Yoshikawa H, Garvey KJ, Ito J. Nucleotide sequence analysis of DNA replication origins of the small Bacillus bacteriophages: evolutionary relationships. Gene 1985; 37:125-130.
17. Pecenkova T, Benes V, Paces J et al. Bacteriophage B103: complete DNA sequence of its genome and relationship to other Bacillus phages. Gene 1997; 199:157-163.
18. Escarmis C, Gomez A, Garcia E et al. Nucleotide sequence at the termini of the DNA of Streptococcus pneumoniae phage Cp-1. Virology 1984; 133:166-171.
19. Caldentey J, Blanco L, Bamford DH et al. In vitro replication of bacteriophage PRD1 DNA. Characterization of the protein-primed initiation site. Nucleic Acids Res 1993; 21:3725-3730.
20. de Jong RN, van der Vliet PC, Brenkman AB. Adenovirus DNA replication: protein priming, jumping back and the role of the DNA binding protein DBP. Curr Top Microbiol Immunol 2003; 272:187-211.
21. Weaver D, Karoonuthaisiri N, Tsai HH et al. Genome plasticity in Streptomyces: identification of 1 Mb TIRs in the S. coelicolor A3(2) chromosome. Mol Microbiol 2004; 51:1535-1550.
22. Yang CC, Huang CH, Li CY et al. The terminal proteins of linear Streptomyces chromosomes and plasmids: a novel class of replication priming proteins. Mol Microbiol 2002; 43:297-305.
23. Dufour E, Rodriguez I, Lazaro JM et al. A conserved insertion in protein-primed DNA polymerases is involved in primer terminus stabilisation. J Mol Biol 2003; 331:781-794.
24. Rybchin VN, Svarchevsky AN. The plasmid prophage N15: a linear DNA with covalently closed ends. Mol Microbiol 1999; 33:895-903.
25. Chaconas G, Stewart PE, Tilly K et al. Telomere resolution in the Lyme disease spirochete. EMBO J 2001; 20:3229-3237.
26. Ravin NV. Mechanisms of replication and telomere resolution of the linear plasmid prophage N15. FEMS Microbiol Lett 2003; 221:1-6.
27. Hertwig S, Klein I, Lurz R et al. PY54, a linear plasmid prophage of Yersinia enterocolitica with covalently closed ends. Mol Microbiol 2003; 48:989-1003.
28. Casjens SR, Gilcrease EB, Huang WM et al. The pKO2 linear plasmid prophage of Klebsiella oxytoca. J Bacteriol 2004; 186:1818-1832.

29. Ravin NV, Kuprianov VV, Gilcrease EB et al. Bidirectional replication from an internal ori site of the linear N15 plasmid prophage. Nucleic Acids Res 2003; 31:6552-6560.
30. Ziegelin G, Tegtmeyer N, Lurz R et al. The repA gene of the linear Yersinia enterocolitica prophage PY54 functions as a circular minimal replicon in Escherichia coli. J Bacteriol 2005; 187:3445-3454.
31. Mardanov AV, Ravin NV. Functional characterization of the repA replication gene of linear plasmid prophage N15. Res Microbiol 2006; 157:176-183.
32. Ravin NV, Strakhova TS, Kuprianov VV. The protelomerase of the phage-plasmid N15 is responsible for its maintenance in linear form. J Mol Biol 2001; 312:899-906.
33. Ravin V, Ravin N, Casjens S et al. Genomic sequence and analysis of the atypical temperate bacteriophage N15. J Mol Biol 2000; 299:53-73.
34. Deneke J, Ziegelin G, Lurz R et al. Phage N15 telomere resolution. Target requirements for recognition and processing by the protelomerase. J Biol Chem 2002; 277:10410-10419.
35. Huang WM, Joss L, Hsieh T et al. Protelomerase uses a topoisomerase IB/Y-recombinase type mechanism to generate DNA hairpin ends. J Mol Biol 2004; 337:77-92.
36. Casjens S, Murphy M, DeLange M et al. Telomeres of the linear chromosomes of Lyme disease spirochaetes: nucleotide sequence and possible exchange with linear plasmid telomeres. Mol Microbiol 1997; 26:581-596.
37. Deneke J, Burgin AB, Wilson SL et al. Catalytic residues of the telomere resolvase ResT: a pattern similar to, but distinct from, tyrosine recombinases and type IB topoisomerases. J Biol Chem 2004; 279:53699-53706.
38. Beaurepaire C, Chaconas G. Mapping of essential replication functions of the linear plasmid lp17 of B. burgdorferi by targeted deletion walking. Mol Microbiol 2005; 57:132-142.
39. Kobryn K, Chaconas G. ResT, a telomere resolvase encoded by the Lyme disease spirochete. Mol Cell 2002; 9:195-201.
40. Bankhead T, Chaconas G. Mixing active-site components: a recipe for the unique enzymatic activity of a telomere resolvase. Proc Natl Acad Sci USA 2004; 101:13768-13773.
41. Casjens S, Palmer N, van Vugt R et al. A bacterial genome in flux: the twelve linear and nine circular extrachromosomal DNAs in an infectious isolate of the Lyme disease spirochete Borrelia burgdorferi. Mol Microbiol 2000; 35:490-516.
42. Huang WM, Robertson M, Aron J et al. Telomere exchange between linear replicons of Borrelia burgdorferi. J Bacteriol 2004; 186:4134-4141.
43. Allardet-Servent A, Michaux-Charachon S, Jumas-Bilak E et al. Presence of one linear and one circular chromosome in the Agrobacterium tumefaciens C58 genome. J Bacteriol 1993; 175:7869-7874.
44. Goodner B, Hinkle G, Gattung S et al. Genome sequence of the plant pathogen and biotechnology agent Agrobacterium tumefaciens C58. Science 2001; 294:2323-2328.
45. Wood DW, Setubal JC, Kaul R et al. The genome of the natural genetic engineer Agrobacterium tumefaciens C58. Science 2001; 294:2317-2323.
46. Blakely GW, Davidson AO, Sherratt DJ. Sequential strand exchange by XerC and XerD during site-specific recombination at dif. J Biol Chem 2000; 275:9930-9936.
47. Cui T, Moro-oka N, Ohsumi K et al. Escherichia coli with a linear genome. EMBO Rep 2007; 8:181-187.
48. Fukuhara H, Sor F, Drissi R et al. Linear mitochondrial DNAs of yeasts: frequency of occurrence and general features. Mol Cell Biol 1993; 13:2309-2314.
49. Garcia AD, Moss B. Repression of vaccinia virus Holliday junction resolvase inhibits processing of viral DNA into unit-length genomes. J Virol 2001; 75:6460-6471.
50. Maringele L, Lydall D. The PAL-mechanism of chromosome maintenance: causes and consequences. Cell Cycle 2005; 4:747-751.
51. Delaroque N, Muller DG, Bothe G et al. The complete DNA sequence of the Ectocarpus siliculosus Virus EsV-1 genome. Virology 2001; 287:112-132.
52. Wilson WH, Schroeder DC, Allen MJ et al. Complete genome sequence and lytic phase transcription profile of a Coccolithovirus. Science 2005; 309:1090-1092.
53. Casjens S, Hendrix R. Bacteriophage roles in bacterial chromosome evolution. In: Higgins P, ed. The Bacterial Chromosome. Washington, D.C.: ASM Press; 2005:39-52.
54. Huang CH, Chen CY, Tsai HH et al. Linear plasmid SLP2 of Streptomyces lividans is a composite replicon. Mol Microbiol 2003; 47:1563-1576.
55. Bentley SD, Brown S, Murphy LD et al. SCP1, a 356,023 bp linear plasmid adapted to the ecology and developmental biology of its host, Streptomyces coelicolor A3(2). Mol Microbiol 2004; 51:1615-1628.

Chapter 12

Mitochondrial Telomeres:
An Evolutionary Paradigm for the Emergence of Telomeric Structures and Their Replication Strategies

Jozef Nosek* and Ľubomír Tomáška

Abstract

Linear DNA genomes are sporadically found among viruses, bacteria and organelles. In contrast, virtually all eukaryotic species harbor in their nuclei chromosomes consisting of linear DNA molecules that terminate with specific structures termed telomeres, indicating that this genomic or chromosomal form may, under specific conditions, provide a selective advantage. As the molecular form of eukaryotic chromosomes and their telomeric structures does not seem to be related to any linear genome known in free living prokaryotes, linear chromosomes in eukaryotic nuclei may represent evolutionary innovation. This raises the question of how linear chromosomes and primordial pathways for the maintenance of their terminal structures emerged in eukaryotes. In this chapter we review what we have learned from studies on linear DNA genomes and their terminal structures in yeast mitochondria. We briefly outline how linear DNA genomes might have emerged in organelles and, based on parallels between the mitochondrial and nuclear systems, suggest a scenario for emergence of linear chromosomes in the nuclei of early eukaryotes.

Introduction

The linear DNA genome in mitochondria was first reported in 1968 by Syuama and Miura.[1] A number of studies during the following decades revealed that the occurrence of species harboring linear mitochondrial genomes is unexpectedly high. These organisms belong to taxonomically distant taxa such as jakobid, ciliate and apicomplexan protists, algae, oomycete fungi, yeasts and even several metazoan species from the phylum Cnidaria. The list of linear mitochondrial genomes may be complemented by linear DNA plasmids isolated from the mitochondria of plants, slime molds, filamentous fungi and yeasts (for review see refs. 2, 3). Studies in yeasts were particularly important for the development of concepts regarding emergence of linear DNA genomes, evolution of their terminal structures and maintenance pathways. Species with linear mitochondrial genomes occur almost randomly on the phylogenetic tree and closely related species or even different strains of the same organism may contain different mitochondrial genome forms (Fig. 1). This indicates that linear mitochondrial genomes emerged independently in different lineages, presumably via simple molecular mechanism(s). This conclusion is further supported by differences in their telomeric structures and/or sequence motifs. In addition, the molecular diversity of mitochondrial telomeres indicates that fortuitous emergence of linear DNA genomes was accompanied by applying different solutions to the end-replication problem (for review see refs. 2, 3). This can be illustrated

*Corresponding Author: Jozef Nosek—Departments of Biochemistry and Genetics, Faculty of Natural Sciences, Comenius University, Mlynska dolina CH-1, 842 15 Bratislava, Slovakia. Email: nosek@fns.uniba.sk

Origin and Evolution of Telomeres, edited by Jozef Nosek and Ľubomír Tomáška.

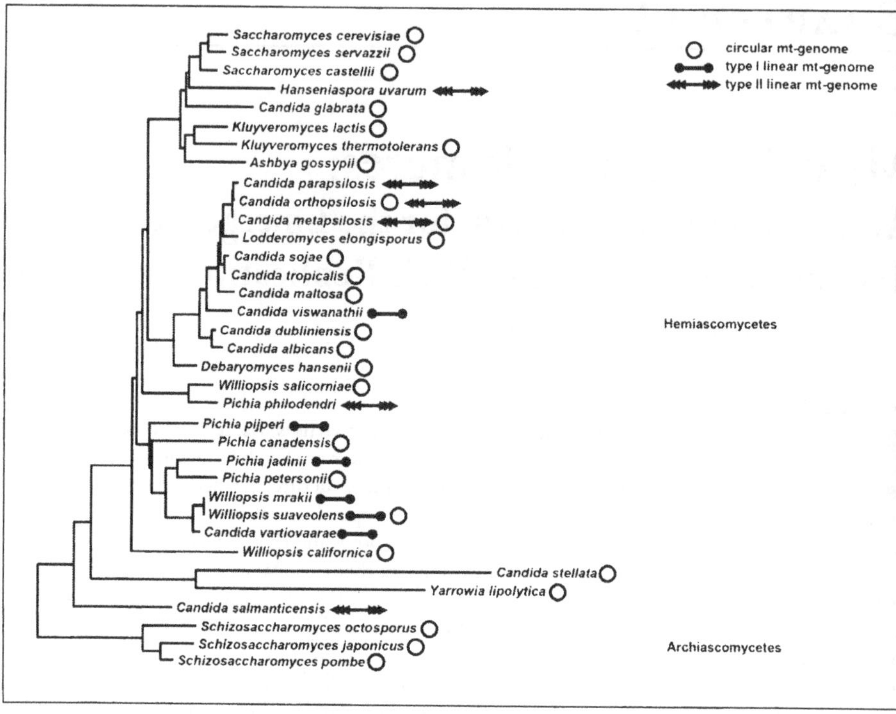

Figure 1. Phylogenetic tree of yeast species indicating the occurrence of linear mitochondrial genomes (for details see refs. 2-5,7,9,12,20,21,45,46; M. Valach and J. Nosek, unpublished). The tree was calculated from the sequences of the D1/D2 region of the 26S ribosomal RNA gene using the Neighbor-Joining method. Note that strains of *W. suaveolens, C. orthopsilosis* and *C. metapsilosis* exhibit intraspecific variability in the form of the organellar genome. Mitochondrial telomeres of type I and type II linear genomes are represented by telomeric hairpins (t-palindromes) and arrays of tandem repeats, respectively. T-circles were detected in *C. parapsilosis, C. orthopsilosis, C. metapsilosis, P. philodendra* and *C. salmanticensis.*

by three structurally different types of mitochondrial telomeres identified in yeasts.[4-6] Type I linear mitochondrial genomes found in species belonging to the *Williopsis-Pichia* clade[7] terminate with covalently closed single-stranded hairpin loops similar to the termini of poxviral DNA, *Borrelia* chromosomes and telomeric palindromes (t-palindromes) at the ends of nuclear chromosomes of yeast *tlc1 rad52 exo1* mutants.[8] The termini of type II linear genomes (e.g., *Candida parapsilosis*[5]) consist of arrays of tandem repeats remotely resembling the ends of typical nuclear chromosomes. The third type of linear genome found in yeast mitochondria is represented by the linear DNA plasmid pPK2 from *Pichia kluyveri*,[6] which has a terminal protein covalently attached to the 5′ ends of the linear DNA, similar to that seen at the termini of adenoviral DNA and linear plasmids and chromosomes in *Streptomyces*.

A Natural Telomerase-Independent System Occurring in Yeast Mitochondria

In mid 1980s, we entered the field of telomere biology as undergraduate students in the laboratory of Ladislav Kovac (Department of Bioenergetics, Institute of Animal Physiology, Slovak Academy of Sciences, Ivanka pri Dunaji, Czechoslovakia). Our goal was to investigate how the linear mitochondrial genome, discovered by Ladislav Kovac, Jaga Lazowska and Piotr P. Slonimski[9] in *Candida rhagii* SR23 (now taxonomically reclassified as *C. parapsilosis*), solves the end-replication

Table 1. *Analogy between nuclear and mitochondrial telomeres*

	Human Nuclear Telomere	*C. parapsilosis* Mitochondrial Telomere
Telomere architecture		
tandem repeat motif	6 bp	738 bp
single-stranded overhang	3′	5′
t-loops	+	+
t-circles	+	+
telomeric ssDNA binding protein	hPot1	mtTBP
Telomere maintenance		
telomerase	+	not detected
ALT pathway(s)	+	+

problem. At this time a major breakthrough in the field of telomere biology occurred, when Elizabeth Blackburn and Carol Greider discovered telomerase, a special nucleoprotein enzyme with reverse transcriptase activity copying its own template to the 3′ ends of the chromosomal termini, in nuclear extracts of *Tetrahymena*.[10,11]

Although telomerase is considered a typical feature of telomere maintenance in eukaryotes, certain species, and importantly also a significant fraction of human cancers, lack its activity. In these cells the problems associated with the ends of linear DNA molecules are solved by alternative means. An intriguing example of alternative is chromosome-end maintenance via telomere-associated retrotransposons (t-posons), which apparently replaced telomerase in fruit flies (Pardue and DeBaryshe, this volume).

As there is no counterpart of telomerase in yeast mitochondria with a linear DNA genome, the mitochondrial system provided us with a unique playground for (i) analysis of the nature of telomerase-independent pathways of telomere maintenance and mechanisms employed to solve telomere-associated problems, and (ii) understanding how linear chromosomes emerged during evolution.

In the following years, in close collaboration with Hiroshi Fukuhara (Institute Curie, Orsay, France) and Jack D. Griffith (University of North Carolina, Chapel Hill), we characterized the structure of mitochondrial telomeres of *C. parapsilosis* in more detail and showed that they display essentially the same structural features as typical telomeres of nuclear eukaryotic chromosomes (Table 1, Fig. 2), including terminal arrays of tandem repeats, single-stranded overhangs[5,12] protected by a specific telomere-binding protein[13-15] and higher order structures[16] resembling the t-loops of mammalian cells.[17] In 2000, we published a paper describing extragenomic circular molecules derived exclusively from the telomeric sequence (telomeric circles, t-circles) and proposed their active role in the maintenance of mitochondrial telomeres.[18] Subsequently, we demonstrated that t-circles replicate via a rolling-circle strategy generating long arrays of telomeric tandem repeats.[19] As a result of these findings, we concluded that mitochondrial telomeres are maintained by a novel t-circle-dependent pathway. Isolation of mutant strains lacking t-circles and harboring a circularized derivative of the mitochondrial genome further supported this idea and led to the suggestion that telomere maintenance via t-circles may represent the main, or even the only, mechanism operating in mitochondria containing a type II linear genome.[19-21]

On the Origin of Linear Chromosomes

Structurally different telomeres found in phylogenetically distant linear DNA genomes such as animal viruses, bacteriophages, plasmids, organelles and bacterial and nuclear eukaryotic chromosomes illustrate the range of successful replication strategies that evolved to evade the terminal erosion caused by the end-replication problem.[22-24] The most prominent mechanism operating in

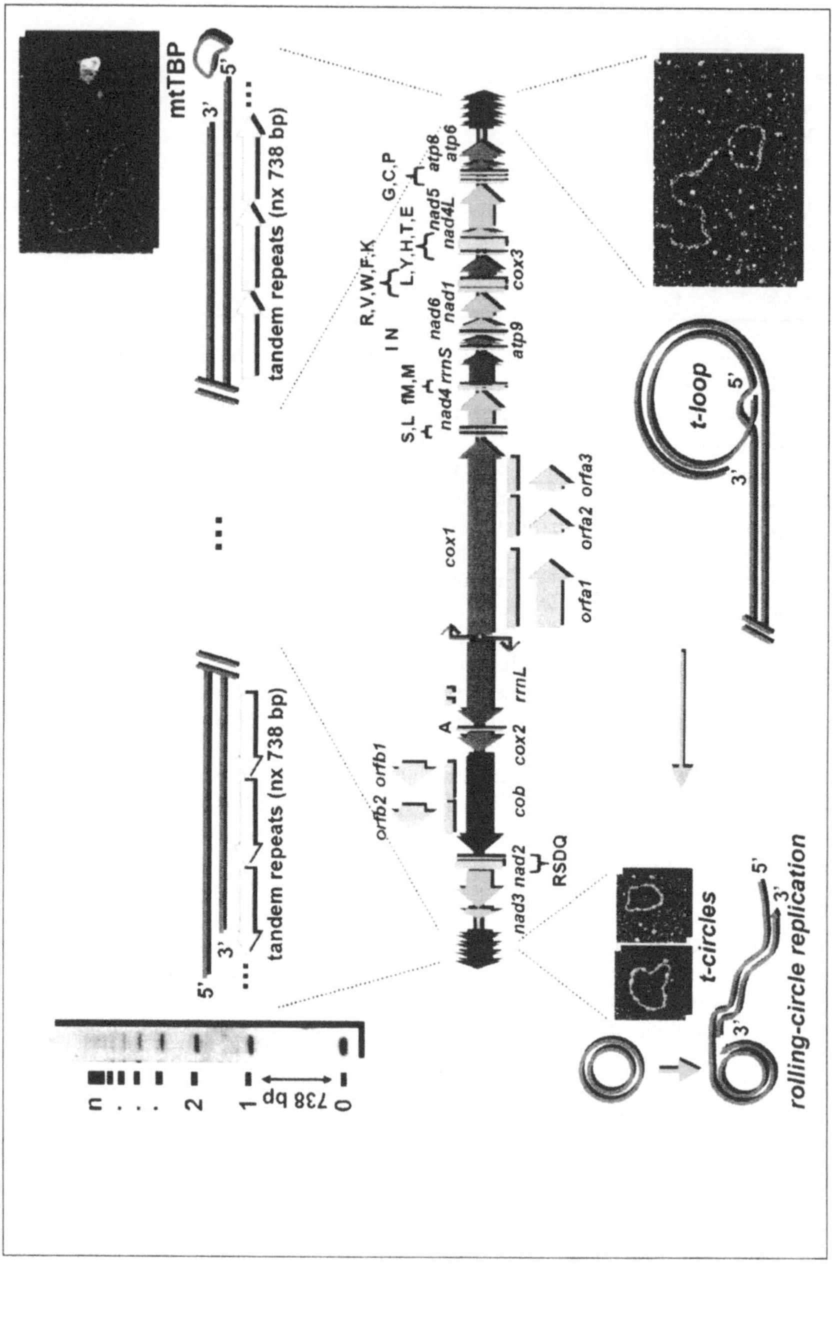

Figure 2. Figure legend on next page.

Figure 2, previous page. The yeast *C. parapsilosis* represents a unique model system for the study of telomerase-independent mechanisms of telomere maintenance. Mitochondria in this species harbor a linear mitochondrial genome (30,923 bp) encoding a standard set of mitochondrial genes. The linear molecules terminate with arrays of tandem repeats (nx 738 bp) and single-stranded 5′ overhang at both ends. The very ends of these linear molecules are protected by either mitochondrial telomere binding protein (mtTBP) and/or by formation of t-loop structures. In addition, the mitochondria of *C. parapsilosis* contain series (nx 738 bp) of circular DNA molecules derived exclusively from the sequence of the mitochondrial telomere termed t-circles, which amplify via rolling circle replication mechanism and seem to play a key role in the maintenance of the linear mtDNA.[5,12-16,18,19]

the majority of eukaryotes is represented by telomerase (Brault et al, this volume). As the enzyme was identified in major eukaryotic lineages including protists, fungi, plants and animals, it might have been recruited for telomere maintenance relatively early in the evolution of eukaryotes. Moreover, the origin of the enzyme can be traced back to the world of RNA.[25,26] On the other hand, this does not necessarily imply that the enzyme maintained the ends of linear chromosomes in the first eukaryotes. Rather, as recently suggested, telomerase might have been recruited later and replaced and/or complemented primordial telomere maintenance strategies.[27-29]

The intriguing question is how and why linear chromosomes emerged in evolution. One possible scenario includes an accidental linearization of an originally circular chromosome accompanied by the formation of specific terminal structures that stabilized the linear DNA molecules. In addition, we suggested that an invasion of selfish genetic element(s), such as plasmids or retrotransposons that integrated into an ancestral circular genome, forced conversion toward a linear form and, at the same time, provided the means for stabilization and replication of its termini.[29,30] Collisions of circular genomes with linear DNA plasmids resulting in their linearization are well documented in the chromosomes of *Streptomyces*[31] as well as in the mitochondrial genomes of several species[32,33] and the extension of telomeres by retrotransposition is known from telomere maintenance in *Drosophila* (Pardue and DeBaryshe, this volume). In addition, dysfunctional telomeres can serve as a substrate for endonuclease-independent retrotransposition of LINE-1 elements.[34] These examples illustrate that telomeres can be considered as structural and functional modules that are transferable between different replicons.

Suitable candidates for primordial telomeres that can be classified as selfish elements are telomeric palindromes. In addition to solving the end replication problem, palindrome insertion into a circular genome accompanied by the activity of a specific enzyme able to resolve the double-stranded palindrome into terminal covalently closed single-stranded hairpins provides a simple way for linearization and, in the case of several insertions, for genome segmentation into multiple chromosomes.

Recently, we proposed that t-circles may represent another candidate for a selfish genetic element involved in the formation of a linearized genome with telomeric tandem arrays on its ends. Widespread occurrence and horizontal transfer of rolling-circle-dependent replicons among prokaryotic and eukaryotic species supports the feasibility of such a scenario. Moreover, the t-circles and rolling-circle dependent pathway were observed in various nuclear telomere maintenance systems including telomerase-negative human and rodent cell lines, plants and yeasts (reviewed in ref. 35). This raises the possibility that t-circles not only represent a general feature of telomeres composed of arrays of tandem repeats, but may be considered as molecular fossils from an era preceding telomerase recruitment.

Cases of interconversion between linear and circular DNA forms seen in yeast mitochondrial genomes,[7,21] bacterial chromosomes[31] and nuclear chromosomes of fission yeast mutants[36] provide the possibility to experimentally address questions concerning chromosomal linearity.

While no specific advantage associated with linear chromosomes was reported in bacteria, competition experiments with isogenic *C. metapsilosis* mutants differing in their mitochondrial genome form revealed that the presence of telomeres and/or the linearity per se can, at least in certain circumstances, provide a specific growth advantage.[21] The nature of this advantage in mitochondria remains unknown. However, the strong preference for linear chromosomes in eukaryotic nuclei seems to be clearer. Although circular or ring chromosomes can be tolerated during vegetative growth of eukaryotic cells, they are associated with genetic anomalies and exhibit problems during meiosis.[37] This indicates that sexual reproduction, which occurs in most eukaryotes, depends on chromosomal linearity. Hence, benefits of sex functioning as a ratchet led to preference for the linear chromosomal form among eukaryotes. Therefore, the selective pressure toward linearization must have been associated with emergence of robust and redundant mechanisms for maintenance of terminal structures. At the same time, the evolutionary success of telomerase over alternative, presumably ancient, telomere maintenance mechanisms may be due to its auxiliary activities that enhance cell survival independently of the synthesis of telomeric repeats (Brault et al, this volume).

On the Origin of T-Circles

If the t-circle-dependent pathway belongs to a primordial means of telomere maintenance, the question of what their evolutionary origin is remains. On one hand, there are a number of yeast species with linear mitochondrial genomes possessing terminal tandem repeats maintained with the assistance of telomeric circles. On the other hand, the sequences of mitochondrial t-circles are extremely diverse and there may be only some common structural features that enable their propagation. How these (very different) t-circles emerged in the corresponding phylogenetic branches?

We propose three possibilities to explain (i) how t-circles appear, (ii) why their sequences are very different, and (iii) how and where they can integrate into the main genome. The most trivial possibility is that similarly as in the case of rDNA cluster[38] extrachromosomal circular DNA molecules result from intramolecular recombination between two or more head-to-tail repeats present within the genome. Naturally, only small subset of such circles capable of rolling-circle synthesis would be stably maintained and further amplified. Eventually, the linear tandem arrays generated by this mechanism re-integrate into the circular genome leading to formation of linearized chromosome.

Another mechanism is based on similarities between mitochondrial t-circles and *rho-* derivatives of mitochondrial DNA in hypersuppressive (HS) *petite* yeast strains pointing to their common origin. HS strains of baker's yeast are frequently isolated as spontaneous respiratory-deficient mutants or after treatment of cells with DNA intercalating agents such as ethidium bromide. HS cells harbor amplified subgenomic mtDNA fragments which are able to out-compete the wild-type mitochondrial genome without any apparent advantage for its host.[39,40] Hence, HS genomes can be classified as selfish genetic elements. In contrast to baker's yeast, in a *petite*-negative, strictly aerobic species such as *C. parapsilosis* they would be unable to eliminate the wild-type mtDNA from cells and might have been forced to co-exist with the main genome. Therefore, it is possible to hypothesize that mitochondrial t-circles originated from HS derivative that integrated into an ancestral, presumably circular mitochondrial genome and subsequently became essential for maintenance of its linearized form.

Third scenario can be based on the ability of some types of reverse transcriptases (RT) (e.g., R2Bm) to initiate end-to-end intramolecular reverse transcription resulting in extrachromosomal circular DNA molecules.[41,42] Mitochondria contain numeroous RT (e.g., derived from retroposons such as *Mauriceville* retroplasmid of *Neurospora crassa*[43,44]) that can employ endogenous RNA molecules as substrates for the above activity. The RNA molecules may not necessarily be full-length, but can be degradation products. Furthermore, some RTs are capable of non-templated extension of existing RNA molecules and thus generate templates with very high sequence variability. These templates then may lead to generation of circular DNA molecules. Naturally, most of these pieces of circular DNA will be discarded because they are unable to replicate. However, although at extremely low frequency (in addition to the above because in strictly aerobic yeast linearization must occur in a way that does not compromise the coding capacity of mtDNA), a DNA circle may appear that has the appropriate structural features allowing it to be replicated (e.g., via rolling-circle strategy) and invade the main mitochondrial chromosome (via homologous recombination, since t is derived from an endogenous mitochondrial RNA molecule) leading to its linearization. This possibility is testable. For example, a R2Bm RT (encoded by a *Bombyx mori* retrotransposon) may be overexpressed from an inducible promoter on a nuclear plasmid in *Saccharomyces cerevisiae* and targeted to mitochondria. If the scenario is correct, there should be a higher frequency of *petites* under inducible conditions, and physical mapping of the mtDNA in these mutant strains should reveal (at least in some instances) linear molecules with tandem repeats at the ends.

The RT-assisted emergence of t-circles may not be limited to mitochondrial telomeres, but may be at the heart of the origin of the first nuclear telomere. One possibility is that t-circles generated in the mitochondrial compartment escaped from the organelle into the nucleus, where it linearized the genome and established a t-circle dependent mechanism of telomere maintenance. Alternatively, nuclear t-circles evolved independently and telomerase is the remnant of this event: as a RT it produced the first t-circle by end-to-end RT activity and then it adopted the ability to maintain nuclear telomeres through extension of the 3′ overhang. Thus, the emergence of nuclear telomeres is inseparable from both reverse transcriptase (pre-telomerase) and t-circles, which resulted from its RT activity. The question of whether telomerase or t-circles evolved first would then become irrelevant.

Conclusion

Phylogenetically independent cases of yeast species with linear mitochondrial genomes having specific molecular architecture at their termini, the ability to form of higher order structures and the t-circle-dependent pathway implicated in their maintenance illustrate repeated emergence of linear DNA genomes with terminal tandem arrays in the absence of telomerase. This provides a paradigm for the evolutionary origin of linear chromosomes and their telomeres in early eukaryotes. Naturally, the parallels between nuclear and mitochondrial systems may have limits. Nevertheless, t-palindromes, t-circles, t-loops, t-posons and telomerases found in the nuclei of diverse species together with certain subtelomeric sequences may be considered as molecular fossils from an early phase of eukaryotic chromosome evolution. T-palindromes and t-circles may represent selfish genetic elements that integrated into a presumably circular genome and forced its conversion into the linear form. At the same time, they provided a means for its stable maintenance. Primordial telomeres were then able to form t-loops and were also suitable substrates for telomerase recruitment. Consequently, these structures and strategies for their maintenance were out-competed or concealed by the telomerase-dependent mode of telomere synthesis, which operates in most modern eukaryotes. Importantly, the molecular fossils from early phases of telomere evolution can still co-exist with telomerase or be selectively re-activated when a cell is depleted of telomerase activity, providing a back up system for telomere maintenance. Both, telomerase-dependent and alternative molecular mechanisms represent a source of redundancy ensuring robustness of the system.

Acknowledgements

We thank Ladislav Kovac (Comenius University, Bratislava, Slovakia), Hiroshi Fukuhara (Institut Curie, Orsay, France) and Jack D. Griffith (University of North Carolina, Chapel Hill, NC, USA) for continuous support and helpful discussions. The work in our laboratory was supported by grants from the Howard Hughes Medical Institute (55005622), the Fogarty International Research Collaboration Award (2-R03-TW005654-04A1), the Slovak grant agencies VEGA (1/2331/05, 1/3247/06) and APVT (20-001604, LPP-0164-06).

References

1. Suyama Y, Miura K. Size and structural variations of mitochondrial DNA. Proc Natl Acad Sci USA 1968; 60:235-242.
2. Nosek J, Tomaska L. Mitochondrial telomeres: Alternative solutions to the end-replication problem. In: Krupp G, Parwaresch R, eds. Telomeres, telomerases and cancer. New York: Kluwer Academic/Plenum Publishers, 2002:396-417.
3. Nosek J, Tomaska L, Fukuhara H et al. Linear mitochondrial genomes: 30 years down the line. Trends Genet 1998; 14(5):184-188.

4. Dinouel N, Drissi R, Miyakawa I et al. Linear mitochondrial DNAs of yeasts: closed-loop structure of the termini and possible linear-circular conversion mechanisms. Mol Cell Biol 1993; 13(4):2315-2323.
5. Nosek J, Dinouel N, Kovac L et al. Linear mitochondrial DNAs from yeasts: telomeres with large tandem repetitions. Mol Gen Genet 1995; 247(1):61-72.
6. Blaisonneau J, Nosek J, Fukuhara H. Linear DNA plasmid pPK2 of Pichia kluyveri: distinction between cytoplasmic and mitochondrial linear plasmids in yeasts. Yeast 1999; 15(9):781-791.
7. Fukuhara H, Sor F, Drissi R et al. Linear mitochondrial DNAs of yeasts: frequency of occurrence and general features. Mol Cell Biol 1993; 13(4):2309-2314.
8. Maringele L, Lydall D. Telomerase- and recombination-independent immortalization of budding yeast. Genes Dev 2004; 18(21):2663-2675.
9. Kovac L, Lazowska J, Slonimski PP. A yeast with linear molecules of mitochondrial DNA. Mol Gen Genet 1984; 197(3):420-424.
10. Greider CW, Blackburn EH. Identification of a specific telomere terminal transferase activity in Tetrahymena extracts. Cell 1985; 43(2 Pt 1):405-413.
11. Greider CW, Blackburn EH. The telomere terminal transferase of Tetrahymena is a ribonucleoprotein enzyme with two kinds of primer specificity. Cell 1987; 51(6):887-898.
12. Nosek J, Novotna M, Hlavatovicova Z et al. Complete DNA sequence of the linear mitochondrial genome of the pathogenic yeast Candida parapsilosis. Mol Genet Genomics 2004; 272(2):173-180.
13. Tomaska L, Nosek J, Fukuhara H. Identification of a putative mitochondrial telomere-binding protein of the yeast Candida parapsilosis. J Biol Chem 1997; 272(5):3049-3056.
14. Nosek J, Tomaska L, Pagacova B et al. Mitochondrial telomere-binding protein from Candida parapsilosis suggests an evolutionary adaptation of a nonspecific single-stranded DNA-binding protein. J Biol Chem 1999; 274(13):8850-8857.
15. Tomaska L, Makhov AM, Nosek J et al. Electron microscopic analysis supports a dual role for the mitochondrial telomere-binding protein of Candida parapsilosis. J Mol Biol 2001; 305(1):61-69.
16. Tomaska L, Makhov AM, Griffith JD et al. t-loops in yeast mitochondria. Mitochondrion 2002; 1:455-459.
17. Griffith JD, Comeau L, Rosenfield S et al. Mammalian telomeres end in a large duplex loop. Cell 1999; 97(4):503-514.
18. Tomaska L, Nosek J, Makhov AM et al. Extragenomic double-stranded DNA circles in yeast with linear mitochondrial genomes: potential involvement in telomere maintenance. Nucleic Acids Res 2000; 28(22):4479-4487.
19. Nosek J, Rycovska A, Makhov AM et al. Amplification of telomeric arrays via rolling-circle mechanism. J Biol Chem 2005; 280(11):10840-10845.
20. Rycovska A, Valach M, Tomaska L et al. Linear versus circular mitochondrial genomes: intraspecies variability of mitochondrial genome architecture in Candida parapsilosis. Microbiology 2004; 150 (Pt 5):1571-1580.
21. Kosa P, Valach M, Tomaska L et al. Complete DNA sequences of the mitochondrial genomes of the pathogenic yeasts Candida orthopsilosis and Candida metapsilosis: insight into the evolution of linear DNA genomes from mitochondrial telomere mutants. Nucleic Acids Res 2006; 34(8):2472-2481.
22. Olovnikov AM. Principle of marginotomy in template synthesis of polynucleotides. Dokl Akad Nauk SSSR 1971; 201(6):1496-1499.
23. Watson JD. Origin of concatemeric T7 DNA. Nat New Biol 1972; 239(94):197-201.
24. Olovnikov AM. A theory of marginotomy. The incomplete copying of template margin in enzymic synthesis of polynucleotides and biological significance of the phenomenon. J Theor Biol 1973; 41(1):181-190.
25. Eickbush TH. Telomerase and retrotransposons: which came first? Science 1997; 277(5328):911-912.
26. Nakamura TM, Cech TR. Reversing time: origin of telomerase. Cell 1998; 92(5):587-590.
27. de Lange T. T-loops and the origin of telomeres. Nat Rev Mol Cell Biol 2004; 5(4):323-329.
28. Fajkus J, Sykorova E, Leitch AR. Telomeres in evolution and evolution of telomeres. Chromosome Res 2005; 13(5):469-479.
29. Nosek J, Kosa P, Tomaska L. On the origin of telomeres: a glimpse at the pre-telomerase world. Bioessays 2006; 28(2):182-190.
30. Nosek J, Tomaska L. Mitochondrial genome diversity: evolution of the molecular architecture and replication strategy. Curr Genet 2003; 44(2):73-84.
31. Chen CW, Huang CH, Lee HH et al. Once the circle has been broken: dynamics and evolution of Streptomyces chromosomes. Trends Genet 2002; 18(10):522-529.
32. Schardl CL, Lonsdale DM, Pring DR et al. Linearization of maize mitochondrial chromosomes by recombination with linear episomes. Nature 1984; 310:292-296.
33. Sakurai R, Sasaki N, Takano H et al. In vivo conformation of mitochondrial DNA revealed by pulsed-field gel electrophoresis in the true slime mold, Physarum polycephalum. DNA Res 2000; 7(2):83-91.

34. Morrish TA, Garcia-Perez JL, Stamato TD et al. Endonuclease-independent LINE-1 retrotransposition at mammalian telomeres. Nature 2007; 446(7132):208-212.
35. Tomaska L, McEachern MJ, Nosek J. Alternatives to telomerase: keeping linear chromosomes via telomeric circles. FEBS Lett 2004; 567(1):142-146.
36. Nakamura TM, Cooper JP, Cech TR. Two modes of survival of fission yeast without telomerase. Science 1998; 282(5388):493-496.
37. Ishikawa F, Naito T. Why do we have linear chromosomes? A matter of Adam and Eve. Mutat Res 1999; 434(2):99-107.
38. Gangloff S, Zou H, Rothstein R. Gene conversion plays the major role in controlling the stability of large tandem repeats in yeast. EMBO J 1996; 15(7):1715-1725.
39. MacAlpine DM, Kolesar J, Okamoto K et al. Replication and preferential inheritance of hypersuppressive petite mitochondrial DNA. EMBO J 2001; 20(7):1807-1817.
40. Zweifel SG, Fangman WL. A nuclear mutation reversing a biased transmission of yeast mitochondrial DNA. Genetics 1991; 128(2):241-249.
41. Bibillo A, Eickbush TH. End-to-end template jumping by the reverse transcriptase encoded by the R2 retrotransposon. J Biol Chem 2004; 279(15):14945-14953.
42. Weiner A. SINEs and LINEs: Troublemakers, saboteurs, benefactors, ancestors. In: RNA World, 3rd ed, Cold Spring Harbor, NY, Cold Spring Harbor Laboratory Press, 2006.
43. Chen B, Lambowitz AM. De novo and DNA primer-mediated initiation of cDNA synthesis by the mauriceville retroplasmid reverse transcriptase involve recognition of a 3' CCA sequence. J Mol Biol 1997; 271(3):311-332.
44. Kennell JC, Wang H, Lambowitz AM. The Mauriceville plasmid of Neurospora spp. uses novel mechanisms for initiating reverse transcription in vivo. Mol Cell Biol 1994; 14(5):3094-3107.
45. Pramateftaki PV, Kouvelis VN, Lanaridis P et al. The mitochondrial genome of the wine yeast Hanseniaspora uvarum: a unique genome organization among yeast/fungal counterparts. FEMS Yeast Res 2006; 6(1):77-90.
46. Wesolowski M, Fukuhara H. Linear mitochondrial deoxyribonucleic acid from the yeast Hansenula mrakii. Mol Cell Biol 1981; 1(5):387-393.

Index

A

B

C

D

E

F

G

H

I

K

L

M

N

P

R

S

T